The Molecular Biology of Neurological Disease

Butterworths International Medical Reviews

Neurology 9

Published titles

1 **Clinical Neurophysiology**
Erik Stålberg and Robert R. Young

2 **Movement Disorders**
C. David Marsden and Stanley Fahn

3 **Cerebral Vascular Disease**
Michael J. G. Harrison and Mark L. Dyken

4 **Peripheral Nerve Disorders**
Arthur K. Asbury and R. W. Gilliatt

5 **The Epilepsies**
Roger J. Porter and Paolo L. Morselli

6 **Multiple Sclerosis**
W. I. McDonald and Donald H. Silberberg

7 **Movement Disorders 2**
C. David Marsden and Stanley Fahn

8 **Infections of the Nervous System**
Peter G. E. Kennedy and Richard T. Johnson

The Molecular Biology of Neurological Disease

Edited by

Roger N. Rosenberg, MD
Professor of Neurology and Physiology; Chairman, Department of Neurology, University of Texas Southwestern Medical Center, Southwestern Medical School, Dallas, Texas, USA

and

A. E. Harding, MD, MRCP
Reader in Clinical Neurology at the Institute of Neurology; Consultant Neurologist at the National Hospitals for Nervous Diseases, London, UK

Butterworths
London Boston Singapore Sydney Toronto Wellington

First published, 1988

British Library Cataloguing in Publication Data

The Molecular biology of neurological disease.
– (Butterworths international medical
reviews. Neurology; ISSN 0206-0137).
1. Nervous system–Diseases 2. Pathology,
Molecular
I. Rosenberg, Roger N. II. Harding, A. E.
616.8 RC347

ISBN 0-407-02400-X

Library of Congress Cataloging-in-Publication Data

The Molecular biology of neurological disease/edited by Roger N.
Rosenberg and A. E. Harding.
p. cm. – (Butterworths international medical reviews.
Neurology)
Includes bibliographies and index.
ISBN 0-407-02400-X
1. Nervous system–Disease–Pathophysiology. 2. Molecular
neurobiology. I. Rosenberg, Roger N. II. Harding, A. E.
III. Series.
[DNLM: 1. Gene Expression Regulation. 2. Molecular Biology.
3. Nervous System Disease–familial. WL 100 M7186]
RC346.M639 1988
616.8′047–dc19
DNLM/DLC 87-35478
for Library of Congress CIP

Photoset by Butterworths Litho Preparation Department
Printed and bound by Robert Hartnoll (1985) Ltd, Bodmin, Cornwall

Foreword

For almost a quarter of a century (1951–1975), subjects of topical interest were written about in the periodic volumes of our predecessor, *Modern Trends in Neurology*. Although both that series and its highly regarded editor, Dr Denis Williams, are now retired, the legacy continues in the present Butterworths series in Neurology. As was the case with *Modern Trends*, the current volumes are intended for use by physicians who grapple with the problems of neurological disorders on a daily basis, be they neurologists, neurologists in training, or those in related fields such as neurosurgery, internal medicine, psychiatry, and rehabilitation medicine.

Our purpose is to produce annually a monograph on a topic in clinical neurology in which progress through research has brought about new concepts of patient management. The subject of each monograph is selected by the Series Editors using two criteria: first, that there has been significant advance in knowledge in that area and, second, that such advances have been incorporated into new ways of managing patients with the disorders in question.

This has been the guiding spirit behind each volume, and we expect it to continue. In effect we emphasize research, both in the clinic and in the experimental laboratory, but principally to the extent that it changes our collective attitudes and practices in caring for those who are neurologically afflicted.

C. D. Marsden
A. K. Asbury
Series Editors

Preface

The 'guiding spirit' behind each of the volumes in the Butterworths series of neurological reviews is to focus on topics in which progress through research has altered the management of patients with neurological disease. In a field festooned with verbal hyperbole, it seems feeble to state that advances in molecular biology will revolutionize the practice of medicine, including neurology, over the next few years. The revolution has begun, and it is important that those who look after neurological patients are aware of it.

This book is designed to review recent advances in our understanding of the molecular mechanisms of neurological disease, and immediate and future applications of molecular biological techniques to clinical practice. We hope that it will be of value to neuroscientists in general, but particularly to established clinicians and those in training.

We wish to thank the authors for their excellent contributions, and two more tangible guiding spirits, David Marsden and Art Asbury, for asking us to edit this volume.

Roger N. Rosenberg
Anita Harding

Contributors

Alastair Compston, PhD, FRCP
Professor of Neurology, University of Wales College of Medicine, Cardiff, UK

Fred Gilbert, MD
Associate Professor of Medical Genetics and Pediatrics, Division of Medical Genetics, Mt Sinai School of Medicine, New York, USA

T. Conrad Gilliam
Departments of Psychiatry and Neurology, College of Physicians and Surgeons, Columbia University, New York, USA

Michel Goedert, MD, PhD
Medical Research Council Laboratory of Molecular Biology, Hills Road, Cambridge, UK

Sue Griffin, PhD
Professor, Departments of Pediatrics and Anatomy, University of Arkansas for Medical Sciences, Little Rock, Arkansas, USA

Abraham Grossman
Research Assistant Professor of Neurology, University of Texas Southwestern Medical Center, Dallas, Texas, USA

James F. Gusella, PhD
Genetics Unit, Massachusetts General Hospital, Boston, Massachusetts, USA

William E. Hahn, PhD
Professor, Department of Cellular and Structural Biology, University of Colorado School of Medicine, Denver, Colorado, USA

A. E. Harding, MD, MRCP
Reader in Clinical Neurology at the Institute of Neurology; Consultant Neurologist at the National Hospitals for Nervous Diseases, London, UK

Peter S. Harper
Professor and Consultant in Medical Genetics, Institute of Medical Genetics, University of Wales College of Medicine, Cardiff, UK

Ian J. Holt, BSc
Research Assistant, Department of Clinical Neurology, Institute of Neurology, London, UK

Ruth F. Itzhaki, MSc, PhD, MA
Honorary Reader in Cell Biology, UMIST, Manchester, UK

Albee Messing, VMD, PhD
Assistant Professor of Neuropathology, Department of Pathobiological Sciences, School of Veterinary Medicine, University of Wisconsin, Madison, Wisconsin, USA

Girish Modi, MB, BCh, MSc, PhD
Assistant Professor, Department of Neurology, University of Natal, Durban, South Africa

Marcelle R. Morrison, PhD
Associate Professor, Departments of Neurology and Biochemistry, University of Texas Southwestern Medical Center, Dallas, Texas, USA

Gregory P. Owens, PhD
Postdoctoral Fellow, Department of Cellular and Structural Biology, University of Colorado School of Medicine, Denver, Colorado, USA

David Pleasure, MD
Professor of Neurology, University of Pennsylvania, Philadelphia, Pennsylvania, USA

Roger N. Rosenberg, MD
Professor of Neurology and Physiology and Chairman, Department of Neurology, University of Texas Southwestern Medical Center, Dallas, Texas, USA

Robert S. Sparkes, MD
Professor of Medicine, UCLA School of Medicine, Los Angeles, California, USA

R. J. Swingler, BSc, MRCP (UK)
Lecturer in Neurology, University of Edinburgh, Edinburgh, UK

Kenneth L. Tyler, MD
Assistant Professor, Departments of Neurology and Microbiology and Molecular Genetics, Harvard Medical School; Assistant Neurologist, Massachusetts General Hospital, Harvard Medical School, Boston, Massachusetts, USA

Frank S. Walsh, BSc, PhD
Wellcome Trust Senior Lecturer, Department of Neurochemistry, Institute of Neurology, Queen Square, London, UK

Contents

1
Molecular genetics and neurological disease: basic principles and methods

A. E. Harding and Roger N. Rosenberg

INTRODUCTION

Prenatal diagnosis of a genetic disease, α-thalassaemia, was first performed using molecular genetic techniques 23 years after the description of the double helical structure of deoxyribonucleic acid (DNA) by Watson and Crick in 1953 (Watson and Crick, 1953; Kan, Golbus and Dozy, 1976). Subsequently, rapid progress has been made in elucidating the molecular basis of inherited disease. At least 3000 disorders exhibiting mendelian inheritance are known to occur in man; 936 of these had been localized to specific chromosomal regions, many using molecular genetic techniques, by 1986 as compared to 300 in 1975 (McKusick, 1986). The molecular genetics of a number of inherited neurological disorders will be discussed in detail in the Chapters 10 to 18. The 'new genetics' also has many applications in other neurological and neurobiological problems, including oncology, virology, and development; selected topics pertaining to these are dealt with in Chapters 2–9.

For the clinician or scientist without a strong background in cell biology or genetics, understanding recombinant DNA technology is hampered not only by lack of familiarity with the techniques and their application, but also by the jargon which has been generated by this rapidly expanding field. It is hoped that this chapter will provide a general introduction and glossary to the more specific topics covered by the rest of the book. Many advances in neurogenetics have been made using molecular genetic techniques in genetic linkage studies; this topic will be discussed in some detail, partly for illustrative purposes but also because of its immediate clinical relevance.

PRINCIPLES OF MENDELIAN INHERITANCE

Every human cell contains a pair of each of 22 chromosomes called the autosomes. In addition there are either two X chromosomes or one X and one Y chromosome, depending on the sex of the individual. During cell division the process of mitosis results in two daughter cells which are diploid, that is, each contains an identical chromosomal complement to the parent cell. Gametogenesis involves the more complex process of meiosis, during which each pair of chromosomes exchanges genetic material, resulting in gametes which are haploid, containing only half the number of chromosomes of the parent cell.

Defective genes on the autosomes (chromosomes 1–22) may be inherited as dominant or recessive traits. An individual affected by an autosomal dominant disorder has a 50% chance of transmitting it to offspring (*Figure 1.1*). In this case, a heterozygous gene carrier manifests the disease despite the presence of a normal corresponding gene (allele) on the other half of the chromosome pair.

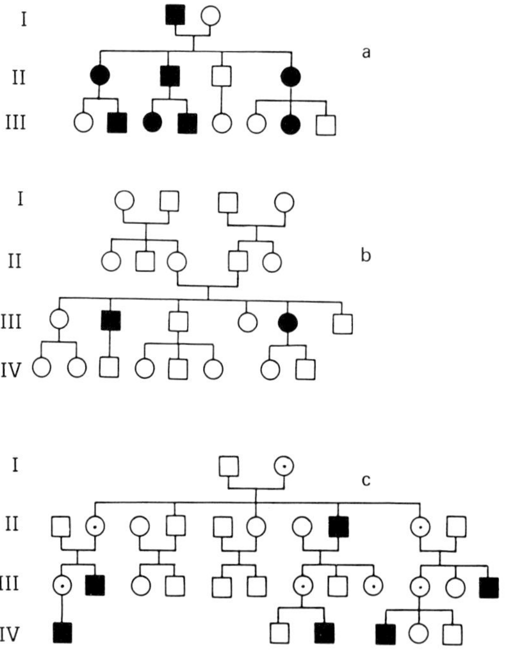

Figure 1.1 Pedigrees illustrating (*a*) autosomal dominant; (*b*) autosomal recessive; and (*c*) X-linked inheritance. Square = male; circle = female; filled symbol = affected; open symbol = unaffected; ⊙ in (*c*) indicates obligate carrier

This is not so in autosomal recessive inheritance, where it is necessary for both alleles to be abnormal in order for the disease to be expressed (*see Figure 1.1*), that is, the affected individual is homozygous. We are all heterozygous for one or two recessive genes; the commonest autosomal recessive disorder in the United Kingdom is cystic fibrosis, and about 5% of the population carry the cystic fibrosis gene (Harper, 1981). If two heterozygotes for the same autosomal recessive gene mate, on average one in four of their children will be affected, two out of four will be carriers and one out of four will not carry the gene. It should be obvious that, in populations where family size is small, the majority of individuals with autosomal recessive disorders do not have affected sibs, and often the clinician does not suspect the presence of a genetically determined disorder. Autosomal recessive disorders are more common amongst the offspring of consanguineous parents, for example cousins, as these are more likely to share autosomal recessive genes in common than unrelated members of the population.

Defective genes on the X chromosomes show a distinctive pattern of inheritance in which males are most severely affected, and females carrying the gene may be moderately or mildly affected or clinically normal. The variation in expression of an X-linked disorder in females is due to the process of lyonization, during which the expression of one X chromosome is suppressed randomly in each cell. The distinction between X-linked recessive (in which female carriers are normal) and X-linked dominant (female carriers affected) disorders is rather artificial, although these terms are often used. An important feature of X-linked inheritance is that male-to-male transmission never occurs (*see Figure 1.1*), but all the female offspring of affected males inherit the abnormal gene.

STRUCTURE AND FUNCTION OF NUCLEIC ACIDS

Both DNA and ribonucleic acid (RNA) consist of sequences of nucleotides, each of which contain a pentose sugar, a phosphate group, and a nitrogenous base. The last may be either pyrimidines (uracil, cytosine, and thymine), or purines (adenine and guanine). DNA contains adenine (A), guanine (G), thymine (T), and cytosine (C) (*Figure 1.2*). In RNA uracil (U) replaces thymine. RNA is usually single-stranded, but DNA is generally double-stranded, forming an antiparallel double helix; G and C always pair together by means of hydrogen bonds and the same applies to A and T. Compact mammalian chromosomal DNA is supercoiled around proteins called histones. The human genome, that is all the chromosomal DNA, consists of about 3 $\times 10^9$ base pairs (bp) of DNA.

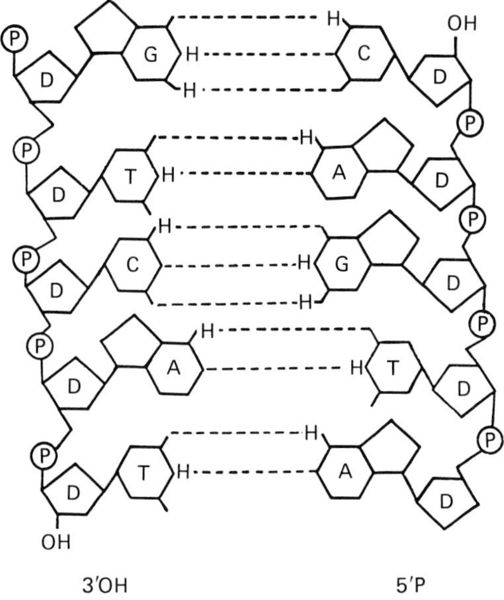

Figure 1.2 The structure of deoxyribonucleic acid. P = phosphate; D = deoxyribose; H = hydrogen; A = adenine; C = cytosine; G = guanine; T = thymine; OH = hydroxyl group. Dotted lines indicate hydrogen bonds between paired nucleotides

Only about 1% of genomic DNA forms structural genes consisting of unique sequences (single copy sequences). Nearly half of human DNA consists of repetitive short sequences which are either not transcribed or, less commonly, encode for small high abundance proteins such as histones. The function of the remaining 50%, some of which codes for *introns* (non-coding parts of genes which are spliced out after transcription), is also not fully understood. Introns may regulate mRNA stability, thereby increasing the level of the translated gene product. They may exert a major influence on the overall size of mRNA transcripts. For example, the ovalbumin gene, representing some 7000 bases, has only 1859 bases which encode for *exons* and thus for structural protein. Introns represent the majority of bases in this gene, with approximately 5000 bases present in seven introns (Alberts *et al.*, 1983).

The majority of structural proteins and enzymes appear to be encoded by single copy sequences of DNA. Genes coding for similar functions are sometimes located as compact groups on the same chromosome, such as the immunoglobulin heavy chain gene cluster on chromosome 14. Subunits of multimeric proteins may be encoded by genes scattered throughout the genome, for example the hexosaminidase enzyme complex which is composed of subunits coded for by genes on chromosomes 5 and 15 (McKusick, 1986). Genes may also occur as dispersed families representing multiple copies on several different chromosomes, as has been described for the argininosuccinate synthase and actin genes (Beaudet *et al.*, 1982; McKusick, 1986). It should be stressed that not all of these are transcribed; many may be defective, non-transcribed copies, the so-called pseudogenes. Isogenic copies (different genes coding for the same function) may have tissue specificity. Multiple gene copies can also be expressed on double minute chromosomes, multiple copies of miniature chromosomes which appear to amplify gene action (Schimke *et al.*, 1978; Baskin, Rosenberg and Dev, 1981).

There are also DNA sequences which initiate or stop *transcription*, the synthesis of messenger RNA (mRNA) complementary to single-stranded DNA sequences catalysed by the enzyme RNA polymerase. This enzyme traverses the DNA template strand from the 3′ to the 5′ end, and synthesizes RNA starting at its 5′ end. The promotor, a specific DNA sequence upstream from the coding region, signals where RNA synthesis is to begin; when RNA polymerase reaches a second DNA sequence, the termination signal, it dissociates from the DNA and releases the RNA chain. A tail of up to 200 adenylic acids may be added to the 3′ end of this chain, and it is capped with methylated guanosine residues at the 5′ end.

In eukaryocytes (higher organisms containing nuclei), nearly all primary RNA transcripts contain introns which are spliced out, leaving a molecule consisting largely of exons; mRNA then migrates from the nucleus to ribosomes in the cytoplasm. The 'genetic code' dictates that any amino acid is encoded by a triplet of nucleotide bases (a codon), for example, CGG for tryptophan. As most amino acids may be encoded by more than one codon, for example, UUU or UUC for phenylalanine, the code is said to be degenerate. Transfer (t)RNAs, each containing a specific codon, successively pair with the mRNA and add amino acids to form a polypeptide chain. This process is known as *translation*. The ribosomes associated with one molecule of mRNA are referred to as a *polysome*. Many proteins undergo post-translational modification, such as methylation, glycosylation, phosphorylation or cleavage, before they serve any biological function. Large polypeptides may be cleaved to produce a number of active peptides, and this process can vary between different tissues. This is clearly an oversimplified summary of the complex processes of transcription and translation which have been reviewed recently by Alberts *et al.* (1983) in a highly readable account of cell biology.

THE MOLECULAR BASIS OF GENETIC VARIATION

DNA *replication* requires separation of the two strands of the double helix; each strand then acts as a template for the formation of complementary sequences, a process catalysed by DNA polymerase. As well as successively adding complementary nucleotides to the growing chain, this enzyme has the capacity to read and correct errors, allowing a high degree of precision in replication. Mutation occurs when this mechanism fails, either by substituting a single incorrect nucleotide (point mutation), or by skipping or adding a number of bases (deletion or insertion). The former may lead to an amino acid substitution in a protein, such as the change from UUA (phenylalanine) to UUG (leucine), or premature termination of transcription if, for example, UGG (tryptophan) changes to UGA (stop codon). Deletions of one or more bases may result in frame shift mutation, with misreading of the genetic code downstream from that site because the mRNA is not read in its correct sequence. If mutations occur in germ cells, they may be inherited by progeny, but mutations also occur in somatic cells and these are increasingly recognized as being important in oncogenesis.

Mutations changing codon signals occur frequently and potentially can affect the primary amino acid sequence of many proteins, including those of the nervous system. It has been estimated that 70 000–100 000 proteins are expressed in a brain cell from about 3×10^5 genes (Rosenberg, 1983). Mutation in structural genes can cause cell death as a result of impaired synthesis of an essential protein; more rarely the change is genetically advantageous. Variation in the primary amino acid sequence of brain proteins resulting from mutation may give rise to genetic diversity in the normal population. Enzyme activity, receptor binding affinity, or structural protein functions may be slightly altered, albeit still within the normal range. This background heterogeneity is probably responsible for the marked variation in expression which is characteristic of autosomal dominant disorders affecting the central nervous system. In other words, expression of the disease mutation is altered by modifying genes.

Genetic variation at a molecular level may be abnormal and result in disease. However, much of the normal variation which occurs in humans is caused by variation in DNA sequences, and these variable (polymorphic) sequences are inherited. This normal variation may be externally apparent, such as the diversity in hair or eye colour observed between different individuals. Further evidence of normal genetic heterogeneity can be obtained at a subclinical level, for example in the different blood group systems, and enzyme polymorphism.

Harris (1966) showed that humans exhibit marked variation in enzyme activity. Enzyme polymorphism is the basis for adverse reactions to drugs in some instances, for example the development of peripheral neuropathy with isoniazid therapy in slow acetylators. Dopamine β-hydroxylase (DBH) and catechol-O-methyltransferase (COMT) both express activity through two alleles, one inherited from each parent. DBH^L-DBH^L and $COMT^L$-$COMT^L$ result in a low level of enzyme activity; DBH^H-DBH^H and $COMT^H$-$COMT^H$ result in high activity; and with allelic heterozygosity, seen in 50% of the population, intermediate levels of activity occur (Weinshilboum, 1983).

On statistical grounds, variation in DNA sequence is more likely to occur in non-coding regions of the genome. This type of polymorphism often has no functional significance. It can be detected by restriction endonucleases, a class of enzymes possessed by microorganisms. Type II restriction endonucleases recognize specific sequences of the DNA molecule and cleave it at these sites and nowhere else (*Figure 1.3*). A single base change may lead to loss or gain of a specific recognition site. Thus cleavage of genomic DNA by any restriction endonuclease will result in fragments which vary in length between individuals. These can be detected by hybridization with a radioactively labelled probe complementary to the polymorphic DNA sequence being studied. These polymorphisms are inherited, and are known as *restriction fragment length polymorphisms* (RFLPs).

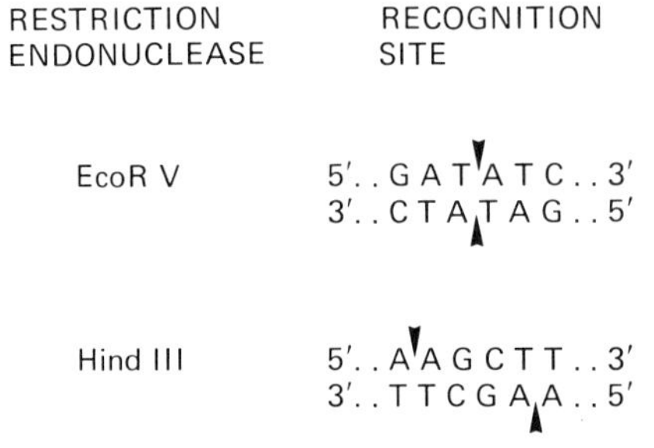

RESTRICTION ENDONUCLEASE	RECOGNITION SITE
EcoR V	5′. . G A T▼A T C . . 3′ 3′. . C T A▲T A G . . 5′
Hind III	5′. . A▼A G C T T . . 3′ 3′. . T T C G A▲A . . 5′

Figure 1.3 EcoR V and Hind III are type II restriction endonucleases (from *Escherichia coli* and *H. influenzae*, respectively) which recognize DNA sequences consisting of the six bases shown and cleave the DNA at sites indicated by the arrows. Cleavage of double-stranded DNA by EcoR V yields blunt-ended fragments, whereas these have 'sticky' ends after cleavage with Hind III

RFLPs are extremely useful as genetic markers in genetic linkage studies, as will be explained subsequently. The techniques used for detecting them (*Figure 1.4*) are relatively simple (Old, 1986). DNA is extracted from leukocytes and incubated with a given restriction endonuclease to produce a large number (about 10^6) of DNA fragments, usually ranging up to 20 kilobases (kb) in length. These are electrophoresed on an agarose gel, which separates the fragments by length. Small fragments migrate faster than large ones. For ease of handling and further analysis the DNA fragments are transferred to nitrocellulose or nylon membrane filters by a process developed by Southern and now known as Southern blotting (Southern, 1975). The gel is placed on filter paper which hangs over the sides of a glass plate into a bath of sodium chloride/sodium citrate solution. The nitrocellulose or nylon filter is placed on top of the gel and more filter paper, a stack of paper towels, and a weight are placed on top of this. The solution is drawn up through the gel by the filter paper wicks, and the DNA fragments are deposited on the filter. It should be explained at this point that the transfer of RNA and proteins from gels to filters is usually referred to rather whimsically as Northern and Western blotting, respectively! (Davis, Dibner and Battey, 1986).

After blotting, the DNA on the filter is permanently bound to it by baking or ultraviolet irradiation; the blot can then be stored virtually indefinitely. When

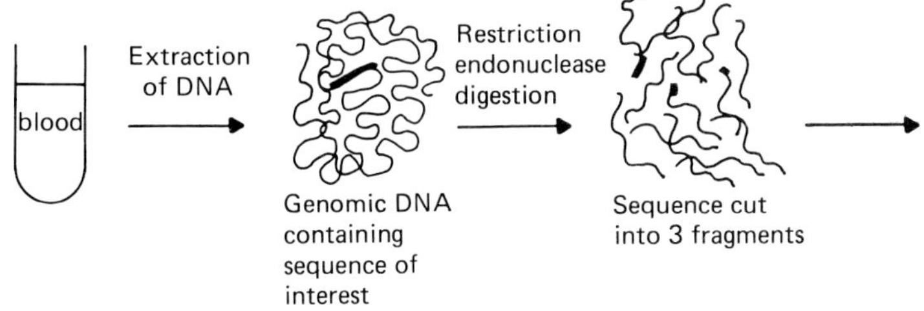

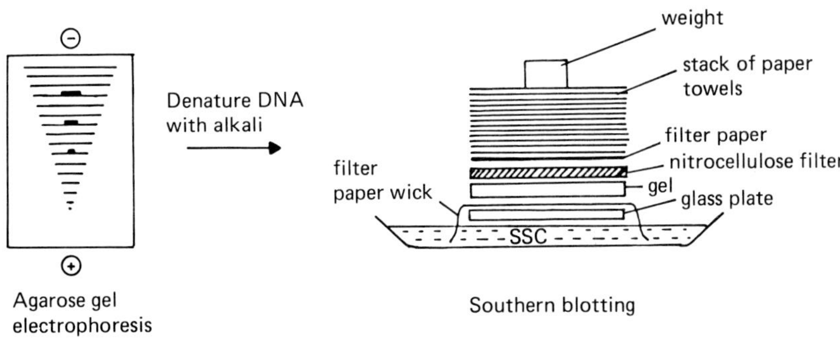

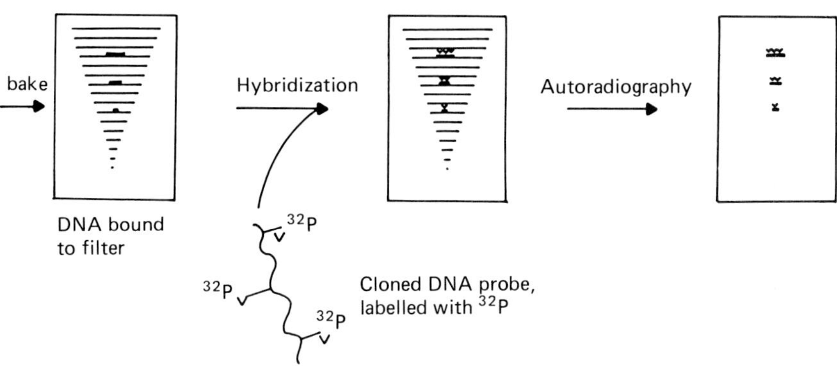

Figure 1.4 A diagrammatic summary of the techniques used for analysing restriction fragment length polymorphisms (RFLPs). DNA is extracted from peripheral blood leukocytes and digested with a given restriction endonuclease. In this example, the genomic DNA sequence of interest contains two restriction sites and is therefore cut into three fragments of different lengths. The genomic DNA fragments (about 10^6) are separated by agarose gel electrophoresis, shorter fragments migrating faster than long ones. They are then denatured with alkali and transferred to a nitrocellulose filter by Southern blotting. After baking the DNA is permanently bound to the filter. A DNA probe, complementary to the genomic sequence of interest, is labelled with radioactive phosphorus and hybridized to the DNA on the filter. After autoradiography the position of the three fragments can be visualized. (From Harding (1988) by courtesy of the publisher)

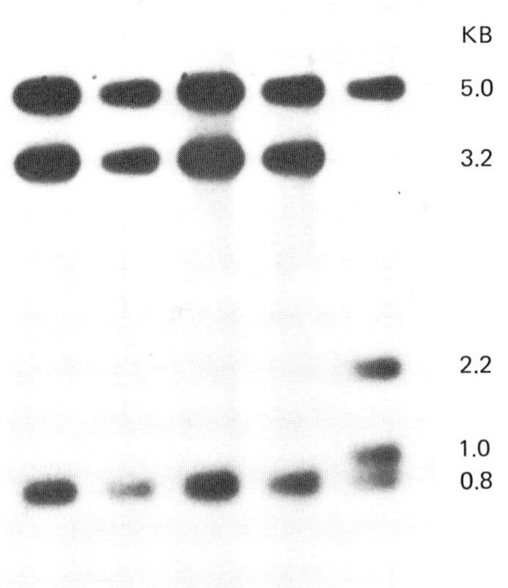

Figure 1.5 Autoradiograph showing DNA fragments, produced by digesting genomic DNA with a restriction enzyme, hybridized to a radioactively labelled DNA probe. In the four lanes on the left, the probe hybridizes to three fragments 5.0, 3.2, and 0.8 kb in length. In the lane on the right, an additional restriction site (RFLP) has resulted in cleavage of the 3.2 kb fragment into two of 2.2 and 1.0 kb

required, the filter is placed in a hybridization solution containing the denatured labelled DNA probe complementary to the sequence of interest. The location of the hybridized fragments is detected by autoradiography (*Figure 1.5*).

GENE CLONING, GENE LIBRARIES, AND GENE PROBES

The discovery of restriction endonucleases was one of the most critical developments in recombinant DNA technology. As well as detecting polymorphisms, they enable investigators to cut DNA into manageable fragments, and also to insert these fragments into prokaryocyte vectors such as bacteriophage (more commonly called phage, a virus which infects bacteria), cosmids (plasmid DNA packaged into phage), or plasmids.

Plasmids are the most frequently used vector. They are naturally occurring small circular DNA molecules which are found in the cytoplasm of bacteria and replicate independently of the host chromosomal DNA. Plasmids can be cleaved by restriction endonucleases and foreign DNA, for example a DNA fragment containing a human gene sequence (cut out of genomic DNA with the same enzyme) can be inserted using the enzyme DNA ligase. This hybrid DNA molecule is called a recombinant. Bacteria (usually *Escherichia coli*) infected with recombinants can be cultured, allowing the production of multiple copies of the inserted DNA fragment (*Figure 1.6*). This procedure is known as DNA cloning (Glover, 1985), and it has numerous applications, not least of which is the

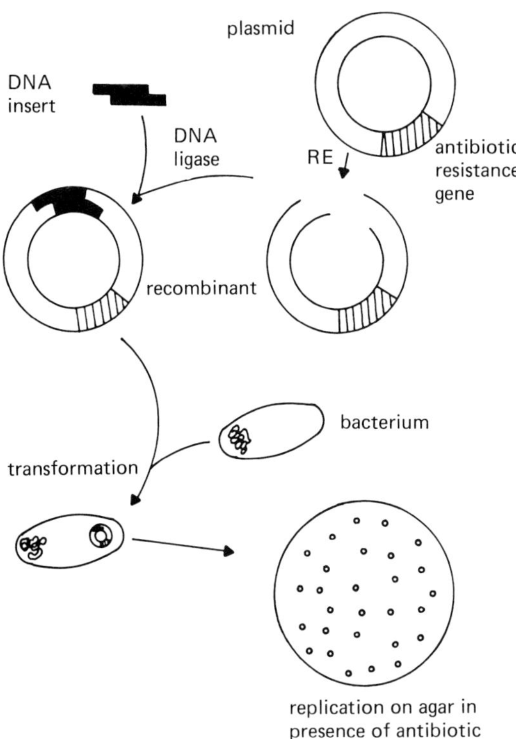

Figure 1.6 Cloning of a sticky-ended DNA fragment in a plasmid vector. The plasmid contains an antibiotic resistance gene. In this example it is cleaved by the same restriction site as that used to generate the inserted foreign fragment; plasmid and insert are ligated using DNA ligase to form a recombinant. This is inserted into competent bacteria, for example, *Escherichia coli*, which are grown on agar plates. The presence of the antibiotic to which the plasmid is resistant selects in favour of growth of bacteria containing the recombinant

large-scale production of therapeutic products, for example insulin from expression of the human insulin gene in bacteria.

Sequences from the entire human genome can be cloned in a given vector, usually lambda (λ) (phage) or cosmid, producing a human gene library. A more specific genomic library can be constructed by using sequences of DNA from a single chromosome, isolated by flow sorting (Lebo *et al.*, 1984); this is particularly useful if probes are to be used for linkage studies (*see below*, p. 11). It is also possible to construct libraries containing the DNA sequences complementary to mRNA transcribed in different tissues such as muscle or brain. mRNA is extracted from the relevant tissue and complementary DNA (cDNA) to the mRNA sequences is synthesized using the enzyme reverse transcriptase. The cDNA is then cloned into a vector, producing a cDNA library which differs from a genomic library in that it only contains sequences of DNA that are transcribed in the tissue studied (*Figure 1.7*). A third type of library, an expression library, is produced by inserting foreign DNA fragments into a beautifully engineered phage vector, the survival of which is dependent on expression of the insert. Clones of interest can be

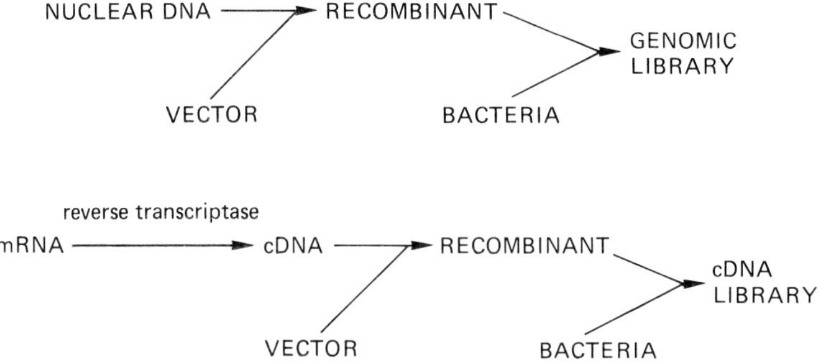

Figure 1.7 The production of genomic and cDNA libraries

detected using antibodies directed against the protein which they synthesize (Davis, Dibner and Battey, 1986).

Gene libraries may be used to generate gene probes, an essential tool in the molecular analysis of inherited diseases. A probe is a fragment of DNA which detects complementary sequences. Probes may be produced synthetically, the nucleotide sequence inferred from the amino acid composition of the gene product if this is known. More commonly, small fragments of genomic or cDNA are cloned and isolated from libraries for use as probes.

In order to identify complementary sequences on a nitrocellulose filter or section of tissue which hybridize to a probe, it is usually labelled with P^{32} by means of either a process called *nick translation* or *oligolabelling*. In the former, the enzyme DNase I randomly removes nucleotides from ('nicks') the DNA molecule and DNA polymerase then inserts labelled nucleotides into the gaps (Rigby *et al.*, 1984). The latter labels DNA to a very high specific activity and employs the Klenow fragment of DNA polymerase to incorporate radiolabelled nucleotides into the probe (Feinberg and Vogelstein, 1983). The DNA to be analysed is made single-stranded (denatured) by heating or alkaline treatment so that the probe can hybridize to it.

GENE MAPPING, GENE TRACKING AND NEUROLOGICAL DISEASE

In many autosomal recessive disorders, such as the aminoacidurias, leukodystrophies and gangliosidoses, the underlying metabolic defect is at least partly understood, as there is often accumulation of a metabolite caused by a specific enzyme deficiency (Stanbury *et al.*, 1983). This is not true of the majority of autosomal dominant disorders, including diseases affecting the nervous system such as Huntington's disease, neurofibromatosis, the hereditary ataxias, and myotonic dystrophy, in which there is no metabolic clue or storage product to indicate their molecular basis. Nevertheless, it is theoretically possible to locate defective dominant genes to a specific chromosomal region by means of linkage analysis.

If two independent genes are located on the same chromosome, they are said to be linked; the closer they are, the tighter the linkage, and the smaller the likelihood that they will be separated as a result of the exchange of genetic material that occurs between homologous chromosomes during meiosis (crossing-over). Depending on the disease studied, the demonstration of close genetic linkage between a mutant gene and a genetic marker may allow preclinical or antenatal detection of gene carriers, and makes it easier to locate the defective gene itself in order to study the abnormal product and its deleterious effects.

Genetic linkage studies in neurological disease

The genetic markers used in linkage studies are the variable sequences of DNA (polymorphisms) referred to earlier which are inherited and occur in the normal population. These may be recognized by their products, such as blood group antigens or serum proteins, or by analysis of the DNA itself using RFLPs. The advantage of using DNA markers is that there are more of them than protein or other markers, and the number of recognized RFLPs is increasing rapidly.

Linkage of autosomal dominant neurological disorders to blood group markers has been demonstrated in myotonic dystrophy (to the ABO secretor and Lutheran blood group loci), tuberous sclerosis (to the ABO blood group locus), and hereditary motor and sensory neuropathy (HMSN) type I (to the Duffy blood group locus) (Harper *et al.*, 1972; Bird, Ott and Giblett, 1982; Fryer *et al.*, 1987). It should be noted (*Figure 1.8*) that the marker loci are linked to the disease gene in

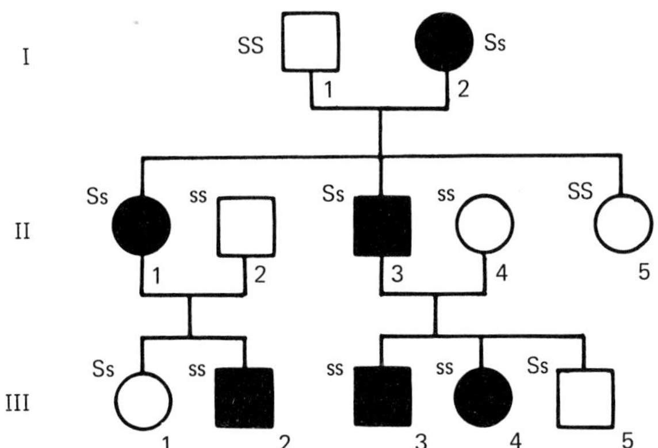

Figure 1.8 Pedigree of a family with myotonic dystrophy (symbols as in *Figure 1.1*) illustrating linkage to the ABO secretor locus. SS and Ss indicate homozygous and heterozygous secretors, respectively (these cannot be distinguished), and ss homozygous non-secretors. I.2 has myotonic dystrophy and must have passed the s gene, together with myotonic dystrophy, to II.1 and II.3, as they both have children with the disease who are non-secretors. The unaffected members of generation III are both secretors, having inherited the S gene from their unaffected grandparent

question, as opposed to any association existing between a specific marker type and the disease. In the pedigree shown in *Figure 1.8*, the myotonic dystrophy gene can be tracked by means of the ss (non-secretor) phenotype, but it segregates with the Ss (secretor) gene in other families. Since the observation of linkage to the secretor locus, the myotonic dystrophy gene has been mapped more accurately to the

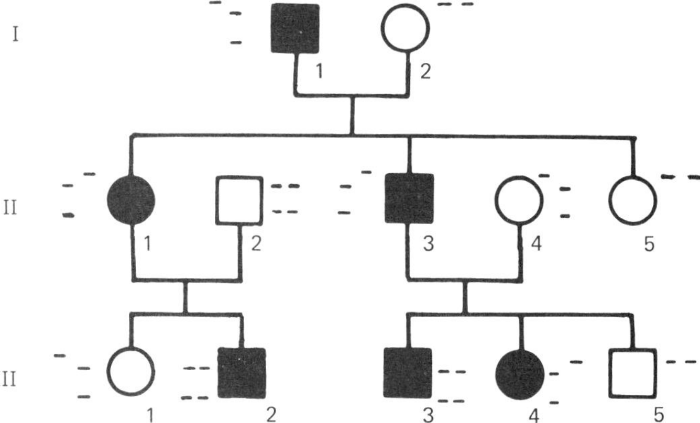

Figure 1.9 Pedigree of a family with an autosomal dominant disorder illustrating linkage to an RFLP. I.2 is homozygous for an anonymous DNA fragment 8.3 kb in length (indicated by two horizontal lines above pedigree symbol). I.1 has the 8.3 kb fragment on one chromosome, but polymorphism has led to an extra restriction site on the other chromosome, giving two fragments of 6 (upper line) and 2.3 (lower line) kb representing this sequence. I.1 has transmitted the disease and the sequence yielding the 6 and 2.3 kb fragments to II.1 and II.3, whereas II.5 is unaffected and has inherited the 8.3 kb fragment from I.1. The disease gene can also be tracked in generation III using the 6/2.3 kb fragment pattern which segregates with the disease

centromeric region of chromosome 19 using RFLPs (*see* Harper, Chapter 13 of this volume). Linkage between an RFLP and an autosomal dominant disorder is illustrated in *Figure 1.9*.

The Duffy blood group locus was known to be on chromosome 1 when linkage between it and the HMSN type I locus was demonstrated; further attempts to locate the gene accurately are in progress using DNA markers such as the sequence encoding antithrombin III which is also on chromosome 1 near the Duffy locus. The study of HMSN I illustrates a further use of linkage analysis in that it can be used to investigate the possibility of genetic heterogeneity. The majority of families with dominantly inherited HMSN I do not show linkage to Duffy, suggesting that the disorder is caused by more than one gene from different chromosomal regions (Bird *et al.*, 1983; Dyck *et al.*, 1983; Harding, unpublished data). In addition, DNA probes have been used to localize the gene causing X-linked HMSN to part of the long arm of the X chromosome (Fischbeck *et al.*, 1986).

Linkage studies using RFLPs have also shown that dominantly inherited manic depressive psychosis is genetically heterogeneous. In a large Amish pedigree the defective gene was localized to the short arm of chromosome 11 (Egeland *et al.*, 1987), but linkage to markers in this region was excluded in Icelandic and North American families (Detera-Wadleigh *et al.*, 1987; Hodgkinson *et al.*, 1987). This is not the case in Huntington's disease (*see* Gilliam and Gusella, Chapter 12 of this volume), where linkage studies in more than 20 families indicate that even if the mutation itself is heterogeneous, the disease locus is always on the short arm of chromosome 4.

The likelihood of linkage between two genetic loci is assessed statistically using lod scores (Ott, 1986). Lod stands for the 'log of the odds'; linkage between two loci at a given distance is usually considered to be proven or excluded if the lod score is greater than +3 or −2 respectively. A lod score of 3 means that the odds in favour of linkage are 1000 to 1. Lod scores are calculated for a number of theoretical genetic distances between the marker and disease loci. If two loci are not linked, one would expect that, on average, they would be transmitted together or separately in equal proportions. The 'recombination fraction' would therefore be 0.5. If a lod score is highest at a recombination fraction (θ) of 0.5, the likelihood of the two loci being linked is low. If, however, the maximum lod score is 3 or greater at, for example, $\theta = 0.05$, this indicates that the two loci are linked and that recombination will occur, on average, during 1 out of 20 meioses.

The distance between two loci is defined genetically in terms of potential recombination, and genetic distance is measured in centimorgans (cM). Crossing-over between two loci 1 cM apart will occur in 1% of meioses. Genetic distance is not directly comparable to physical distance, as some parts of the genome appear to be more prone to recombination than others, but a distance of 1 cM is roughly equal to 1 million bp, or 1000 kilobases (kb; 1 kb = 1000 bp).

In autosomal dominant disorders it is highly desirable to study large multigeneration families, in order to track the defective gene by analysing the maximum number of potentially informative meioses. It is possible to combine data from more than one family for the purposes of calculating lod scores, but this approach assumes that there is no question of genetic heterogeneity and this is rarely the case in neurological disorders.

A further vital requirement for gene tracking is that key individuals in the pedigree are informative for the marker studied. For example, in *Figure 1.9*, if I.1 was homozygous for the 8.3 kb fragment, it would be impossible to assess which fragment was segregating with the disease gene. Thus the best genetic markers for linkage studies are highly polymorphic, that is, the gene frequency for each allele approaches 50%, maximizing the chances of any family member being heterozygous and therefore informative.

It has been estimated that only about 200 polymorphic DNA markers, evenly spaced 10–20 million bp apart, are needed to map the entire human genome, theoretically allowing chromosomal localization of the majority of mendelian disorders (Botstein *et al.*, 1980). This number is almost certainly an underestimate, and the hypothesis presupposes that there are suitable large families available for study.

Candidate genes

Given the existence of appropriate families, how does one start to track the mutant gene? In some neurological disorders, linkage studies can be initiated using 'candidate' chromosomal regions or genes. Dystonia can be produced experimentally by injection of the N-terminal fragments of adrenocorticotropin which is derived from the pro-opiomelanocortin (POMC) peptide precursor. The POMC gene was studied as a candidate gene for autosomal dominant torsion dystonia, with negative results (Kramer *et al.*, 1985). Even more striking 'candidates' are not always the correct ones; for example the Wilson's disease gene is on a different chromosome from that encoding caeruloplasmin (*see* Chapter 18).

The long arm of chromosome 21 has been suggested as a candidate region for the familial Alzheimer's disease gene because of the association with Down's syndrome (trisomy 21), and linkage studies using RFLPs mapping to 21q suggest that this hypothesis is correct (St George-Hyslop *et al.*, 1987). This is of particular interest in view of the observation that the gene encoding the amyloid plaque core protein precursor maps to the same region (Tanzi *et al.*, 1987a). However, establishing linkage between a candidate gene and a disease locus does not always imply that the mutation directly involves the candidate. Subsequent linkage studies in familial Alzheimer's disease using probes detecting RFLPs in the amyloid protein gene sequence itself have shown a number of crossovers, indicating that this is not the site of the primary disease mutation (van Broeckhoven *et al.*, 1987; Tanzi *et al.*, 1987b).

Similarly, Seizinger *et al.* (1987) have shown that the von Recklinghausen's neurofibromatosis (VRNF) locus is linked to that for the nerve growth factor receptor (NGFR) gene on chromosome 17. The observation of several recombinant events between the two loci in this study suggests that NGFR is not the disease gene, although it is possible that mutation of a related gene located close to the NGFR locus could affect its function.

The observations of Seizinger, Martuza and Gusella (1986), that there is loss of chromosome 22 in acoustic neuromas, provided a candidate chromosomal region for the gene giving rise to bilateral acoustic neurofibromatosis (Rouleau *et al.*, 1987). Candidate chromosomal regions may also be suggested by the occurrence of an autosomal or X-linked disease in a patient with a chromosomal anomaly such as a deletion or translocation. Such patients have been of crucial importance in the hunt for the Duchenne muscular dystrophy gene (*see* Harper, Chapter 13 of this volume).

If there is no clue whatsoever as to the chromosomal localization of a mendelian disorder, it is possible to look for linkage between the disease locus and a randomly selected series of genetic markers. In this way, Gusella *et al.* (1983) were able to map the Huntington's disease gene to the short arm of chromosome 4 using anonymous DNA probes from a human gene library. They were fortunate to find the right marker amongst the first 12 probes tested! This approach is laborious, expensive, and not for the fainthearted. Nevertheless, it has also been productive in localizing the gene for VRNF to the pericentromeric region of chromosome 17 (Barker *et al.*, 1987; Seizinger *et al.*, 1987), and a number of groups have used random linkage analysis to exclude various regions of the genome from containing some genes causing neurological disease. For example, the Friedreich's ataxia locus has been excluded from about 50% of the human genome using linkage analysis with RFLPs and protein markers (Chamberlain, personal communication). It should be mentioned in this context that it is more difficult to establish linkage in autosomal recessive diseases than in dominant ones because of the limitations imposed by family structure.

RFLP mapping

Another issue which needs to be resolved either before or after establishing linkage to an RFLP, is how anonymous RFLPs from gene libraries are mapped to specific chromosomal regions. Chromosomal localization is not necessary for gene tracking and its practical application in genetic counselling or prenatal diagnosis; linkage between the myotonic dystrophy and secretor loci could be used in clinical practice before it was known that these genes are on chromosome 19 (*see* Harper, Chapter 13 of this volume).

RFLPs and other markers can be mapped by a number of methods. The commonest uses somatic cell hybrids containing both human and, for example, mouse chromosomes, in which the human chromosomal content is randomly lost. The anonymous DNA sequence can be hybridized to a fixed chromosomal preparation from such cells on glass slides (*in situ* hybridization) and mapped to a chromosome by a proces of elimination, assuming that the human chromosomal content of the hybrids is known (Buckle and Craig, 1986). For regional localization within a particular chromosome, the DNA markers can be hybridized to sequences from cell lines carrying deletions of that chromosome, and therefore mapped within or outside the deleted region. Alternatively, linkage studies between the 'new' marker and other loci known to be on the relevant chromosome can be performed in large reference families, in order to establish the genetic distances between them. The first two of these techniques were used by Gusella's group in mapping the G8 marker to the short arm of chromosome 4 (Gilliam and Gusella, Chapter 12 of this volume).

WHAT NEXT AFTER GENETIC LINKAGE?

One of the ultimate aims of genetic linkage studies is to identify the defective gene itself, and its abnormal product, in order to gain understanding of disease pathophysiology and possibly develop a rational approach to therapy. Progressing from the identification of a linked genetic marker to isolating the mutant gene is a formidable task. Even closely linked RFLPs are usually more than a million bp away from the disease gene.

One way of trying to identify the gene is by chromosome 'walking' (or as it is sometimes more appropriately called, 'crawling'). This involves generating overlapping small clones in the direction of the gene locus, but this is extremely laborious as it is difficult to take steps of more than 20–30 kb at a time. The introduction of pulsed field gradient electrophoresis is promising, as this technique allows resolution of DNA fragments up to 2000 kb in length. These are produced by using restriction enzymes which cleave human DNA very infrequently. This makes it possible to construct maps of large adjacent regions of the genome (van Ommen and Verkerk, 1986).

One question which often arises is how to recognize the gene once the relevant sequence has been cloned. The approach taken by Monaco *et al.* (1986) in the investigation of Duchenne muscular dystrophy (DMD) is promising in this respect (*see also* Harper, Chapter 13 of this volume). During molecular analysis of 57 boys with DMD who had deletions of the short arm of the X chromosome, Kunkel *et al.* (1986) cloned more than 200 kb of DNA from the vicinity of the DMD gene. Monaco *et al.* (1986) used these clones to search for regions which were transcribed by hybridizing them with DNA from other species, including mouse, hamster, cow, and chicken, on the basis that expressed exons tend to show evolutionary conservation. Two probes hybridized to DNA from all mammals tested, and one of these to a 16 kb transcript present in fetal muscle RNA. This transcript appears to be derived from eight exons; the complete mRNA might correspond to a gene covering one to two million bp. This elegant series of experiments illustrates the potentially complex molecular basis of at least some inherited neurological diseases, but there is no reason why a similar approach could not be taken in genetic disorders affecting the central nervous system such as Huntington's disease and the hereditary ataxias.

Once a gene has been cloned, its sequence can be determined by a number of different methods, the most frequently used being the dideoxy method of Sanger (Sanger, Nicklen and Coulson, 1977; Davies, 1982). This involves using single-stranded DNA as a template to synthesize a complementary strand, incorporating dideoxynucleotides in place of normal nucleotides. The former lack the 3′ hydroxy groups and chain synthesis is terminated when they are incorporated. Four reactions containing template, DNA polymerase, the four normal nucleotides (one of which is radioactively labelled), and one of the four dideoxynucleotides, are set up and then each subjected to gel electrophoresis and autoradiography. The chains in all four reactions will end with one of the chain terminating dideoxynucleotides, so the sequence of the DNA can be read directly from the autoradiograph.

Using this technique it has been possible to analyse large sequences of DNA such as mitochondrial DNA (16.5 kb), and the Epstein Barr virus genome which is 170 kb in length (Emery, 1984). The prospect of systematically sequencing the entire human genome is actively being explored at present (Newmark, 1986).

THE CLINICAL APPLICATION OF LINKED DNA MARKERS AND GENE SPECIFIC PROBES

The identification of a genetic marker linked to a disease locus has important clinical applications in presymptomatic and prenatal detection of gene carriers. Once linkage has been established, it is often relatively easy to generate more closely linked DNA markers, thus enhancing the accuracy of gene tracking which is so important in genetic counselling. The ethical and practical problems of using linked genetic markers in a late onset autosomal dominant disorder such as Huntington's disease are mentioned by Gilliam and Gusella (Chapter 12 of this volume), and have been reviewed by Harper and Sarfarazi (1985) and Crauford and Harris (1986). The clinical application of linkage studies in the muscular dystrophies is described by Harper in Chapter 13 of this volume.

Prenatal diagnosis

In dominant, recessive, and X-linked diseases the availability of chorionic biopsy allows the early detection of an affected fetus, the degree of accuracy depending on the genetic distance between the marker and the disease loci. Prenatal diagnosis of a number of enzyme defects using amniotic cells has been available for several years, but this involves a second trimester abortion if the fetus is abnormal.

On the other hand, chorionic villi, which are derived from the trophoblastic layer of the developing blastocyst, can be sampled between 8 and 12 weeks' gestation. A flexible cannula is inserted through the cervix into the uterus under ultrasonic control. A small amount of chorionic tissue is aspirated and this is separated from any contaminating maternal tissue using a dissecting microscope. The size of the biopsy varies considerably, but the average yield of DNA is about 35 μg, which is ample for DNA analysis using Southern blotting (Old, 1986).

The use of this technique means that the parents of an affected fetus can opt for a first-trimester abortion, which is physically and psychologically less hazardous to the mother. This advantage probably outweighs the currently higher risk of chorionic villus sampling (CVS) as compared to amniocentesis. The risk of CVS inducing miscarriage is about 5%, and that of amniocentesis 1%, but it is important to take the natural incidence of spontaneous abortion during the first trimester into account when comparing these figures (Old, 1986).

The diagnosis of genetic diseases using linked RFLPs is obviously an indirect method and has a number of disadvantages, not the least of which is the inaccuracy generated by the possibility of recombination during meiosis. It also necessitates studying several family members in order to establish how the genetic marker is segregating with the disease locus, and these individuals are not always available for study, particularly in a late-onset life-limiting disorder such as Huntington's disease (Harper and Sarfarazi, 1985).

Gene specific probes

In some diseases, direct detection of a gene defect is possible using DNA analysis. In the Portuguese type of familial amyloid polyneuropathy (FAP), an autosomal dominant disorder, the amyloid deposits are derived from an abnormal prealbumin, or transthyretin (TTR), in which there is a substitution of methionine

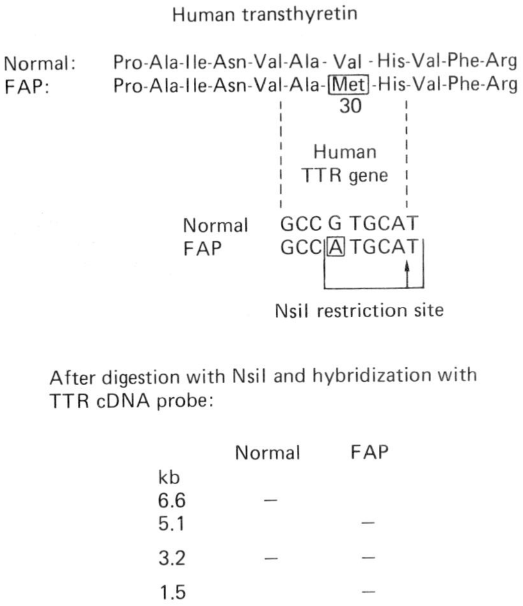

Figure 1.10 The abnormal transthyretin (TTR) found in patients with familial amyloid polyneuropathy contains a substitution of methionine for valine at position 30. This corresponds to a single base change of A for G in the TTR gene, changing the codon GUG to AUG in mRNA. The base change results in a new restriction site for the restriction endonuclease NsiI which cleaves the normal 6.6 kb restriction fragment into two, 5.1 and 1.5 kb in length. (From Harding (1988), by courtesy of the publisher)

for valine at position 30 (Saraiva *et al.*, 1984). This corresponds to a single base change in the gene for TTR, changing the codon GUG to AUG. The nucleotide substitution results in new restriction sites for the restriction endonucleases Nsi I and Bal I which can be detected using a cDNA probe for the TTR gene (*Figure 1.10*), making it possible to diagnose FAP presymptomatically and prenatally (Ide *et al.*, 1986). These observations illustrate the use of a gene specific probe.

Gene specific probes are also available for the hypoxanthine guanine phosphoribosyl transferase locus on the X chromosome. Deficiencies of this enzyme, which may be produced by at least 16 different mutations, give rise to the Lesch–Nyhan syndrome (Wilson *et al.*, 1986). The use of gene specific probes does not necessarily involve detection of a mutation by a specific restriction endonuclease; some types of β-thalassaemia caused by deletions of part of the β-globin gene can be detected by hybridization with a DNA probe for the normal gene sequence resulting in restriction fragments of different lengths compared to the normal pattern (Old, 1986).

It is probable that many genetic disorders are caused by point mutations or small deletions which do not give rise to loss of restriction endonuclease recognition sites, or enough loss of genetic material to detect the deletion using the type of DNA probe (usually > 5 kb in length) employed in Southern blotting analysis. If the exact sequences of the normal and mutant genes are known, it is possible to synthesize complementary oligonucleotides approximately 20 bases in length. If these are labelled and exposed to human DNA samples under very strict conditions, only the perfectly matched sequences will hybridize, as mismatched hybrids exhibit less thermal stability than perfectly matched ones. These oligonucleotide probes have been used in the diagnosis of some β-thalassaemias and α-1 antitrypsin deficiency (Thein and Wallace, 1986).

GENE THERAPY

Recombinant DNA technology is already used in the manufacture of therapeutic products, such as opioid peptides, which have relevance to clinical neurology (Thompson, 1984). Another rather indirect example is the availability of genetically engineered human growth hormone (GH) which has assumed great importance since it was recognized that the administration of GH from pooled postmortem pituitary tissue carried a risk of transmitting Creutzfeldt–Jakob disease to the deficient recipients (Powell-Jackson *et al.*, 1985).

The prospect of replacing defective genes in patients is futuristic but already theoretically possible (Emery, 1984). There are numerous technical problems, but it is likely that this approach to treatment will become available in the next decade or so, albeit in a limited number of diseases. The main difficulties arise not from cloning the gene, but introducing it into the patient in such a way that it will be expressed in the appropriate tissue.

Foreign genes could be transferred to a human host by viral vectors, such as the retroviruses, but there would be no guarantee that the foreign gene would not be incorporated into other important genes and this could clearly be hazardous to the recipient. If DNA sequences are microinjected into germ cells (germ cell gene therapy), there is also random incorporation of them into the genome, and the genes may not be expressed in the target tissue unless accompanied by a tissue specific promotor. Nevertheless, Palmiter *et al.* (1982) produced transgenic 'supermice' by microinjecting the rat growth hormone gene into mouse ova soon after fertilization (*see* Messing, Chapter 9 of this volume).

It is technically possible to apply the same techniques to somatic cells (somatic cell gene therapy). For example, bone marrow stems cells from patients with haemoglobinopathies could be microinjected with globin genes and replaced, with the hope that the transformed cells would multiply, express the gene, and eventually replace the mutant cells. There are obvious problems in using this approach in neurological disorders. However, in rare instances, such as the Lesch–Nyhan syndrome, neurological dysfunction results from a generalized metabolic defect but is not associated with gross structural abnormalities in the brain. In these circumstances replacing the deficient enzyme might be beneficial. The hypoxanthine-guanine phosphoribosyl transferase gene, linked to retroviral or promotor sequences, has been transferred to mouse bone marrow stem cells with subsequent expression of the gene (Stein and Morrison, 1985). It remains to be seen whether this type of approach will be successful in patients with the Lesch–Nyhan syndrome, or any other neurological disease.

Comings (1980) predicted 7 years ago that recombinant DNA techniques would have a major impact on clinical medicine and basic research, referring to their application as 'the new genetics'. The new genetics have entered the arenas of clinical neurology and neurobiology and are providing a precise means of investigating inherited neurological diseases. More generally, these techniques are already enhancing our understanding of the development and function of the normal human nervous system.

References

ALBERTS, B., BRAY, D., LEWIS, J., RAFF, M., ROBERTS, K. and WATSON, J. D. (1983) *Molecular Biology of the Cell.* New York: Garland

BARKER, D., WRIGHT, E., NGUYEN, K., CANNON, L., FAIN, P., GOLDCAR, D. *et al.* (1987) Gene for von Recklinghausen neurofibromatosis is on the pericentromeric region of chromosome 17. *Science,* **236,** 1100–1102

BASKIN, F., ROSENBERG, R. N. and DEV, V. (1981) Correlation of double-minute chromosomes with unstable multidrug cross-resistance in uptake of neuroblastoma cells. *Proceedings of the National Academy of Sciences USA,* **78,** 3654–3658

BEAUDET, A. L., SU, T.-S., O'BRIEN, W. E., D'EUSTACHI, O. P., BARKER, P. E. and RUDDLE, F. H. (1982) Dispersion of argininosuccinate-synthetase-like human genes to multiple autosomes and the X chromosome. *Cell,* **30,** 287–293

BIRD, T. D., OTT, J. and GIBLETT, E. R. (1982) Evidence for linkage of Charcot–Marie–Tooth neuropathy to the Duffy locus on chromosome 1. *American Journal of Human Genetics,* **34,** 388–394

BIRD, T. D., OTT, J., GIBLETT, E. R., CHANCE, P. F., SUMI, S. M. and KRAFT, G. H. (1983) Genetic linkage evidence for genetic heterogeneity in Charcot–Marie–Tooth neuropathy (HMSN type I). *Annals of Neurology,* **14,** 679–684

BOTSTEIN, D., WHITE, R. L., SKOLNICK, M. and DAVIS, R. W. (1980) Construction of a genetic linkage map in man using restriction fragment length polymorphisms. *American Journal of Human Genetics,* **32,** 314–331

BUCKLE, V. J. and CRAIG, I. W. (1986) *In situ* hybridization. In *Human Genetic Diseases: A Practical Approach,* edited by K. E. Davies, pp. 85–100. Oxford: IRL Press

COMINGS, D. (1980) Editorial. *American Journal of Human Genetics,* **32,** 453

CRAUFORD, D. I. O. and HARRIS, R. (1986) Ethics of predictive testing for Huntington's chorea: the need for more information. *British Medical Journal,* **293,** 249–251

DAVIES, R. W. (1982) DNA sequencing. In: *Gel Electrophoresis of Nucleic Acids: A Practical Approach,* edited by D. Rickwood and B. Haines, pp. 117–172. Oxford: IRL Press

DAVIS, L. G., DIBNER, M. D. and BATTEY, J. F. (1986) *Basic Methods in Molecular Biology.* Amsterdam: Elsevier

DETERA-WADLEIGH, S. D., BERRETTINI, W. H., GOLDIN, L. R., BOORMAN, D., ANDERSON, S. and GERSHON, E. S. (1987) Close linkage of c-Harvey-*ras*-1 and the insulin gene to affective disorder is ruled out in three North American pedigrees. *Nature,* **325,** 806–808

DYCK, P. J., OTT, J., MOORE, S. B., SWANSON, C. J. and LAMBERT, E. H. (1983) Linkage evidence for genetic heterogeneity among kinships with hereditary motor and sensory neuropathy, type 1. *Mayo Clinic Proceedings,* **58,** 430–435

EGELAND, J. A., GARHARD, D. S., PAULS, D. L., SUSSEX, J. N., KIDD, K. K., ALLEN, C. R. *et al.* (1987) Bipolar affective disorders linked to DNA markers on chromosome 11. *Nature,* **325,** 783–787

EMERY, A. E. H. (1984) *An Introduction to Recombinant DNA.* Chichester: John Wiley

FEINBERG, A. P. and VOGELSTEIN, B. (1983) A technique for radiolabelling DNA restriction endonuclease fragments to high specific activity. *Analytical Biochemistry,* **132,** 6–13

FISCHBECK, K. H., AR-RUSHDI, N., PERICAK-VANCE, M., ROZEAR, M., ROSES, A. D. and FRYNS, J. P. (1986) X-linked neuropathy: gene localization with DNA probes. *Annals of Neurology,* **20,** 527–532

FRYER, A. E., CONNOR, J. M., POVEY, S., YATES, J. R. W., CHALMERS, A., FRASER, I. *et al.* (1987) Evidence that the gene for tuberous sclerosis is on chromosome 9. *Lancet,* **i,** 659–670

GLOVER, D. M. (1985) (ed) *DNA Cloning: A Practical Approach,* Vol. 1. Oxford: IRL Press

GUSELLA, J. F., WEXLER, N. S., CONNEALLY, P. M., NAYLOR, S. L., ANDERSON, M. A., TANZI, R. E. *et al.* (1983) A polymorphic DNA marker genetically linked to Huntington's disease. *Nature,* **306,** 234–238

HARDING, A. E. (1988) Molecular genetics and neurological disease. In: *Recent Advances in Clinical Neurology 5,* edited by C. Kennard. Edinburgh: Churchill Livingstone

HARPER, P. S. (1981) *Practical Genetic Counselling.* Bristol: John Wright

HARFER, P. S. and SARFARAZI, M. (1985) Genetic prediction and family structure in Huntington's chorea. *British Medical Journal,* **290,** 1929–1931

HARPER, P. S., RIVAS, M. L., BIAS, W. B., HUTCHINSON, J. R., DYKEN, P. R. and McKUSICK, V. A. (1972) Genetic linkage confirmed between the locus for myotonic dystrophy and the ABH-secretion and Lutheran blood groups. *American Journal of Human Genetics,* **55,** 213–224

HARRIS, H. (1966) Enzyme polymorphism in man. *Proceedings of the Royal Society of London,* **164,** 298–310

HODGKINSON, S., SHERRINGTON, R., GURLING, H., MARCHBANKS, R., REEDERS, S., MALLET, J. *et al.* (1987) Molecular genetic evidence for heterogeneity in manic depression. *Nature,* **325,** 805–806

IDE, M., MITA, S., IKEGAWA, S., MAEDA, S., SHIMADA, K. and ARAKI, S. (1986) Identification of carriers of mutant prealbumin gene associated with familial amyloidotic polyneuropathy type I by Southern blot procedures: study of six pedigrees in the Arao district of Japan. *Human Genetics,* **73,** 281–285

KAN, Y. W., GOLBUS, M. S. and DOZY, A. M. (1976) Prenatal diagnosis of α-thalassemia. Clinical application of molecular hybridization. *New England Journal of Medicine,* **295,** 1165–1167

KRAMER, P., BREAKEFIELD, X., BRESSMAN, S., OZELIUS, L., MOSKOWTZ, C., TANZI, R. E. *et al.* (1985) Linkage analysis in family with dominantly inherited torsion dystonia: exclusion of proopiomelanocortin gene and regions of chromosomes 4 and 21. *American Journal of Human Genetics,* **37,** A163

KUNKEL, L. M., HEJTMANCIK, J. F., CASKEY, C. T., SPEER, A., MONACO, A. P., MIDDLESWORTH, W. *et al.* (1986) Analysis of deletions in DNA from patients with Becker and Duchenne muscular dystrophy. *Nature,* **322,** 73–77

LEBO, R. V., GORIN, F., FLETTERICK, R. J., KAO, F. T., CHEUNG, M. C., BRUCE, B. D. *et al.* (1984) High resolution chromosome sorting and DNA spot-blot analysis assign McArdle's syndrome to chromosome 11. *Science,* **225,** 57–59

McKUSICK, V. A. (1986) *Mendelian Inheritance in Man: Catalogs of Autosomal Dominant, Autosomal Recessive, and X-linked Phenotypes.* 7th edition, Baltimore: Johns Hopkins University Press

MONACO, A. P., NEVE, R. L., COLLETTI-FEENER, C., BERTELSON, C. J., KURNIT, D. M. and KUNKEL, L. M. (1986) Isolation of candidate cDNAs for portions of the Duchenne muscular dystrophy gene. *Nature,* **323,** 646–650

NEWMARK, P. (1986) Is megasequencing madness? *Nature,* **323,** 291

OLD, J. M. (1986) Fetal DNA analysis. In *Human Genetic Diseases: A Practical Approach,* edited by K. E. Davies, pp. 1–17. Oxford: IRL Press

OTT, J. (1986) A short guide to linkage analysis. In *Human Genetic Diseases: A Practical Approach,* edited by K. E. Davies, pp. 19–32. Oxford: IRL Press

PALMITER, R. D., BRINSTER, R. L., HAMMER, R. E., TRUMBAUER, M. E., ROSENFELD, M. G., BIRNBERG, N. C. *et al.* (1982) Dramatic growth of mice that develop from eggs microinjected with metallothionein-growth hormone fusion genes. *Nature,* **300,** 611–615

POWELL-JACKSON, J., KENNEDY, P., WHITCOMBE, E. M., WELLER, R. O., PREECE, M. A. and NEWSOM-DAVIS, J. (1985) Creutzfeldt–Jacob disease after administration of human growth hormone. *Lancet,* **ii,** 244–246

RIGBY, P. W. J., DIECKMANN, M., RHODES, C. and BERG, P. (1984) Labelling deoxyribonucleic acid to high specific activity *in vitro* by nick translation with DNA polymerase 1. *Journal of Molecular Biology,* **113,** 237–251

ROSENBERG, R. N. (1983) Recombinant DNA and neurological disease: the coming of a new age. *Neurology,* **33,** 622–625

ROULEAU, G. A., WERTELECKI, W., HAINES, J. L., HOBBS, W. J., TROFATTER, J. A., SIEZINGER, B. R. *et al.* (1987) Genetic linkage of bilateral acoustic neurofibromatosis to a DNA marker on chromosome 22. *Nature,* **329,** 246–248

ST GEORGE-HYSLOP, P. H., TANZI, R. E., POLINSKY, R. J., HAINES, J. L., NEE, L., WATKINS, P. C. *et al.* (1987) The genetic defect causing familial Alzheimer's disease maps on chromosome 21. *Science,* **235,** 885–890

SANGER, F., NICKLEN, S. and COULSON, A. R. (1977) DNA sequencing with chain-terminating inhibitors. *Proceedings of the National Academy of Sciences USA,* **74,** 5463–5467

SARAIVA, M. J. M., BIRKEN, S., COSTA, P. P. and GOODMAN, D. S. (1984) Amyloid fibril protein in familial amyloidotic polyneuropathy, Portuguese type: definition of molecular abnormality in transthyretin (prealbumin). *Journal of Clinical Investigation,* **74,** 104–119

SCHIMKE, R. T., KAUFMAN, R. D., ALT, F. W. and KELLENS, R. F. (1978) Gene amplification and drug resistance in cultured murine cells. *Science,* **202,** 1051–1055

SEIZINGER, B., MARTUZA, R. L. and GUSELLA, J. F. (1986) Loss of genes on chromosome 22 in tumorigenesis of human acoustic neuroma. *Nature,* **322,** 644–647

SEIZINGER, B. R., ROULEAU, G. A., OZELIUS, L. J., LANE, A. H., FARYNIARZ, A. G., CHAO, M. V. *et al.* (1987) Genetic linkage of von Recklinghausen neurofibromatosis to the nerve growth factor receptor gene. *Cell,* **49,** 589–594

SOUTHERN, E. M. (1975) Detection of specific sequences among DNA fragments separated by gel electrophoresis. *Journal of Molecular Biology,* **98,** 503–517

STANBURY, J. B., WYNGAARDEN, J. B., FREDRICKSON, D. S., GOLDSTEIN, J. L. and BROWN, M. S. (1983) (eds) *The Metabolic Basis of Inherited Disease,* 5th edn. New York: McGraw-Hill

STEIN, S. A. and MORRISON, M. R. (1985) The molecular biology of Lesch–Nyhan syndrome. *Trends in Neurosciences,* **8,** 148–150

TANZI, R. E., GUSELLA, J. F., WATKINS, P. C., BRUNS, G. A. P., ST GEORGE-HYSLOP, P., VAN KEUREN, M. L. *et al.* (1987a) Amyloid β-protein gene: cDNA, mRNA distribution, and genetic linkage near the Alzheimer locus. *Science,* **235,** 880–884

TANZI, R. E., ST GEORGE-HYSLOP, P. H., HAINES, J. L., POLINSKY, R. J., NEE, L., FONCIN, J.-F. *et al.* (1987b) The genetic defect in familial Alzheimer's disease is not tightly linked to the amyloid β-protein gene. *Nature,* **329,** 156–157

THEIN, S. L. and WALLACE, R. B. (1986) The use of synthetic oligonucleotides as specific hybridization probes in the diagnosis of genetic disorders. In: *Human Genetic Diseases: A Practical Approach,* edited by K. E. Davies, pp. 33–50. Oxford: IRL Press

THOMPSON, J. W. (1984) Opioid peptides. *British Medical Journal,* **288,** 259–261

VAN BROECKHOVEN, C., GENTHE, A. M., VANDENBERGHE, A., HORSTHEMKE, B., BACKHOVENS, H., RAEYMAEKERS, P., VAN HUL, W. *et al.* (1987) Failure of familial Alzheimer's disease to segregate with the A4-amyloid gene in several European families. *Nature,* **329,** 153–155

VAN OMMEN, G. J. B. and VERKERK, J. M. H. (1986) Restriction analysis of chromosomal DNA in a size range up to two million base pairs by pulsed field gradient electrophoresis. In: *Human Genetic Diseases: A Practical Approach,* edited by K. E. Davies, pp. 113–133. Oxford: IRL Press

WATSON, J. D. and CRICK, F. H. C. (1953) Molecular structure of nucleic acids – a structure for deoxyribose nucleic acid. *Nature,* **171,** 737–738

WEINSHILBOUM, R. M. (1983) Biochemical genetics of catecholamines in humans. *Mayo Clinic Proceedings,* **58,** 319–330

WILSON, J. M., STOUT, J. T., PALELLA, T. D., DAVIDSON, B. L., KELLEY, W. N. and CASKEY, C. T. (1986) A molecular survey of hypoxanthine–guanine phosphoribosyl transferase deficiency in man. *Journal of Clinical Investigation,* **77,** 188–195

2
Genes expressed in the brain: evolutionary and developmental considerations

William E. Hahn and Gregory P. Owens

INTRODUCTION

In this chapter several aspects of genetic expression in the brain are discussed, mostly in broad interpretative terms rather than in factual and descriptive detail. The first of the more general topics we address is the extent to which the genome is expressed in the brain.

THE COMPLEXITY OF GENE EXPRESSION IN THE BRAIN

It is evident from measurements of the sequence complexity of messenger RNA (mRNA) that a substantial portion of genetic information in mammals, as well as invertebrate animals, is apparently required for development and function of the brain (Bantle and Hahn, 1976; Chikaraishi, 1979; Van Ness, Maxwell and Hahn, 1979; Kaplan, 1986; Hahn *et al.*, 1986). Many of the genes expressed in the brain are expressed in a variety of other organs (Hahn and Chaudhari, 1984) but quantitative differences in expression of many of these 'shared' genes are evident (Milner and Sutcliffe, 1983). In other words, the relative abundance of a given messenger RNA species can differ markedly among various tissues and organs.

Of greater interest regarding functions unique to the brain are measurements that indicate the presence of a wide variety of putatively brain specific mRNAs (Milner and Sutcliffe, 1983; Hahn and Chaudhari, 1984). Presumably these mRNAs encode for proteins that have presently evolved such that they are of specific adaptive value in the development and function of the brain. The suggestion that the expression of many genes might be restricted to the brain first came from comparative measurements on the sequence complexity of nuclear RNAs obtained from various mammalian organs (Hahn and Laird, 1971; Grouse, Chilton and McCarthy, 1972). While these initial measurements were underestimates of the linear sequence complexity of the RNA in question, they nonetheless showed that more of the genome is expressed in brain than in other complex organs such as the liver and kidney. Subsequent investigations showed that the greater sequence complexity of brain nuclear RNA was also reflected in the diversity of the mRNA population (Bantle and Hahn, 1976). The complexity of polysomal RNA from mouse and rat brain, as measured by saturation hybridization of single copy genomic DNA, is in the range of 2.3–2.9 $\times$ 10^8 nucleotides, of which 1.1–1.8 $\times$ 10^8 nucleotides are attributable to the polyadenylated fraction (Van Ness, Maxwell and Hahn, 1979; Chikaraishi, 1979). These values are regarded as estimates, as the determination of sequence complexity, either from saturation hybridization of single copy genomic sequences or from hybridization kinetics of copy DNA (cDNA) transcribed from mRNA, is not precise (Van Ness and Hahn, 1980; 1982). However, certain refinements in technique and the fact that there is fair consistency between different investigators strengthens the conclusion that these complexity values are fairly reliable estimates (Kaplan, 1986).

Conversion of the linear sequence complexity of RNA into the number of different mRNA species is an uncertain step owing to the fact that mRNAs are highly heterogeneous in length. The average size of messenger RNA as determined by electrophoretic mobility and sedimentation in density gradients is around 1500–2000 nucleotides (Bantle and Hahn, 1976; Meyuhas and Perry, 1979). Most of the sequence complexity is contained in the fraction of mRNA that comprises the rare or infrequent copy class, and these species make up less than half of the mass of total mRNA (Young, Birnie and Paul, 1976; Hahn, Van Ness and Chaudhari, 1982). Some experiments with cDNA probes indicate that less abundant mRNAs are on average longer than abundant species (Meyuhas and Perry, 1979; Milner and Sutcliffe, 1983); mRNAs of 500 to nearly 10 000 nucleotides have been observed in the brain (Rutishauser and Goridis, 1986; Adelman *et al.*, 1987; Owens and Hahn, unpublished data). It is also possible that a wide variety of the mRNA species that are restricted to the brain may, on average, be larger than mRNAs found in other organs. Perhaps many of the genes necessary for brain development and behaviour capabilities encode large polyproteins (Sutcliffe and Milner, 1983). Whatever the case, the number of individual messenger species can be roughly estimated by dividing the total sequence complexity of polysomal RNA by the number of nucleotides in mRNA molecules of average size. Assuming an average size of around 5000 nucleotides for mRNA molecules the number of species is around 50 000 for total polysomal RNA and about 30 000 for the polyadenylated fraction of polysomal RNA (Hahn and Chaudhari, 1984; Milner, 1986). Additional molecular variety can result from heterogeneity of certain mRNAs in which the same basic coding sequence is present but different 5′ untranslated sequence and use of alternative polyadenylation signals result in a polymorphic family of mRNAs specified by the same gene (Sutcliffe, McKinnon and Tsau, 1986).

The correspondence between linear sequence complexity of mRNA and proteins is conjectural. In many instances the protein coding region of mRNA molecules has been found to be considerably shorter than the 3′ untranslated sequence (Kuwano *et al.*, 1984; Milner *et al.*, 1985). The 3′ untranslated region may be specified by both single copy as well as repetitious sequences in the genome, and thus comprise a considerable portion of the total sequence complexity. Hence the complexity of the code sequence of brain mRNA is unclear. This aspect would not decrease the predicted number of proteins as based upon an estimated number of different mRNAs, but would simply reduce the average size (linear amino acid sequence complexity) of the protein molecules.

It should be noted that the number of protein species detectable by two-dimensional electrophoresis in cultured neural cells is much less than predicted on the basis of the complexity of the mRNA populations (Schubert, Brass and

Dumas, 1986). Other points to consider for perspective on the mRNA population in relationship to the predicted number of protein species include differential processing of primary transcripts and alternate pathways of post-translational processing of proteins. These processes increase the diversity of functional peptides and polypeptides beyond that suggested simply from linear sequence complexity of mRNAs (Rosenfeld, Amara and Evans, 1984; *see below*).

The approximate number of gene products that are restricted to brain is not clearly established, although hybridization experiments with poly(A)$^+$ mRNA from other complex organs suggest half or more are specific to the brain (Hahn, Van Ness and Chaudhari, 1982; Milner and Sutcliffe, 1983). Numerous unidentified cloned cDNAs have been shown to represent mRNA that can be detected in the brain but not in other organs by RNA blot assays (Milner and Sutcliffe, 1983; Hahn and Chaudhari, 1984). High resolution hybridization assays substantiate the presence of brain restricted mRNAs (Hahn *et al.,* 1986). The frequency and apparent ease of identifying clones in brain cDNA libraries corresponding to brain restricted mRNAs, and experiments with fractionated cDNA probes, point to the probable existence of a wide array of proteins restricted to the brain. Presumably many of these proteins function in specific developmental and physiological processes that are unique to this organ.

To what extent the sequence complexity of polysomal RNA represents the inherent complexity of neurons and glial cells as opposed to differential distribution of mRNAs within a diverse population of cells in the brain is unclear (Kaplan and Finch, 1982; Takahashi, 1984). Some mRNAs apparently have restricted cellular distribution as reflected at the protein level, for example as shown by antibody probes in the visual cortex (Arimatsu, Naegele and Barnstaple, 1987). A number of mRNAs encoding regulatory polypeptides appear to be restricted to certain hypothalamic neurons (Mason *et al.,* 1986). Cloned cDNA representing mRNAs present in forebrain but absent in the cerebellum have been isolated (Rhyner *et al.,* 1986). As more sequences are examined it should become clear whether a large number of mRNAs are limited in distribution within the brain, as would be anticipated from differences in cellular attributes such as morphology, connectivity, neurotransmitters, receptors, etc. Regarding abundant class mRNAs, most of these appear to be present throughout the brain. When several relatively abundant anonymous mRNAs were screened by RNA blot and *in situ* hybridization, none was found to be anatomically restricted, although they were present in greater abundance in some regions than in others (Branks and Wilson, 1986; Wood *et al.,* 1986; Uhl and Sasek, 1986). This observation would be expected if these mRNAs are globally abundant because they are present in almost all neurons and/or glia.

Messenger RNAs that are relatively rare (abundance range of 10^{-5}–10^{-7} of total mRNA mass) are not detected by commonly used screening methods. These mRNAs are more likely to show cellular or anatomical restriction because they are probably produced in only a fraction of the cells of the brain. To date, little has been done to ascertain the extent to which these sequences are restricted to certain anatomical regions or cell types in the brain. In one study about 5% of 200 cDNA clones representing moderately rare mRNAs were found to correspond to mRNAs that were detectable in the forebrain but not in the cerebellum (Rhyner *et al.,* 1986). If this study provides any basis for forecasting future observations then it would appear that very few mRNAs are truly restricted with respect to their anatomical distribution. If this is the case then the majority of rare class mRNAs are present in most cells on average in the range of 0.1–10 copies per cell. Whether such mRNAs are physiologically significant with respect to cellular differentiation and specificity of function is unclear.

It should be noted that some mRNAs may be controlled at the post-transcriptional level (Wold *et al.*, 1978, Tobin, 1979; Carneiro and Schibler, 1984). That is, an mRNA sequence may be detectable in nuclear RNA but not in the polysomal or cytoplasmic RNA (Hahn and Chaudhari, 1984). Thus, in using *in situ* hybridization techniques it is important to determine if cellular localization is nuclear, cytoplasmic, or both. In probing mRNA purified from different brain regions it is necessary to determine the purity of the mRNA preparation (absence of nuclear RNA).

Differential expression of genes adds to molecular complexity

Sequence analysis of genes and/or cDNAs of mRNAs for several neuropolypeptides and neuroendocrine peptides has revealed several mechanisms by which diversity can be generated from a single linear DNA sequence (for review *see* Schwartz and Costa, 1986; Lynch and Snyder, 1986). Many genes are organized into complex transcriptional units (Leff, Rosenfeld and Evans, 1986) from which multiple protein products can be derived due to alternative processing of primary gene products.

Alternative processing of a primary transcript to yield different mRNAs was first demonstrated for the calcitonin gene. During investigations on expression of the calcitonin gene in rat medullary thyroid carcinomas a new mRNA transcript referred to as calcitonin gene-related product (CGRP) was observed (Rosenfeld *et al.*, 1981, 1982). CGRP mRNA shares sequences with calcitonin mRNA and encodes a 16 kD protein that is synthesized in the brain (Amara *et al.*, 1982; Rosenfeld *et al.*, 1983). Sequence analysis of the calcitonin gene showed it to contain six exons (sequences complementary to mRNA). The first three of these are represented in both calcitonin and CGRP mRNAs. Exon 4 encodes the peptide sequence and the 3′ untranslated region of calcitonin mRNA, whereas exons 5 and 6 encode the CGRP protein sequence and 3′ untranslated portion of the CGRP mRNA. These mRNAs are formed by an alternative splicing pathway involving the third exon/intron junction. Biogenesis of these mRNAs is dependent in part upon the use of one or the other of two polyadenylation signals that are widely separated on the primary transcript (Amara, Evans and Rosenfeld, 1984). The synthesis of calcitonin and CGRP is tissue specific, with CGRP being found predominantly in the nervous system and calcitonin in the thyroid.

A second example of alternative splicing of a neuropeptide gene transcript is the bovine preprotachykinin that contains the regulatory peptide, substance P. Two mRNAs, α-PPT and β-PPT, for preprotachykinin have been identified that are expressed in different regions of the nervous system. The larger β-PPT mRNA contains an additional sequence that codes for a decapeptide termed substance K due to its homology with the amphibian peptide kassinin (Nawa *et al.*, 1983). The preprotachykinin gene consists of seven exons of which the last six are protein-coding sequences. The sixth exon contains the substance K coding sequence that is excluded in the production of α-PPT mRNA. Both messages contain the substance P coding region present in exon 3 (Nawa, Kotani and Nakanishi, 1984). The alternative splicing mechanism utilized in the expression of the preprotachykinin gene is based upon an exon inclusion/exclusion rather than alternative sites of polyadenylation.

Evolution and molecular complexity

It is clear from evolutionary history that species of greater complexity have emerged through time. This increased complexity is reflected in behavioural plasticity, precision and specialization. Using vertebrate animals as an example it would appear that many of the organ systems in vertebrates are of similar complexity. That is, the basic tissues and various organs such as the liver, kidney, spleen and the like appear to be morphologically and physiologically similar in complexity amongst various species. Animals vary more with respect to the plasticity and specialization of their nervous systems. The extent to which neurological complexity differences are reflected at the molecular genetic level is an interesting though scarcely examined evolutionary aspect. Comparative measurements of mRNA complexity amongst vertebrates as well as invertebrates are very limited. The small amount of data available suggests that early in phylogeny there has been selection for a substantial number of genes that are specifically expressed in brain. For example, mRNA from the squid optic lobe is three times more complex than that of the gill (Kaplan and Gioio, 1986), and in the fruit fly *Drosophila*, 75% of the sequence complexity of the RNA from the whole body is measurable in RNA isolated from the head (Zimmerman, Fouts and Manning, 1980). In the goldfish the sequence complexity of polysomal RNA from brain is somewhat greater than twice that of the kidney (Perrone-Capano *et al.*, 1986). Brain polysomal RNA from the mouse is four to five times more complex than in the liver and kidney (Van Ness, Maxwell and Hahn, 1979; Van Ness and Hahn, 1980). As much of the survival ability of a species is dependent upon behavioural capabilities, both in the way of specialized 'hardwired' and more plastic adaptive aspects, it should not appear at all surprising that a substantial portion of the genome has evolved as a consequence of selection at the behavioural level. In this context the possibility should be considered that there might be a number of brain proteins that are not essential for function but under certain conditions confer adaptive capabilities of positive value.

Although data are very limited, it appears that there is some correspondence between an increase in the complexity of the brain and a greater number of gene products in this organ. In support of this view, the brain polysomal RNA of the goldfish appears to be five to six times less complex than in the mouse brain (Van Ness, Maxwell and Hahn, 1979; Chikaraishi, 1979; Kaplan and Gioio, 1986). Direct comparisons between the sequence complexity of polysomal RNAs from the goldfish and from mammals (rodents) based on saturation hybridization of genomic DNA might be misleading as the sequence complexity of the single copy fraction of the goldfish genome is less well documented than that for the mouse. But assuming that one strand of the haploid single copy fraction of the goldfish genome is about 6.5 times less than that of the mouse (3.3×10^8 versus 2×10^9 nucleotides as estimated from DNA renaturation kinetics), then this comparison points to an increase in molecular genetic complexity of the brain in association with advanced phylogenetic position.

It is commonly thought that only one strand of DNA at a given locus functions as template for RNA synthesis. More recent investigations have shown that certain mRNAs originate by transcription of the complementary strands in the opposite direction. Therefore the complexity of the genome for RNA template purposes is not necessarily restricted to a single strand at a given genetic locus (Henikoff *et al.*, 1986; Adelman *et al.*, 1987).

The complexity variation of nervous systems amongst species is much more extensive when compared at the cellular level. For example there are an estimated 10^5 neurons in a *Drosophila* brain and 6×10^7 and 1×10^{11} neurons in mouse and human brains respectively, yet the genomic diploid DNA content is about 0.25 picograms (pg) in *Drosophila* and 6 pg in the mammals. It is clear that the great expansion in the cellular population of the brain was not accompanied by a comparable linear expansion in the size of the genome. Hence as nervous systems became more complex as a cellular population, the extent to which singularity (uniqueness) of the individual neurons exists appears more a matter of microdifferentiation (differential patterns of gene expression, quantitative differences, etc.) rather than a matter of different gene products restricted to individual cells (Changeux, 1983). This implies that neurons in the mammalian brain of the same type apparently share very similar, if not identical, sets of gene products although the management of gene products at the post-transcriptional and post-translational level may vary amongst the individual cells.

To conclude on matters pertaining to the complexity of gene expression in brain, measurements to date indicate that a large number of gene products have arisen during the course of evolution that pertain to the development and function of the central nervous system. Genes in this category appear to be more numerous in the case of higher animals than in lower ones. Estimates of the sequence complexity of various RNA populations in brains of different species provide perspective on what lies ahead for investigators pursuing the identity, evolution and functional role of a wide variety of brain-specific proteins.

GENE EXPRESSION AND POSTNATAL DEVELOPMENT OF THE BRAIN

The brain undergoes extensive postnatal developmental changes. Functional processes emerge that undoubtedly depend upon the growth of the individual neurons and the extent and nature of their connectivity. Using cloned cDNA probes, a variety of anonymous RNAs have been detected that are developmentally regulated in that they appear just prior to or shortly after birth (Hahn *et al.*, 1983; Lenoir *et al.*, 1986; Miller *et al.*, 1986). Extrapolation from limited observations on cDNA libraries have led some investigators to estimate that somewhere between 5000 and 10 000 new mRNAs appear during the course of postnatal development of the rodent brain (Sutcliffe, McKinnon and Tsau, 1986). Using complex cDNA probes representing polyadenylated messenger RNA from mouse brain, a number of mRNA sequences present in adult were found to be absent in the near-term fetal brain (Chaudhari and Hahn, 1983). Only small, if any, differences between the poly(A)$^+$ mRNA populations in whole adult and newborn brain are evident. Saturation hybridization measurements using single copy genomic DNA showed that the complexity of the poly(A)$^+$ mRNA from the newborn brain was ~10% less than in the adult. Though reproducible, this difference may not be real. If it is, this value amounts to about 10^4 kilobases (kb) or sufficient sequence for perhaps 2000–5000 different mRNAs depending on the average size of the mRNA species. Complexity measurements on neuronal nuclear RNA also point to postnatal activation of a substantial number of genes (Takahashi, 1984). Thus both extrapolation from cDNA cloning experiments and direct complexity measurements using single copy DNA both point to the postnatal appearance of a significant number of mRNAs.

Measurements on both total nuclear RNA and RNA from mouse brain polysomes show much greater changes in the sequence complexity occurring after birth in the fraction of RNA molecules that do not bind to oligo(dT)cellulose; in other words, RNA that lacks tracts of poly(A) (Chaudhari and Hahn, 1983). While it is clear from many determinations that most of the measurable sequence complexity contained in polysomal RNA from mammalian organs is recoverable as poly(A)$^+$ mRNA (Chikaraishi, 1979; Bantle, Courchesne and Couch, 1980), this is not the case for polysomal RNA from brain where only ~60% of the measurable sequence complexity of polysomal RNA can be observed in the fraction of molecules containing poly(A) (Chikaraishi, 1979; Van Ness, Maxwell and Hahn, 1979). The complex non-polyadenylated fraction of polysomal RNA molecules is referred to as poly(A)$^-$ mRNA in part because of complementarity to single copy genomic DNA, and the fact that mRNA is the only known polysomal RNA of high sequence diversity. Furthermore, this fraction of RNA apparently has the chemical attributes required to serve as template in protein synthesis in cell-free systems (Chikaraishi, 1979; Hahn *et al.*, 1983). Poly(A)$^-$ mRNA comprises about 0.5% of the total polysomal RNA, and like poly(A)$^+$ mRNA, it is released as ribonucleoprotein particles when ribosomes are dissociated (Chikaraishi, 1979). As is of course the situation for the bulk of the poly(A)$^+$ mRNAs, it has not been established whether these RNA sequences encode for protein. While there are some known and anonymous mRNAs that lack poly(A) tracts (Houdebine, Gaye and Favre, 1974; Nemer, Graham and Dubroff, 1974; Hunter and Garrels, 1977; Kaufman *et al.*, 1977), it is also not known whether these brain polysome(A)$^-$ RNAs function as mRNA despite the attributes stated above. Ongoing investigations in various laboratories are directed towards cloning of individual sequences representing these poly(A)$^-$ RNAs and the identification of putative coding sequences. From this sequence information peptides can be produced and used to raise antibodies suitable for detection of the corresponding proteins.

Although somewhat speculative, it is probable that many of the mRNA species, either of the polyadenylated or non-polyadenylated type, that are first detectable postnatally are likely to be restricted to the brain. Hence cDNA libraries representing mostly postnatally appearing sequences should be a rich source for identifying proteins that are brain-specific. For example, cDNA clones corresponding to mRNAs that are both brain-specific and developmentally regulated have been identified. One such mRNA is that for synapsin I, a neuron-specific phosphoprotein associated with presynaptic vesicles. Two types of synapsin mRNA species have been identified and they are differentially and temporally regulated (Haas and DeGennaro, 1986). The smaller message of ~4800 nucleotides (nt) is expressed at high levels in the rat cerebellum in concert with granule cell proliferation and synaptogenesis and reaches peak levels at postnatal day 20. The larger 5800 nt mRNA is only detected in the brain between birth and postnatal day 7. The mRNA for myelin-associated glycoprotein (clone 1B236) reaches its highest levels in oligodendrocytes of white matter about 15–25 days postnatally (Higgins *et al.*, 1986; Lai *et al.*, 1987). This expression parallels that for myelin basic protein which peaks at about postnatal day 18 (Zeller *et al.*, 1984). The neural cell adhesion molecules are another example of developmentally regulated molecules for which corresponding cDNAs have been cloned (Murray *et al.*, 1984; Goridis *et al.*, 1985; Rutishauser and Goridis, 1986).

To conclude on the point of postnatal gene activation in developing brain, it is clear from both the studies on polysomal RNA as well as nuclear RNA that a variety of new transcripts appear during the course of postnatal development. The appearance of these transcripts is occurring at a time when the brain is developing the capabilities of receiving and responding to sensory input. The fact that much of the connectivity of the brain develops postnatally, and the observation that various types of inputs to the cells can influence genetic expression, lead to the notion that synaptic input into a given population of neurons may activate or silence gene expression (Grouse, Schrier and Nelson, 1979). Indeed it has been demonstrated that the level of tyrosine hydroxylase mRNA is influenced by synaptic input (Black, Chikaraishi and Lewis, 1985). This mRNA increases in relative concentration when input is elevated, and decreases when input is repressed or blocked as measured by *in situ* hybridization. In contrast, synaptic input seems to downregulate the expression of the acetylcholine receptor in muscle since denervation leads to increased levels of mRNA for the alpha subunit of this receptor (Merlie *et al.*, 1984). The presence of tetrodotoxin can also induce increased receptor content via increased mRNA levels in chicken myotubes (Klarsfeld and Changeux, 1985).

Although investigations presented above indicate the activation of genes during the course of postnatal development of mammalian brain, in the rat cerebellum a decrease in the complexity of the polyadenylated nuclear RNA population occurs as postnatal development is completed (Kaplan, 1986). While this decrease appears to be significant in regard to the measurements that were made, it is not known whether this equates to a reduction in the diversity of the mRNA population. However, this decrease does correlate with a reduction in the rate of protein synthesis in the cerebellum (Dunlop, Van Elden and Lajtha, 1977) and with alterations in the pattern or spectrum of protein synthesized during the second to third week of development (Morrison, Pardue and Griffin, 1981). These observations point towards a variety of genes being 'turned off' once postnatal development has been accomplished. If this is the case, these gene products specify proteins necessary for developmental processes rather than proteins required for function of the adult cerebellum.

Another RNA species, which is not an mRNA, increases in abundance during the early phases of postnatal development (Sutcliffe *et al.*, 1984a). This RNA species accumulates specifically in brain cell cytoplasm (neurons and glia?) and hence has been termed brain cytoplasmic (BC1) RNA (Sutcliffe *et al.*, 1984b). BC1 RNA is about 160 nucleotides in length and is complementary, in part, to a sequence referred to as the brain identifier (ID) sequence, a repetitive element that appears in the genome of the rat about 140 000 times (Sapienza and St Jacques, 1986). The ID sequence has substantial homology with certain tRNA genes and pseudogenes (Lawrence, McDonnell and Ramsey, 1985). The ID element is also represented in many heterogeneous nuclear (hn) RNA molecules, usually as an intronic sequence that in some instances is conserved in a few mRNA species as part of non-coding regions (Owens, Chaudhari and Hahn, 1985; DeChiara and Brosius, 1987). Initially it was thought that transcription of the ID sequence occurred specifically in the brain (Sutcliffe *et al.*, 1984a and 1984b). Measurements on nuclear RNAs from various organs have shown that this repeat is represented in a wide variety of hnRNA molecules and hence its transcription is not brain-specific (Witney and Furano, 1984; Owens, Chaudhari and Hahn, 1985; Lone *et al.*, 1986), although the presence of the polymerase III product, BC1 RNA, apparently is unique to brain. The sequencing of cDNA clones representing BC1 RNA showed that this RNA, beginning at the 5′ end, is composed of a sequence complementary to the repetitive ID sequence followed by a poly(A) tract and concluded by a 3′

sequence of 25 nucleotides (DeChiara and Brosius, 1987). The 3′ region appears to be complementary to a 'single copy' sequence, suggesting that the BC1 RNA is transcribed from a single (or very few) site(s) in the genome, rather than from many sites as suggested from the high genomic copy number of the ID repeat.

While the primary structure of BC1 RNA is defined, knowledge regarding its function, and the significance, if any, of its increase in abundance in the early stages of postnatal development is unknown. Other small cytoplasmically accumulating RNA species have also been observed in other organs and tissues and hence there appears to be some cellular specificity in the presence of these functionally unknown RNAs (Walter and Blobel, 1982; Khandekar *et al.*, 1984).

RECOMBINANT DNA AND THE ISOLATION OF GENES SPECIFYING BRAIN PROTEINS

It is clear that a formidable task lies ahead in identifying the proteins involved in brain development and function. Targets of considerable and obvious interest in sorting out the molecular biology of the nervous system include receptors, ionic channel proteins, proteins that play functional and structural roles at synapses, brain-restricted enzymes that function in phosphorylation and dephosphorylation events and synthesis of neurotransmitters, cell surface molecules that function in process outgrowth and synaptogenesis, and proteins that are precursors of a variety of regulatory peptides. Because of the purity and amplification achieved via cloning, rare proteins can be discovered (Young and Davis, 1983).

The isolation of a protein and its corresponding gene, particularly in the case of rare molecules, is often a difficult task and hence many proteins are identified only via certain assayable activities. Enzymatic activity is known to be preserved in fusion proteins produced in recombinant expression systems (de Wet *et al.*, 1985; Kaufman *et al.*, 1986). Also, there appears to be fairly good potential for using specific ligands as probes to identify recombinants in cDNA expression libraries, particularly where suitable antibody probes cannot be obtained. Recently calmodulin has been used as a probe to screen an expression library representing mouse brain mRNA to identify recombinants specifying calmodulin binding proteins (Sikela and Hahn, 1987). In one instance a recombinant that expresses an apparently brain-specific serine calmodulin/Ca^{2+}-dependent kinase was isolated. The fusion protein produced by this recombinant was shown to bind I^{125}-labelled calmodulin in a Ca^{2+}-dependent manner with an affinity equal to or greater than that of various brain calmodulin binding proteins (Sikela and Hahn, 1987). It has also been shown that bungarotoxin binds to a fusion protein specified by an insert that encodes for part of the α-subunit of the acetylcholine receptor (Barkas *et al.*, 1987). Thus ligands can be used successfully as probes to screen expression libraries in instances where a functional binding site is present in the fusion protein even though the full-length polypeptide is not being produced.

This approach requires that the host cells of the recombinant vector do not produce protein that binds the ligand at high affinity. A wide variety of defined ligands are commercially available and some of these are likely to be useful in isolating cloned cDNAs of the corresponding brain proteins. In instances where the binding site is formed by more than one polypeptide or is dependent upon a particular configuration of the polypeptide within the cell membrane, ligand binding would not be detected in recombinant libraries maintained in bacterial host cells. Nonetheless the use of ligands as probes, in contrast to attempting to isolate rare proteins and prepare antibodies, should facilitate efforts to identify and determine the primary structure of a variety of brain proteins.

Another approach to 'identifying' brain proteins is via molecular cloning and sequencing of cDNA molecules representing different mRNAs. From sequence information peptides can be synthesized and used as antigens in the generation of antibodies to the corresponding brain protein. This approach has already been used repeatedly to 'discover' new proteins in order to determine their distribution and possible function in the brain (Sutcliffe *et al.*, 1983). The purity and amplification that can be achieved through molecular cloning allows for the initial identity of proteins that would otherwise go undetected owing to low abundance and the difficulty of extraction from brain tissue. It is evident that we shall elucidate the primary structure and distribution within the brain of a wide variety of brain-specific proteins well in advance of understanding their functional role.

References

ADELMAN, J., BOND, C., DOUGLASS, J. and HERBERT, E. (1987) Two mammalian genes transcribed from opposite strands of the same DNA locus. *Science,* **235,** 1514–1517

AMARA, S., JONES, V., ROSENFELD, M., ONG, E. and EVANS, R. (1982) Alternative RNA processing in calcitonin gene expression generates mRNAs encoding different polypeptide products. *Nature,* **298,** 240–245

AMARA, S., EVANS, R. and ROSENFELD, M. (1984) Calcitonin/calcitonin gene-related peptide transcription unit: tissue-specific expression involves selective use of alternative polyadenylation sites. *Molecular and Cellular Biology,* **4,** 2151–2160

ARIMATSU, Y., NAEGELE, J. and BARNSTABLE, C. (1987) Molecular markers of neuronal subpopulations in layers 4, 5 and 6 of cat primary visual cortex. *Journal of Neuroscience,* **7,** 1250–1263

BANTLE, J. and HAHN, W. (1976) Complexity and characterization of polyadenylated RNA in mouse brain. *Cell,* **8,** 139–150

BANTLE, J., COURCHESNE, C. and COUCH, M. (1980) Complexity and complexity overlap in mouse liver polyadenylated and nonpolyadenylated messenger RNA fractions. *Biochemical and Biophysical Research Communications,* **95,** 1710–1721

BARKAS, T., MAURON, A., ROTH, B., ALLIOD, C., TZATOS, S. and BALLIVET, M. (1987) Mapping the main immunogenic region and toxin-binding site of the nicotinic acetylcholine receptor. *Science,* **235,** 77–80

BLACK, I., CHIKARAISHI, D. and LEWIS, E. (1985) Trans-synaptic increase in RNA coding for tyrosine hydroxylase in rat sympathetic ganglion. *Brain Research,* **339,** 151–153

BRANKS, P. and WILSON, M. (1986) Patterns of gene expression in the murine brain revealed by *in situ* hybridization of brain specific mRNAs. *Molecular Neurobiology,* **1,** 1–16

BROWN, I. (1978) Postnatal appearance of a short DNA repeat length in neurons of the rat cerebral cortex. *Biochemical Biophysical Research Communications,* **84,** 285–292

CARNEIRO, M. and SCHIBLER, U. (1984) Accumulation of rare and moderately abundant mRNAs in mouse L-cells is mainly post-transcriptionally regulated. *Journal of Molecular Biology,* **178,** 869–880

CHANGEAUX, J.-P. (1983) Concluding remarks: about the 'singularity' of nerve cells and its ontogenesis. *Progress in Brain Research,* **58,** 465–478

CHAUDHARI, N. and HAHN, W. (1983) Genetic expression in the developing brain. *Science,* **220,** 924–929

CHIKARAISHI, D. (1979) Complexity of cytoplasmic polyadenylated and nonpolyadenylated rat brain ribonucleic acids. *Biochemistry,* **18,** 3250–3256

DeCHIARA, T. and BROSIUS, J. (1987) Neural BC1 RNA: cDNA clones reveal nonrepetitive sequence content. *Proceedings of the National Academy of Sciences USA,* **84,** 2624–2628

DE WET, J., WOOD, K., HELINSKI, D. and DeLUCA, M. (1985) Cloning of firefly luciferase cDNA and the expression of active luciferase in *Escherichia coli. Proceedings of the National Academy of Sciences USA,* **82,** 7870–7873

DUNLOP, D., VAN ELDEN, W. and LAJTHA, A. (1977) Developmental effects on protein synthesis rates in regions of the CNS *in vivo* and *in vitro. Journal of Neurochemistry,* **29,** 939–945

GORIDIS, C., HIRN, M., SANTONI, M. *et al.* (1985) Isolation of mouse N-CAM related cDNA: detection and cloning using monoclonal antibodies. *EMBO Journal,* **4,** 631–635

GROUSE, L., CHILTON, M. and McCARTHY, B. (1972) Hybridization of RNA with unique sequences of mouse DNA. *Biochemistry,* **11,** 798–805

GROUSE, L., SCHRIER, B. and NELSON, P. (1979) Effect of visual experiences on gene expression during the development of stimulus specificity in cat brain. *Experimental Neurology,* **64,** 354–364

HAAS, C. and DeGENNARO, L. (1986) Developmental regulation of synapsin gene expression. *Society for Neuroscience Abstracts,* **12,** 2

HAHN, W. and LAIRD, C. (1971) Transcription of nonrepeated DNA in mouse brain. *Science,* **173,** 158–161

HAHN, W., VAN NESS, J. and CHAUDHARI, W. (1982) Overviews of the molecular genetics of mouse brain. In: *Molecular Genetic Neuroscience,* edited by F. Schmitt, S. Bird and F. Bloom, pp. 323–334. New York: Raven Press

HAHN, W., CHAUDHARI, N., BECK, L., WILBER, K. and PEFFLEY, D. (1983) Genetic expression and postnatal development of the brain: some characteristics of nonpolyadenylated mRNAs. *Cold Spring Harbor Symposium on Quantitative Biology,* **48,** 465–475

HAHN, W. and CHAUDHARI, N. (1984) Genetic perspectives on brain development and complexity. In: *Gene Expression and Cell–Cell Interactions in the Developing Nervous System,* edited by J. Lander and P. Nelson, pp. 31–42. New York: Plenum Publishing Corporation

HAHN, W., CHAUDHARI, N., SIKELA, J. and OWENS, G. (1986) Messenger RNA in the brain. In: *Role of RNA and DNA in Brain Function,* edited by A. Giuditta, B. Kaplan and C. Zomzely-Neurath, pp. 10–22. Boston: Martinus Nijhoff

HENIKOFF, S., KEENE, M., FECHTEL, K. and FRISTROM, J. (1986) Gene within a gene: nested *Drosophila* genes encode unrelated proteins on opposite DNA strands. *Cell,* **44,** 33–41

HIGGINS, G., SCHMALE, H., NAUS, C., BATTENGERY, E., BLOOM, F., WILSON, M. and MILNER, R. (1986) Developmental shift in the cellular expression of the brain-specific gene 1B236: localization to oligodendrocytes revealed by *in situ* hybridization. *Society for Neuroscience Abstracts,* **12,** 213

HOUDEBINE, L., GAYE, P. and FAVRE, A. (1974) Lack of poly(A) sequence in half of the messenger RNA coding for ewe and s. casein. *Nucleic Acids Research,* **1,** 413–426

HUNTER, T. and GARRELS, J. (1977) Characterization of the mRNA for alpha, beta and gamma actin. *Cell,* **12,** 767–781

KAPLAN, B. (1982) In: *The Handbook of Neurochemistry,* Vol. 2, edited by A. Lajtha, pp. 1–26. New York: Plenum Press

KAPLAN, B. and FINCH, C. (1982) Sequence complexity of brain nucleic acids. In: *Molecular Approaches to Neurobiology,* edited by I. Brown, pp. 71–98. New York: Academic Press

KAPLAN, B. (1986) Current approaches to the study of gene expression in the adult and developing brain. In: *Basic and Clinical Aspects of Nutrition and Brain Development,* edited by D. Rossin, B. Drujan and B. Haber, p. 76. New York: Alan Liss Inc.

KAPLAN, B. and GIOIO, A. (1986) Diversity of gene expression in goldfish brain. *Comparative Biochemistry and Physiology,* **83B,** 305–308

KAUFMAN, D., McGINNIS, J., KRIEGER, N. and TOBIN, A. (1986) Brain glutamate decarboxylase cloned in lambda gt-11: fusion protein produces gamma-aminobutyric acid. *Science,* **232,** 1138–1140

KAUFMAN, Y., MILCAREK, C., BERISSI, A. and PENMAN, S. (1977) Hela cell poly(A)$^-$ mRNA codes for a subset of poly(A)$^+$ mRNA directed proteins with an actin as a major product. *Proceedings of the National Academy of Sciences USA,* **74,** 4801–4805

KHANDEKAR, P., SAIDAPET, C., KRAUSKOPF, M., ZARRAGA, A., LIN, W., MENDOLA, C. and SEDDIGINI, M. (1984) Co-ordinate control of gene expression. Muscle-specific 7S RNA contains sequences homologous to 3′-untranslated regions of myosin genes and repetitive DNA. *Journal of Molecular Biology,* **180,** 417–435

KLARSFELD, A. and CHANGEUX, J. (1985) Activity regulates the levels of acetylcholine receptor subunit mRNA in cultured chicken myotubes. *Proceedings of the National Academy of Sciences USA,* **82,** 4558–4562

KUWANO, R., USUI, H., MAEDA, T., FUKUI, T., YAMANARI, N., OHTSAKA, E., IKEHARA, M. and TAKAHASHI, Y. (1984) Molecular cloning and the complete nucleotide sequence of cDNA to mRNA for 100-S protein of rat brain. *Nucleic Acids Research,* **12,** 7455–7465

LAI, C., BROWN, M., NAVE, K., NORONHA, B., QUARLES, R., BLOOM, F., MILNER, R. and SUTCLIFFE, J. (1987) Two forms of 1B236/myelin associated glycoprotein, a cell adhesion molecule for postnatal neural development are produced by alternative splicing. *Proceedings of the National Academy of Sciences USA,* **84,** 4337–4341

LAWRENCE, C., McDONNELL, D. and RAMSEY, W. (1985) Analysis of repetitive sequence elements containing tRNA-like sequences. *Nucleic Acids Research,* **13,** 4239–4252

LEFF, S., ROSENFELD, M. and EVANS, R. (1986) Complex transcriptional units: diversity in gene expression by alternative RNA processing. *Annual Review of Biochemistry,* **55,** 1091–1117

LENOIR, P., BATTENBERG, E., KIEL, M., BLOOM, F. and MILNER, R. (1986) Brain-specific gene 1B236 is expressed postnatally in the developing rat brain. *Journal of Neuroscience,* **6,** 522–530

LONE, Y., SIMON, M., KAHN, A. and MARIE, J. (1986) Sequences complementary to the brain-specific 'identifier' sequences exist in L-type pyruvate kinase mRNA (a liver specific messenger) and in transcripts especially abundant in muscle. *Journal of Biological Chemistry,* **261,** 1499–1502

LYNCH, D. and SNYDER, S. (1986) Neuropeptides: multiple molecular forms, metabolic pathways and receptors. *Annual Review of Biochemistry,* **55,** 773–799

MASON, A., HAYFLUTE, J., ZOELLER, R., YOUNG, W., PHILLIPS, H., NIKOLI, K. and SEEBURG, P. (1986) A deletion truncating the gondadotrophin-releasing hormone gene is responsible for hypogonadism in the hpg mouse. *Science,* **234,** 1366–1378

MERLIE, J., ISENBERG, K., RUSSELL, S. and SANES, J. (1984) Denervation supersensitivity in skeletal muscle: analysis with a cloned cDNA probe. *Journal of Cell Biology,* **99,** 332–335

MEYUHAS, O. and PERRY, R. (1979) Relationship between size, stability and abundance of messenger RNA of mouse L cells. *Cell,* **16,** 139–148

MILLER, F., NAUS, C., HIGGINS, G., BLOOM, F. and MILNER, R. (1986) Isolation and characterization of developmentally regulated rat brain messenger RNAs. *Society for Neuroscience Abstracts,* **12,** 213

MILNER, R. and SUTCLIFFE, J. (1983) Gene expression in rat brain. *Nucleic Acids Research,* **11,** 5497–5520

MILNER, R., LAI, C., NAVE, K.-A., LENOIR, D., OGATA, J. and SUTCLIFFE, J. (1985) Nucleotide sequences of two mRNAs for rat brain myelin proteolipid protein. *Cell,* **42,** 931–939

MILNER, R. (1986) Expression of brain-specific proteins. In: *Role of RNA and DNA in Brain Function,* edited by A. Giuditta, B. Kaplan and C. Zomzely-Neurath, pp. 32–41. Boston: Martinus Nijhoff Publishing

MORRISON, M., PARDUE, S. and GRIFFIN, U. (1981) Developmental alterations in the levels of translationally active messenger RNAs in the postnatal rat cerebellum. *Journal of Biological Chemistry,* **256,** 3550–3556

MURRAY, B., HEMPERLEY, J., GALLIN, W., MacGREGOR, J., EDELMAN, G. and CUNNINGHAM, B. (1984) Isolation of cDNA clones for the chicken neural cell adhesion molecule (N-CAM). *Proceedings of the National Academy of Sciences USA,* **81,** 5584–5588

NAWA, H., KOTANI, H. and NAKANISHI, S. (1984) Tissue-specific generation of two preprotachykinin mRNAs from one gene by alternative RNA splicing. *Nature,* **312,** 729–734

NAWA, H., HIROSE, T., TAKASHIMA, H., INAYAMA, S. and NAKANISHI, S. (1983) Nucleotide sequences of cloned cDNAs for two types of bovine brain substance P precursor. *Nature,* **306,** 32–36

NEMER, M., GRAHAM, M. and DUBROFF, L. (1974) Coexistence of nonhistone mRNA species lacking and containing poly(A) in sea urchin embryos. *Journal of Molecular Biology,* **89,** 435–454

OWENS, G., CHAUDHARI, N. and HAHN, W. (1985) Brain 'identifier sequence' is not restricted to brain: similar abundance in nuclear RNA of other organs. *Science,* **229,** 1263–1265

PERRONE-CAPANO, C., GIOIO, A., GUIDITTA, A. and KAPLAN, B. (1986) Complexity of nuclear and polysomal RNA from squid optic lobe and gill. *Journal of Neurochemistry,* **46,** 1517–1521

RHYNER, T., FAUCON-BIGUEL, N., BERRAD, S., BORBELY, A. and MALLET, J. (1986) An efficient approach for the selective isolation of specific transcripts from complex brain mRNA populations. *Journal of Neuroscience Research,* **16,** 167–181

ROSENFELD, M., AMARA, S., ROOS, B., ONG, E. and EVANS, R. (1981) Altered expression of the calcitonin gene associated with RNA polymorphism. *Nature,* **290,** 63–65

ROSENFELD, M., LIN, C., AMARA, S., STOLANSKY, L., ROOS, B., ONG, E. and EVANS, R. (1982) Calcitonin mRNA polymorphism: peptide switching associated with alternative RNA splicing events. *Proceedings of the National Academy of Sciences USA,* **79,** 1717–1721

ROSENFELD, M., MERMOD, J., AMARA, S., SWENSON, L., SAWCHENKO, P., RIVIER, J., VALE, W. and EVANS, R. (1983) Production of a novel neuropeptide encoded by the calcitonin gene with tissue-specific RNA processing. *Nature,* **304,** 129–135

ROSENFELD, M., AMARA, S. and EVANS, R. (1984) Alternative RNA processing: determining neuronal phenotype. *Science,* **225,** 1315–1320

RUTISHAUSER, U. and GORIDIS, C. (1986) NCAM: the molecule and its genetics. *Trends in Genetics,* **2,** 72–75

SAPIENZA, C. and ST JACQUES, B. (1986) 'Brain-specific' transcription and evolution of the identifier sequence. *Nature,* **319,** 418–420

SCHUBERT, D., BRASS, B. and DUMAS, J.-P. (1986) Protein complexity in central nervous system cell lines. *Journal of Neuroscience,* **6,** 2829–2836

SCHWARTZ, J. and COSTA, E. (1986) Hybridization approaches to the study of neuropeptides. *Annual Review of Neuroscience,* **9,** 277–304

SIKELA, J. and HAHN, W. (1987) Screening an expression library with a ligand probe: isolation and sequence of a cDNA corresponding to a brain calmodulin-binding protein. *Proceedings of the National Academy of Sciences USA,* **84,** 3038–3042

SUTCLIFFE, J., MILNER, R., SHINNICK, T. and BLOOM, F. (1983) Identifying the protein products of brain-specific genes with antibodies to chemically synthesized peptides. *Cell,* **33,** 671–682

SUTCLIFFE, J., MILNER, R., GOTTESFELD, J. and REYNOLDS, W. (1984a) Control of neuronal gene expression. *Science,* **225,** 1308–1315

SUTCLIFFE, J., MILNER, R., GOTTESFELD, J. and LERNER, R. (1984b) Identifier sequences are transcribed specifically in brain. *Nature,* **308,** 237–241

SUTCLIFFE, J., McKINNON, R. and TSAU, A.-P. (1986) Gene expression in the mammalian brain. In: *Role of RNA and DNA in Brain Function,* edited by A. Giuditta, B. Kaplan, C. and C. Zomzely-Neurath, pp. 23–31. Boston: Martinus Nijhoff

TAKAHASHI, Y. (1984) Brain nuclei. In: *Handbook of Neurochemistry,* Vol. 7, edited by A. Lajtha, pp. 205–229

TOBIN, A. (1979) Evaluating the contribution of post-transcriptional processing to differential gene expressions. *Developmental Biology,* **68,** 47–58

UHL, G. and SASEK, C. (1986) Somatostatin mRNA: regional variation in hybridization densities in individual neurons. *Journal of Neuroscience,* **6,** 3258–3264

VAN NESS, J., MAXWELL, I. and HAHN, W. (1979) Complex population of nonpolyadenylated messenger RNA in mouse brain. *Cell,* **18,** 1341–1349

VAN NESS, J. and HAHN, W. (1980) Sequence complexity of cDNA transcribed from a diverse mRNA population. *Nucleic Acids Research,* **8,** 4259–4269

VAN NESS, J. and HAHN, W. (1982) Physical parameters affecting the rate and completion of RNA driven hybridization of DNA: new measurements relevant to quantitation based on kinetics. *Nucleic Acids Research,* **10,** 8061–8077

WALTER, P. and BLOBEL, G. (1982) Signal recognition particle contains a 7S RNA essential for protein translocation across the ER. *Nature,* **299,** 691–698

WITNEY, F. and FURANO, A. (1984) Highly repeated DNA families in the rat. *Journal of Biological Chemistry,* **259,** 10481–10492

WOLD, B., KLEIN, W., HOUGH-EVENS, B., BRITTEN, R. and DAVIDSON, E. (1978) Sea urchin embryo mRNA sequence in the nuclear RNA of adult tissue. *Cell,* **14,** 941–950

WOOD, T., FRANTZ, G., MENKES, J. and TOBIN, A. (1986) Regional distribution of messenger RNAs in postmortem human brain. *Journal of Neuroscience Research,* **16,** 311–324

YOUNG, B., BIRNIE, G. and PAUL, J. (1976) Complexity and specificity of polysomal poly(A)$^+$ RNA in mouse tissues. *Biochemistry,* **15,** 2823–2829

YOUNG, R. and DAVIS, R. (1983) Efficient isolation of genes by using antibody probes. *Proceedings of the National Academy of Sciences USA,* **80,** 1194–1198

ZELLER, N., HUNKELER, M., CAMPAGNANI, A., SPRAGUE, J. and LAZZARINI, R. (1984) Characterization of mouse myelin basic protein messenger RNAs with a myelin basic protein cDNA clone. *Proceedings of the National Academy of Sciences USA,* **81,** 18–22

ZIMMERMAN, J., FOUTS, D. and MANNING, J. (1980) Evidence for a complex class of nonadenylated mRNA in *Drosophila. Genetics,* **95,** 673–691

3
In situ hybridization: visualizing brain messenger RNA

Sue Griffin

INTRODUCTION

Visualization of polyadenylated messenger RNAs (poly A^+ mRNAs) in individual cells in mammalian brain is the primary topic of this chapter. The application of *in situ* hybridization techniques, combined with immunohistochemical localization of specific proteins and neuropeptides implicated in diseases such as Alzheimer dementia will also be discussed.

EXPRESSION OF GENES ENCODING BRAIN STRUCTURAL PROTEINS DURING DEVELOPMENT

Studying gene expression in the heterogeneous cell types of the brain has been greatly facilitated by the development of *in situ* hybridization techniques for analysis of the levels of specific mRNAs relative to the levels of total polyadenylated mRNAs (poly A^+) in the same cell type (Griffin, 1987). For example, the developmental regulation of genetic expression of specific abundant structural proteins has been the focus of many neurobiological investigations (Morrison and Griffin, 1986; Shay, 1986). The actins and tubulins, and the proteins associated with them, for example, microtubule associated proteins (MAPs) are particularly important in development since cell division, maturation, and function are dependent upon the synthesis and assembly of these structural proteins (Olmsted and Borisy, 1973; Dustin, 1978; Ginzburg *et al.*, 1983). Specific mRNAs encoding the tubulins, actins, and MAPs are regulated during mammalian brain development (Morrison and Griffin, 1981; Morrison, Pardue and Griffin, 1981; Ginzburg *et al.*, 1986). By cell-free *in vitro* translation of RNAs isolated from developing rat cerebellum, we have shown that the levels of each of the mRNAs encoding these proteins is higher at times of peak cell acquisition, migration, and/or maturation (Morrison, Pardue and Griffin, 1981). Developmental regulation of the tubulins and actins is similar in human brain (Morrison and Griffin, 1981).

At present, the only method for analysing differential regulation of mRNAs in specific cell types in a complex structure like brain is *in situ* hybridization. The levels of mRNA(s) encoding one of the developmentally regulated structural proteins, relative to total poly A^+ RNAs, are regulated at the cell level, depending on cell function. In the developing rat cerebellum, we have shown by *in situ* hybridization of a chick β-tubulin cDNA (kindly donated by D. Cleveland; Cleveland, 1986) that less mature granule cells in the external granular layer have increased β-tubulin mRNA relative to total poly A^+ mRNA than do more mature granule cells in the internal granular layer (Griffin *et al.*, 1985), signifying that cells performing developmental tasks that require tubulin have elevated levels of tubulin encoding mRNAs. These results may be particularly helpful in assessing changes in the levels of structural protein expression in Alzheimer's disease as they may relate to the presence of neurofibrillary tangles (NFTs), paired helical filament protein (PHF), and senile plaque formation (Hirano and Zimmerman, 1962; Grundke-Iqbal *et al.*, 1979; Anderton *et al.*, 1982; Ihara, Abraham and Selkoe, 1983; Rasool *et al.*, 1984; Perry *et al.*, 1985; Haugh *et al.*, 1986; Kosik, Joachim and Selkoe, 1986; Miller *et al.*, 1986; Wood *et al.*, 1986; Guiroy *et al.*, 1987; Iqbal, Grundke-Iqbal and Wisniewski, 1987; Ksiczak-Reding and Yen, 1987).

REGULATION OF NEUROPEPTIDES RELATED TO CELL FUNCTION

Neuropeptides and their encoding mRNAs have been characterized using immunohistochemistry and *in situ* hybridization (Sherman *et al.*, 1984; Wolfson *et al.*, 1985; McCabe *et al.*, 1986; Shivers *et al.*, 1986). Using these techniques on tissues from drug-treated animals we have demonstrated the regulation of mRNAs encoding pro-opiomelanocortin (POMC) at the cell level (Chronwall *et al.*, 1986), (for review of POMC *see* Herbert, 1983). Haloperidol and bromocriptine respectively induced and reduced both the levels of the immunoreactive POMC-derived peptides and their encoding mRNA in pituitary cells. Hormone precursors, such as POMC, and their derivatives have been implicated in neurological diseases, especially Alzheimer's disease (Feldman and Conforti, 1980; Richardson and Schonbrunn, 1981; Sapolsky, Krey and McEwen, 1983; Sapolsky *et al.*, 1984; Serby *et al.*, 1984, 1986). Again as with the developmental regulation of structural proteins, the regulation of mRNAs encoding neuropeptides was shown to be related to the functional demand for the translation product.

The mRNAs encoding structural proteins are highly abundant in their cells of origin, reflecting the relative ease with which they can be detected. However, in order to understand a disease process and the seminal event(s) responsible for its initiation, it may be necessary to determine the relative levels of less abundant or even rare mRNAs in their cells of origin. As with the neuropeptides described above, the vast majority of cells in the brain may have none of the mRNAs that are altered by disease, whereas those that encode the peptide have relatively high levels. Conversely, those that transcribe the mRNA(s) of interest may contain few steady state copy numbers of the mRNA(s).

Somatostatin (SS) has now been added to the list of neuropeptides and putative neurotransmitters implicated in Alzheimer pathology (Emson and Lindvall, 1986). Brain SS is consistently reduced in Alzheimer's disease (Serby *et al.*, 1984; Beal *et al.*, 1985; Bouras, Magistretti and Morrison, 1986; Gomes *et al.*, 1986) as well as in epilepsy (Sloviter, 1987) and Parkinson's disease (Dupont *et al.*, 1982; Whitehouse *et al.*, 1983). Impairment of SS immunoreactive (SS^+) cells in Alzheimer's disease may be related to the presence of neurofibrillary tangles (NFTs) (Roberts *et al.*, 1986). These are composed of paired helical filaments (PHF), and are present in many more neurons in Alzheimer's disease than in age-matched controls. Of SS^+ cells in Alzheimer's disease many were also NFT^+ (Roberts *et al.*, 1986). The NFTs contain epitopes that are immunoreactive to antibodies against immune response-generated factors and cytoskeletal proteins (Grundke-Iqbal *et al.*, 1979; Anderton *et al.*, 1982; Rasool and Selkoe, 1984; Perry *et al.*, 1985; Haugh *et al.*, 1986; Kosik, Joachim and Selkoe, 1986; Miller *et al.*, 1986; Wood *et al.*, 1986; Ksiezak-Reding and Yen, 1987); in particular PHFs are immunoreactive to antibodies to MAPs and neurofilament proteins (Perry *et al.*, 1985), suggesting that synthesis or processing of these proteins may be altered in Alzheimer's disease, perhaps especially in SS^+ cells.

The decrease in the number of SS^+ cells (Roberts *et al.*, 1986) and SS content in Alzheimer's disease brain may be of further importance since SS regulates ACTH (Richardson and Schonbrunn, 1981; Serby *et al.*, 1986). There is evidence suggesting that glucocorticoid levels are elevated in Alzheimer's disease – that is, failure of exogenous dexamethasone to suppress endogenous plasma glucocorticoid levels. The hippocampus is the primary target of glucocorticoids in the brain (Meyer and McEwen, 1986) and, in addition to its role in memory, it is involved in terminating the adrenocortical stress response (Feldman and Conforti, 1980; Sapolsky *et al.*, 1984). Lymphocytes also have glucocorticoid receptors (Gametchu, 1987) that may be the anatomical substrates of glucocorticoid-induced immunosuppression (Snyder and Unanue, 1987) as well as the immunosuppression associated with Alzheimer's disease and Down's syndrome. The hippocampal cells that express the glucocorticoid receptor are at present unidentified, but the loss of them (for example, with age; Sapolsky, Krey and McEwen, 1983; Sapolsky *et al.*, 1984) or their surface receptors results in decreased sensitivity to exogenous corticoids and an elevation of plasma glucocorticoids. This association between SS and glucocorticoids, when assessed at the cellular level may reveal possible mechanisms involved in the memory deficits and immune dysfunctions in Alzheimer's disease (Snyder and Unanue, 1962; Gametchu, 1987).

ALZHEIMER DISEASE GENETICS

The genetic aetiology of familial Alzheimer's disease may be manifested by altered expression or processing of normal translation products such as structural proteins, neuropeptides, or hormones (Iqbal, Grundke-Iqbal and Wisniewski, 1987). Recently, the genetic relationship between Alzheimer's disease and Down's syndrome (Heston and Mastri, 1977; Heyman *et al.*, 1983) has been further supported. There are multiple copies (Delabar *et al.*, 1987) of the β-amyloid (Glenner and Wong, 1984) gene in Down's and Alzheimer's disease (Goldgaber *et al.*, 1987; St George-Hyslop *et al.*, 1987; Tanzi *et al.*, 1987). The β-amyloid mRNA sequence is present in normal adult human brain, but what might have been an expected induction of the encoding mRNA in Alzheimer's disease was not evident, perhaps because of RNA degradation (Tanzi *et al.*, 1987). The failure to provide evidence of gene duplication or mRNA induction in Alzheimer's disease brain does not rule out that possibility, but it does imply that extraneural sources may contribute to amyloid deposits in the brain in Alzheimer's disease. Goldgaber *et al.* (1987) postulate that β-amyloid encoded in neurons forms the PHF that are a prominent neuronal component in Alzheimer's disease brain, while β-amyloid synthesized in microglia and brain macrophages is extruded into the extracellular matrix and becomes part of senile plaques. Tanzi *et al.* (1987) suggest that in non-familial Alzheimer's disease, variations in the gene encoding β-amyloid may result in a β-amyloid protein product that makes the individual susceptible to an environmental factor, the interaction resulting in the disease.

VIRUSES AND ALZHEIMER'S DISEASE

Slow viruses are likely candidates as possible environmental cofactors in Alzheimer's disease. Viral and genetic aetiologies of Alzheimer's disease are not mutually exclusive (McLachlan and Lewis, 1985). The amyloids that characterize neural plaques in two transmissible diseases, scrapie and Creutzfeldt–Jakob disease (Brown *et al.*, 1982; Bolton, McKinley and Prusiner, 1984; Prusiner, 1984; Carp *et al.*, 1985; Roberts, Crow and Polak, 1985) are at present indistinguishable from those in Alzheimer's disease brain. Furthermore, the induction of expression in Alzheimer brain cells (Wietgrefe *et al.*, 1985) of a gene expressed in scrapie-infected brain suggests that viral sequences may be present in Alzheimer's disease brain. The infecting sequence could direct DNAs not normally transcribed to transcribe either the viral sequence or a partially homologous sequence encoding a novel protein which predisposes the host to the development of Alzheimer's disease.

TECHNIQUES OF COMBINING *IN SITU* HYBRIDIZATION AND IMMUNOHISTOCHEMISTRY IN BRAIN TISSUE

The goal of much of our work has been to adapt quantitative *in situ* hybridization techniques combined with immunohistochemistry for use in routinely prepared tissue from the neuropathology laboratory (Aksamit, Sever and Major, 1986; Griffin, 1987; Griffin and Morrison, 1985; Griffin *et al.*, 1985). Using these techniques, proteins and mRNAs that are rare in whole brain or specific brain regions may be revealed as abundant in the relatively few cells that synthesize them. The cells that give rise to the peptidergic, cholinergic, and monoaminergic systems are examples of cells whose projection areas have been studied histologically and are doubtless important in the neuropathology of diseases, including Alzheimer's (Gottfries, 1985; Perry, 1986; Price, 1986). Regardless of the obvious relevance of the regulation of mRNAs encoding neuropeptides and putative neurotransmitters or their synthesizing enzymes in neurological disease, the difficulty of the analytical protocols necessary and the limited availability of suitable samples has hampered such studies. Furthermore, because of the cellular heterogeneity of the brain, it is imperative that studies of the regulation of specific and total mRNAs which are implicated in Alzheimer's disease and other disorders are extended to include analysis at the cellular level.

To quantitate the levels of specific mRNAs relative to total poly A^+ RNA/cell, we suggest assessing the ratio of autoradiographic grains per cell after specific mRNA hybridization to grains per cell after poly A^+ RNA hybridization (Griffin and Morrison, 1985). Determining the ratio of specific mRNA to total poly A^+ mRNA in individual cells of a single type enables one to compare and contrast the relative levels in specific cells in Alzheimer's disease and age-matched controls. The singular assumption in computing this ratio is that the radiolabelled polyuridylate sequence and the cDNA (or cRNA) sequence have equal access to the mRNAs in the cells in a given specimen. Accepting this assumption, the degree of degradation of mRNA, postmortem state, fixation, processing, and permeability of the tissue need not be considered.

The methods we routinely use for combining *in situ* hybridization and immunohistochemistry in tissue processed for routine neuropathological examination have been presented in detail elsewhere (Griffin and Morrison, 1985; Griffin, 1987). The general principles of the methods are dependent upon: preservation of the mRNAs and proteins *in situ*; enhancing cell permeability to allow antibodies and cDNA or cRNA probes access to immunogenic epitopes or complementary mRNA; homology of the probe to the mRNA and specificity of the antibody against the antigen; specific activity of the radiolabelled probe and strength of the antibody; appropriate conditions for maximum hybridization; stringency of hybridization and chemistry; choice of chromagen; and appropriate emulsion exposure time and development. The necessary controls for *in situ* hybridization include treatment of adjacent sections with RNase in dilute salt solution ($< 1\times$ physiological saline concentration, PSC) either before or after hybridization. At low salt concentrations RNase degrades both single and double stranded mRNAs. Treatment before hybridization degrades the available single-stranded mRNAs, while treatment after hybridization degrades single-stranded mRNAs and double-stranded RNA hybrids. All sections are treated after hybridization with the same RNase in high salt ($4\times$ PSC) to degrade any available unhybridized mRNAs; mRNA hybrids remain intact. When specific probes are hybridized, it is appropriate to hybridize irrelevant probes, that is, sequences which are not homologous to mRNAs contained in the cells of interest, in addition to the RNase controls. For immunohistochemistry, incubation of adjacent sections with pre-immune serum and/or irrelevant antibody at the same concentration as the relevant antibody provides adequate control.

In preparation for experiments conducted on routine pathology sections from Alzheimer's disease and age-matched control brain sections, the authors prepared rat brain sections according to routine pathology procedures. The sections were hybridized with ^{3}H poly U, immunoreacted with somatostatin, and exposed to Kodak NTB2 emulsion for 5 days, and stained with haematoxylin and eosin. The presence of immunohistochemical staining in cells does not diminish the radioactive signal. We showed (*Figure 3.1a*) that the number of grains per cell was similar in cells of the same size regardless of the intensity of immunoreactive product. In addition, low-salt RNase treatment does not diminish immunohistochemical labelling. The immunoreactive staining was similarly intense, before or after (*Figure 3.1b*) low-salt RNase treatment.

CONCLUSIONS

Techniques of combined *in situ* hybridization and immunohistochemistry are currently available for use in the human brain as it is routinely processed for neuropathology. The advantages of this approach include: the greater availability of such tissue compared to fresh tissue; the pathological diagnosis will have already been made; and relevant findings may be extended to many neurological diseases. In Alzheimer's disease, with the availability of relevant antibodies and DNA probes, and the establishment of appropriate controls and protocols, the regulation

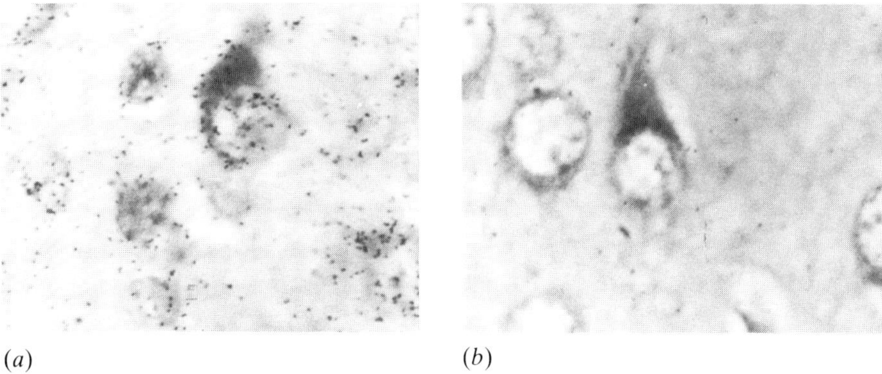

(*a*) (*b*)

Figure 3.1 Sections (5 μm) of routinely fixed rat brain was hybridized *in situ* with ^3H polyuridylate (New England Nuclear) then immunoreacted overnight with somatostatin antibody (DAKO, Santa Barbara, California) and processed, as previously described (Griffin, 1987). Both sections shown were treated after hybridization with RNase A in 4× PSC. The paucity of grains over *Figure 3.1b* illustrates the effects of prehybridization treatment with RNase A in 0.05× PSC compared to no prehybridization treatment in *Figure 3.1a*. The grain density over cells containing immunoreactive product was no different from that over non-immunoreactive cells

of mRNAs in specific cell types can be studied, as intact mRNA is available in routine neuropathology tissue sections for hybridization studies (Morrison *et al.*, 1986). At the cell level, in an equal number of Alzheimer's disease and age-matched control specimens, equivalent amounts of total poly A^+ RNA can be visualized by *in situ* hybridization even in the presence of an immunoreactive product.

Acknowledgements

I would like to thank Ms Penny Struble for secretarial assistance and Dr Chen Ling for technical assistance. Work described here was supported in part by National Institutes of Health grants AG05537 and AI14663.

References

AKSAMIT, A. J., SEVER, J. L. and MAJOR, E. O. (1986) Progressive multifocal leukoencephalopathy: JC virus detection by *in situ* hybridization compared with immunohistochemistry. *Neurology,* **36,** 499–504

ANDERTON, B. H., BREINBURG, D., DOWNES, M. J. *et al.* (1982) Monoclonal antibodies show that neurofibrillary tangles and neurofilaments share antigenic determinants. *Nature,* **298,** 84–86

BEAL, M. F., MAZUREK, M. F., TRAN, V. T., CHATTHA, G., BIRD, E. D. and MARTIN, J. B. (1985) Reduced numbers of somatostatin receptors in the cerebral cortex in Alzheimer's disease. *Science,* **229,** 289–291

BOLTON, D. C., McKINLEY, M. P. and PRUSINER, S. B. (1984) Molecular characteristics of the major scrapie prion protein. *Biochemistry,* **23,** 5898–5906

BOURAS, C., MAGISTRETTI, P. J. and MORRISON, J. H. (1986) An immunohistochemical study of six biologically active peptides in the human brain. *Human Neurobiology,* **5,** 213–226

BROWN, P., SALAZAR, A. M., GIBBS, C. J. JR and GAJDUSEK, D. C. (1982) Alzheimer's disease and transmissible virus dementia (Creutzfeldt–Jakob disease). *Annals of the New York Academy of Sciences,* **396,** 131–143

CARP, R. I., MERZ, P. A., KASCSAK, R. J., MERZ, G. S. and WISNIEWSKI, H. M. (1985) Nature of the scrapie agent: current status of facts and hypotheses. *Journal of General Virology,* **66,** 1357–1368

CHRONWALL, B. M., MILLINGTON, W. R., GRIFFIN, W. S. T., UNNERSTALL, J. R. and O'DONNOHUE, T. L. (1986) Histological evaluation of the dopaminergic regulation of pro-opiomelanocortin gene expression in the rat intermediate lobe of the pituitary using *in situ* hybridization and 3H-thymidine uptake. *Journal of Endocrinology,* **120,** 1201–1211

CLEVELAND, D. W. (1986) Molecular mechanisms controlling tubulin synthesis. In: *Cell and Molecular Biology of the Cytoskeleton,* edited by J. W. Shay, pp. 203–225. New York: Plenum Press

DELABAR, J.-M., GOLDGABER, D., LAMOUR, Y. *et al.* (1987) Beta-amyloid gene duplication in Alzheimer's disease and karyotypically normal Down syndrome. *Science,* **235,** 1390–1392

DUPONT, E., CHRISTENSEN, S. E., HANSEN, A. P., OLIVARIUS, B. F. and ORSKOV, H. (1982) Low cerebrospinal fluid somatostatin in Parkinson disease: an irreversible abnormality. *Neurology,* **32,** 312–314

DUSTIN, P. (1978) *Microtubules.* New York: Springer-Verlag

EMSON, P. C. and LINDVALL, O. (1986) Neuroanatomical aspects of neurotransmitters affected in Alzheimer's disease. *British Medical Bulletin,* **42,** 57–62

FELDMAN, S. and CONFORTI, N. (1980) Participation of the dorsal hippocampus in the glucocorticoid feedback effect on adrenocortical activity. *Neuroendocrinology,* **30,** 52–55

GAMETCHU, B. (1987) Glucocorticoid receptor-like antigen in lymphoma cell membranes: correlation to cell lysis. *Science,* **236,** 456–461

GINZBURG, I., SCHERSON, T., RYBAK, S., KIMHI, Y., NEUMAN, D., SCHWARTZ, M. and LITTAUER, U. Z. (1983) Expression of mRNA for microtubule proteins in the developing nervous system. In: *Molecular Neurobiology,* Cold Spring Harbor Laboratory, **47,** 783–790

GINZBURG, I., TEICHMAN, A., GRIFFIN, W. S. T. and LITTAUER, U. Z. (1986) Differential expression of tubulin mRNA in rat cerebellum as revealed by *in situ* hybridization. *FEBS Letters,* **194,** 161–164

GLENNER, G. G. and WONG, C. W. (1984) Alzheimer's disease: initial report of the purification and characterization of a novel cerebrovascular amyloid protein. *Biochemical and Biophysical Research Communications,* **120,** 885–890

GOLDGABER, D., LERMAN, M. I., McBRIDE, O. W., SAFFIOTTI, U. and GAJDUSEK, D. C. (1987) Characterization and chromosomal localization of a cDNA encoding brain amyloid of Alzheimer's disease. *Science,* **235,** 877–880

GOMEZ, S., DAVOUS, P., RONDOT, P., FAIVRE-BAUMAN, A., VALADE, D. and PUYMIRAT, J. (1986) Somatostatin-like immunoreactivity and acetylcholineterase activities in cerebrospinal fluid of patients with Alzheimer disease and senile dementia of the Alzheimer type. *Psychoneuroendocrinology,* **11,** 69–73

GOTTFRIES, C. G. (1985) Alzheimer's disease and senile dementia: biochemical characteristics and aspects of treatment. *Psychopharmacology,* **86,** 245–252

GRIFFIN, W. S. T. (1987) Methods for hybridization and quantitation of mRNA in individual brain cells. In: *In Situ Hybridization: Applications to Neurology,* edited by J. D. Barchas. New York: Oxford University Press

GRIFFIN, W. S. T. and MORRISON, M. R. (1985) *In situ* hybridization visualization and quantitation of genetic expression in mammalian brain. *Peptides,* **6 (Suppl. 2),** 89–96

GRIFFIN, W. S. T., ALEJOS, M. A., COX, E. J. and MORRISON, M. R. (1985) The differential distribution of beta tubulin mRNAs in individual mammalian brain cells. *Journal of Cellular Biochemistry,* **27,** 205–214

GRUNDKE-IQBAL, I., WISNIEWSKI, H. M., JOHNSON, A. B., TERRY, R. D. and IQBAL, K. (1979) Evidence that Alzheimer neurofibrillary tangles originate from neurotubules. *Lancet,* **i,** 578–580

GUIROY, D. C., MIYAZAKI, M., MULTHAUP, G. *et al.* (1987) Amyloid of neurofibrillary tangles of Guamanian parkinsonism-dementia and Alzheimer disease share identical amino acid sequence. *Proceedings of the National Academy of Sciences USA,* **84,** 2073–2077

HAUGH, M. C., PROBST, A., ULRICH, J., KAHN, J. and ANDERTON, B. H. (1986) Alzheimer neurofibrillary tangles contain phosphorylated and hidden neurofilament epitopes. *Journal of Neurology,* **49,** 1213–1220

HERBERT, E., OATES, E., MARTENS, G., COMB, M., ROSEN, H. and UHLER, M. (1983) Generation of diversity and evolution of opioid peptides. In: *Molecular Neurobiology.* Cold Spring Harbor Laboratory, **47,** 375–384

HESTON, L. L. and MASTRI, A. R. (1977) The genetics of Alzheimer's disease: associations with hematologic malignancy and Down syndrome. *Archives of General Psychiatry,* **34,** 976–981

HEYMAN, A., WILKINSON, W. E., HURWITZ, B. J. *et al.* (1983) Alzheimer's disease: genetic aspects and associated clinical disorders. *Annals of Neurology,* **14,** 597–515

HIRANO, A. and ZIMMERMAN, H. M. (1962) Alzheimer's neurofibrillary changes. *Archives of Neurology,* **7,** 227–242

IHARA, Y., ABRAHAM, C. and SELKOE, D. J. (1983) Antibodies to paired helical filaments in Alzheimer's disease do not recognize normal brain proteins. *Nature,* **304,** 727–730

IQBAL, K., GRUNDKE-IQBAL, I. and WISNIEWSKI, H. M. (1987) Alzheimer's disease, microtubule and neurofilament proteins, and axoplasmic flow. *Lancet,* **i,** 102

KOSIK, K. S., JOACHIM, C. L. and SELKOE, D. J. (1986) Microtubule-associated protein tau (τ) is a major antigenic component of paired helical filaments in Alzheimer disease. *Proceedings of the National Academy of Sciences USA,* **83,** 4044–4048

KSIEZAK-REDING, H. and YEN, S.-H. (1987) Two monoclonal antibodies recognize Alzheimer's neurofibrillary tangles, neurofilament, and microtubule-associated proteins. *Journal of Neurochemistry,* **48,** 455–462

McCABE, J. T., MORRELL, J. I., IVELL, R., SCHMALE, H., RICHTER, D. and PFAFF, D. W. (1986) *In situ* hybridization technique to localize rRNA and mRNA in mammalian neurons. *Journal of Histochemistry and Cytochemistry,* **34,** 45–50

McLACHLAN, D. R. C. and LEWIS, P. N. (1985) Alzheimer's disease: errors in gene expression. *Canadian Journal of Neurological Science,* **12,** 1–5

MEYER, J. S. and McEWEN, B. S. (1982) Evidence for glucocorticoid target cells in the rat optic nerve. Physicochemical characterization of cytosol binding sites. *Journal of Neurochemistry,* **39,** 435–442

MILLER, C. C. J., BRION, J.-P., CALVERT, R. *et al.* (1986) Alzheimer's paired helical filaments share epitopes with neurofilament side arms. *EMBO Journal,* **5,** 269–276

MORRISON, M. R. and GRIFFIN, W. S. T. (1986) Regulation of tubulin expression in brain. In: *Cell and Molecular Biology of the Cytoskeleton,* edited by J. W. Shay, pp. 227–256. New York: Plenum Press

MORRISON, M. R. and GRIFFIN, W. S. T. (1981) The isolation and *in vitro* translation of undegraded messenger RNAs from human postmortem brain. *Analytical Biochemistry,* **113,** 318–324

MORRISON, M. R., GRIFFIN, W. S. T. and WHITE, C. L. (1986) Brain messenger RNA in Alzheimer's disease. In: *Role of RNA and DNA in Brain Function,* edited by A. Guiditta, B. Kaplan and C. Zomzely-Neurath, pp. 142–159. Boston: Nijhoff

MORRISON, M. R., PARDUE, S. and GRIFFIN, W. S. T. (1981) Developmental alterations in the levels of translationally active messenger RNAs in the postnatal rat cerebellum. *Journal of Biological Chemistry,* **256,** 3550–3556

OLMSTED, J. B. and BORISY, G. G. (1973) Microtubules. *Annual Review of Biochemistry,* **42,** 507–540

PERRY, E. K. (1986) The cholinergic hypothesis – ten years on. *British Medical Bulletin,* **42,** 63–69

PERRY, G., RIZZUTO, N., AUTILIO-GAMBETTI, L. and GAMBETTI, P. (1985) Paired helical filaments from Alzheimer's disease patients contain cytoskeletal components. *Proceedings of the National Academy of Sciences USA,* **82,** 3916–3920

PRICE, D. L. (1986) New perspectives on Alzheimer's disease. *Annual Review of Neuroscience,* **9,** 489–512

PRUSINER, S. B. (1984) Some speculations about prions, amyloid, and Alzheimer's disease. *New England Journal of Medicine,* **310,** 661–663

RASOOL, C. G. and SELKOE, D. J. (1984) Alzheimer's disease: exposure of neurofilament immunoreactivity in SDS-insoluble paired helical filaments. *Brain Research,* **322,** 194–198

RASOOL, C. G., ABRAHAM, C., ANDERTON, B. H., HAUGH, M., KAHN, J. and SELKOE, D. J. (1984) Alzheimer's disease: immunoreactivity of neurofibrillary tangles with anti-neurofilament and anti-paired helical filament antibodies. *Brain Research,* **310,** 249–260

RICHARDSON, I. and SCHONBRUNN, A. (1981) Inhibition of adrenocorticotropin secretion by somatostatin in pituitary cells in culture. *Endocrinology,* **108,** 281–290

ROBERTS, G. W., CROW, T. J. and POLAK, J. M. (1985) Location of neuronal tangles in somatostatin neurones in Alzheimer's disease. *Nature,* **314,** 92–94

ROBERTS, G. W., LOFTHOUSE, R., BROWN, R., CROW, T. J., BARRY, R. A. and PRUSINER, S. B. (1986) Prion-protein immunoreactivity in human transmissible dementias. *New England Journal of Medicine,* **315,** 1231–1232

SAPOLSKY, R. M., KREY, L. C., McEWEN, B. S. and RAINBOW, T. C. (1984) Do vasopressin-related peptides induce hippocampal corticosterone receptors? Implications for aging. *Journal of Neuroscience,* **4,** 1479–1485

SAPOLSKY, R. M., KREY, L. C. and McEWEN, B. S. (1983) Corticosterone receptors decline in a site-specific manner in the aged rat brain. *Brain Research,* **289,** 235–240

SERBY, M., RICHARDSON, S. B., RYPMA, B., TWENTE, S. and ROTROSEN, J. P. (1986) Somatostatin regulation of the CRF-ACTH-cortisol axis. *Biology and Psychiatry,* **21,** 971–974

SERBY, M., RICHARDSON, S. B., TWENTE, S., SIEKIERSKI, J., CORWIN, J. and ROTROSEN, J. (1984) CSF somatostatin in Alzheimer's disease. *Neurobiology and Aging,* **5,** 187–189

SHAY, J. W. (1986) *Cell and Molecular Biology of the Cytoskeleton.* New York: Plenum Press

SHERMAN, T. G., WATSON, S. J., HERBERT, E. and AKIL, H. (1984) The co-expression of dynorphin and vasopressin: an *in situ* hybridization and dot blot analysis of mRNAs during stimulation. *Society for Neuroscience Abstracts,* **10,** 359

SHIVERS, B. D., HARLAN, R. E., PFAFF, D. W. and SCHACHTER, B. S. (1986) Combination of immunocytochemistry and *in situ* hybridization in the same tissue section of rat pituitary. *Journal of Histochemistry and Cytochemistry,* **34,** 39–43

SLOVITER, R. S. (1987) Decreased hippocampal inhibition and a selective loss of interneurons in experimental epilepsy. *Science,* **235,** 73–76

SNYDER, D. S. and UNANUE, E. R. (1982) Corticosteroids inhibit murine macrophage Ia expression and interleukin 1 production. *Journal of Immunology,* **129,** 1803–1805

ST GEORGE-HYSLOP, P. H., TANZI, R. E., POLINSKY, R. J. *et al.* (1987) The genetic defect causing familial Alzheimer's disease maps on chromosome 21. *Science,* **235,** 885–890

TANZI, R. E., GUSELLA, J. F., WATKINS, P. C. *et al.* (1987) Amyloid B protein gene: cDNA, mRNA distribution and genetic linkage near the Alzheimer locus. *Science,* **235,** 880–884

WHITEHOUSE, P. J., HEDREEN, J. C., WHITE, C. L. and PRICE, D. L. (1983) Basal forebrain neurons in the dementia of Parkinson disease. *Annals of Neurology,* **13,** 243–248

WIETGREFE, S., ZUPANCIC, M., HAASE, A. (1985) Cloning of a gene whose expression is increased in scrapie and in senile plaques in human brain. *Science,* **230,** 1177–1179

WOLFSON, B., MANNING, R. W., DAVID, L. G., ARENTZEN, R. and BALDINO, F. JR (1985) Co-localization of corticotropin releasing factor and vasopressin mRNA in neurones after adrenalectomy. *Nature,* **315,** 59

WOOD, J. G., MIRRA, S. S., POLLOCK, N. J. and BINDER, L. I. (1986) Neurofibrillary tangles of Alzheimer disease share antigenic determinants with the axonal microtubule-associated protein tau (τ). *Proceedings of the National Academy of Sciences USA,* **83,** 4040–4043

4
Molecular biology of chemical neurotransmission

Michel Goedert

INTRODUCTION

The general nature of chemical synaptic neurotransmission implies that the search for neurotransmitters and the elucidation of their properties constitute an important aspect of contemporary neurobiological research. Over recent years the techniques of molecular biology have revitalized this field and have greatly enriched our understanding of neurotransmission in the vertebrate nervous system. Recombinant DNA technology has already allowed the determination of the primary amino acid sequence of a number of classical neurotransmitter-synthesizing enzymes and of peptidergic neurotransmitter candidate precursors, thereby increasing the known number of these substances. Most importantly, it has also resulted in the determination of the primary structure of some neurotransmitter receptors, thereby allowing for the first time a detailed understanding of signal transmission in the nervous system. The elucidation of the role of acetylcholine at the neuromuscular junction marked the beginning of chemical neurotransmission, and the classification into nicotinic and muscarinic acetylcholine receptors represented the first case of multiple receptors; moreover, the nicotinic acetylcholine receptor from fish electroplaques was the first neurotransmitter receptor to be characterized biochemically. The latter also became the first neurotransmitter receptor of known amino acid sequence and the recent molecular cloning and sequencing of several muscarinic receptors confers a structural basis to the pharmacological classification first proposed by Dale (1914). The emergence of a ligand-gated ion channel receptor gene superfamily and of a gene superfamily of receptor genes whose products function through GTP-binding proteins has led to an important generalization in this rapidly expanding field.

NEUROTRANSMITTERS

Classical neurotransmitters

Classical neurotransmitters are small organic molecules (*Table 4.1*) synthesized through a series of enzymatic reactions. The genes for several of the rate-limiting

Table 4.1 Classical neurotransmitters

Acetylcholine
Adrenaline
Aspartic acid
Dopamine
Gamma-aminobutyric acid
Glutamic acid
Glycine
Histamine
Noradrenaline
Serotonin

synthesizing enzymes have been cloned and at least part of the corresponding amino acid sequences determined. This is the case for tyrosine hydroxylase (Grima *et al.*, 1985, 1987), tryptophan hydroxylase (Darmon *et al.*, 1986), choline acetyltransferase (Itoh *et al.*, 1986) and glutamic acid decarboxylase (Kaufman *et al.*, 1986; Julien *et al.*, 1987). Rat tyrosine hydroxylase consists of 498 amino acids, whereas its human counterpart is 497 amino acids in length and 90% homologous to the rat protein (Grima *et al.*, 1985, 1987). Tyrosine hydroxylase shares extensive sequence homology with phenylalanine hydroxylase and tryptophan hydroxylase, suggesting that the three proteins have evolved from a common ancestor. Tyrosine hydroxylase appears to be composed of two domains: the carboxy-terminal half contains the catalytic region, whereas the amino-terminus contains several consensus phosphorylation sites. Interestingly, three different mRNAs for tyrosine hydroxylase are generated from a single gene through alternative splicing; the sequences differ only in their amino-termini, suggesting the possibility of differential regulation with intact catalytic function.

Beyond the determination of amino acid sequences and the possibility of detailed structure/activity studies, these cloned genes are especially useful for investigating a possible connection between genetic lesions in neurotransmitter-synthesizing enzymes and certain neurological and psychiatric disorders. Catecholamine metabolism is thought to play an important role in the pathogenesis of manic-depressive illness (van Praag, 1982). The gene for human tyrosine hydroxylase has been mapped to the tip of the short arm of chromosome 11 (Powell *et al.*, 1984). Interestingly, recent results point to a close linkage between an autosomal dominant form of manic-depressive psychosis in the Old Order Amish population of south-eastern Pennsylvania and the cellular oncogene Harvey-*ras* 1 which also maps to the short arm of chromosome 11 (Egeland *et al.*, 1987) and is closely linked to the tyrosine hydroxylase locus (Moss *et al.*, 1986). However, bipolar affective disorders appear to be heterogeneous, since two other pedigree studies present no evidence for linkage to cellular Harvey-*ras* 1 (Detera-Wadleigh *et al.*, 1987; Hodgkinson *et al.*, 1987).

The action of classical neurotransmitters is rapidly terminated through selective re-uptake mechanisms and enzymatic inactivation. Acetylcholinesterase, which is responsible for the inactivation of acetylcholine, constitutes the prototype of such an enzyme. In the synaptic cleft it is found as a collagen-like structural subunit linked to the catalytic subunits by disulphide bonds. The catalytic subunit of acetylcholinesterase from *Torpedo californica* electroplaques has been cloned and sequenced (Schumacher *et al.*, 1986); it is a 575 amino acid protein that shows a surprising homology with the carboxy-terminal half of bovine thyroglobulin, the significance of which is unknown.

Neuropeptides

Over the past 15 years the discovery of neuropeptides has constituted one major development in our understanding of chemical neurotransmission. To the ten classical neurotransmitters one has to add now some 40 neuropeptides, each with a potential chemical messenger function (*Table 4.2*).

Table 4.2 Neuropeptides

Adrenocorticotropin	Melanocyte-stimulating hormones
Angiotensins	Motilin
Atrial natriuretic peptides	Neurokinins
β-Endorphin	Neuromedins
Calcitonin gene-related peptides	Neuropeptide Y
Carnosine	Neurotensin
Cholecystokinin	Oxytocin
Corticotropin-releasing factor	Pancreastatin
δ Sleep-inducing peptide	Peptide histidine-isoleucine
Dynorphins	Proctolin
Enkephalins	Prolactin
FMRF-amide	Prolactin-inhibiting factor
Galanin	Secretin
Gastrin-releasing peptide	Somatostatins
Glucagon	Substance P
Growth hormone	Thyroid-stimulating hormone
Growth hormone-releasing hormone	Thyrotropin-releasing hormone
Hydra-head activator peptide	Vasoactive intestinal polypeptide
Insulin	Vasopressin
Luteinizing hormone-releasing hormone	

Molecular biology continues to make substantial contributions in this area, mainly through the discovery of new neuropeptides and the uncovering of unexpected biosynthetic relationships between these substances. The large number of neuropeptides and the fact that the amino acid sequences of at least some are conserved over long evolutionary periods, and appear to predate classical neurotransmitters, suggest important physiological functions.

Whereas the function of classical neurotransmitters is quite well understood at least in the peripheral nervous system, this cannot be said of neuropeptides. Expression or overexpression of peptide precursors in transgenic animals will permit a closer understanding of their physiological functions in the developing and mature nervous system. An encouraging step in this direction has already been made (Hammer *et al.*, 1985). The proof of the function of neuropeptides will depend on a complete blockade of their biological activities. At present this can be done to a limited extent through antibody administration (Bannon *et al.*, 1983; Goedert, Hunter and Ninkovic, 1984), and will no doubt be possible in a more complete manner through peptide receptor antagonists. However, it would be ideal to study animals with a heritable inactivating mutation in a neuropeptide gene. The fact that organisms with advanced genetics, such as *Caenorhabditis elegans* and *Drosophila melanogaster* produce a variety of neuropeptides implies that these organisms may provide the first complete understanding of neuropeptide function.

Neuropeptides are invariably synthesized as parts of larger precursor molecules. Most of the domains encompassing bioactive peptides are flanked by pairs of basic amino acids that constitute cleavage sites for trypsin-like enzymes, but single basic amino acids are occasionally used. Many neuropeptides are further modified, with carboxy-terminal amidation constituting the most common post-translational modification. In these cases the peptide sequence in the precursor is followed by a single glycine residue which functions as the amide donor. In precursors with one identified bioactive peptide, the peptide coding sequence represents only a small fraction of that for the precursor. Such precursors (for instance preprotachykinin B and preprocholecystokinin (Deschenes *et al.*, 1984; Gubler *et al.*, 1984; Kotani *et al.*, 1986) are monofunctional as opposed to the polyproteins that contain more than one biologically active peptide. However, it is possible that monofunctional peptide precursors contain additional bioactive peptide sequences, as they often consist of cryptic domains flanked by pairs of basic amino acids.

Polyproteins can be grouped into three categories: those carrying a duplication of an identical sequence; those carrying related sequences; and those resulting in peptides with different biological functions. To the first class belong precursors, such as preprothyrotropin-releasing hormone which carries several repeats of the same neuropeptide sequence (Lechan *et al.*, 1986). The second class includes precursors, such as preproenkephalin A, preproenkephalin B and preprovasoactive intestinal polypeptide (Comb *et al.*, 1982; Gubler *et al.*, 1982; Noda *et al.*, 1982;

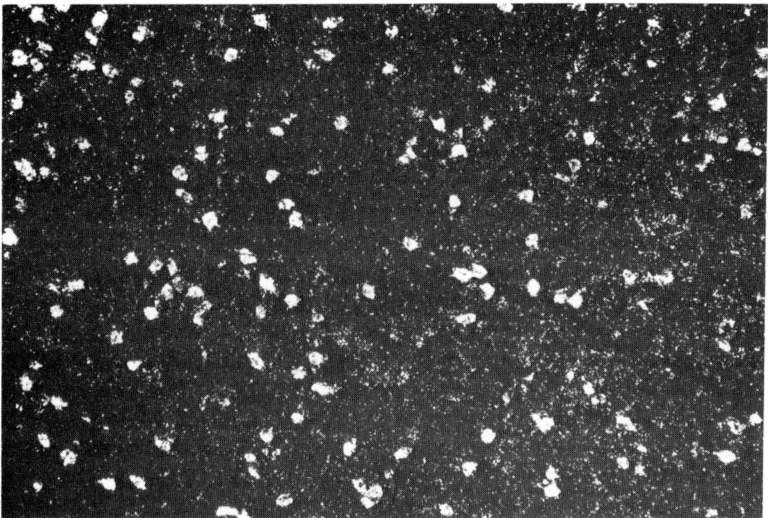

Figure 4.1 Cellular localization of preproenkephalin A mRNA in the bovine caudate nucleus. Tissue sections were incubated in the presence of a radioactively labelled preproenkephalin A cDNA clone that hybridized specifically to the corresponding mRNA molecules; the cells producing preproenkephalin A mRNA were then visualized using standard autoradiographic techniques

Kakidani *et al.*, 1982; Bloom *et al.*, 1983; Itoh *et al.*, 1983) (*Figure 4.1*). The third class comprises neuropeptides with unrelated sequences and unique biological activities. Prepro-opiomelanocortin contains the sequences of β-endorphin, adrenocorticotropin and melanocyte-stimulating hormones (Nakanishi *et al.*, 1979), whereas the precursors for oxytocin and vasopressin each contain a neurophysin in addition to the neuropeptide (Land *et al.*, 1982, 1983). There exist neuropeptide precursors that do not belong to a single polyprotein group, but share characteristics of all three classes. The most extreme case is the precursor for FMRFamide from *Aplysia californica* (Taussig and Scheller, 1986). The 597 amino acid precursor contains 28 copies of the tetrapeptide FMRFamide, a single copy of FLRFamide and three regions sandwiched between pairs of basic amino acids that are homologous to vertebrate corticotropin-releasing factor, α-melanocyte-stimulating hormone and corticotropin-like intermediate lobe peptide.

The generation of multiple bioactive peptides from a single precursor represents a powerful means for increasing the diversity of neuropeptides. It can be further increased through the tissue-specific processing of precursors. The best characterized example is the different processing of pro-opiomelanocortin in the anterior and the intermediate lobe of the pituitary gland (Eipper and Mains, 1980). In the anterior lobe the initial product is cleaved into adrenocorticotropin and β-lipotropin, with some of the latter being cleaved to release β-endorphin. In contrast, the intermediate lobe does not produce adrenocorticotropin or β-lipotropin as final products, as adrenocorticotropin is cleaved into α-melanocyte-stimulating hormone and corticotropin-like intermediate lobe peptide, and β-lipotropin is split into β-endorphin and α-lipotropin.

A further means to increase the diversity of biologically active peptides is through alternative RNA splicing (*see also* Hahn and Owens, Chapter 2 of this volume), with the calcitonin/α-calcitonin gene-related peptide (CGRP) gene representing the best known example (Rosenfeld *et al.*, 1983). In the rat, it consists of six exons; the first three used by both calcitonin and α-CGRP mRNAs are a 5′ non-coding first exon and two exons encoding the region common to both mRNAs and protein precursors. The fourth exon encodes the calcitonin coding and 3′ non-coding sequences, whereas the fifth and sixth exons encode the α-CGRP coding and 3′ non-coding sequences. Transcription proceeds through the entire gene, followed by cleavage and polyadenylation occurring at the end of the fifth exon for calcitonin mRNA, and the sixth exon for α-CGRP mRNA. Thus, calcitonin mRNA is produced by splicing of the first three to the fourth exon, whereas α-CGRP mRNA consists of the first three exons fused to the fifth and sixth.

The alternative production of these mRNAs is regulated in a tissue-specific manner, with the thyroid gland producing mostly calcitonin and very little α-CGRP, whereas the peripheral and central nervous systems produce α-CGRP and no calcitonin. The exact mechanisms involved in selective RNA splicing and choice of polyadenylation sites are unknown (Amara, Evans and Rosenfeld, 1984). It has been proposed that neurones produce a factor that determines the use of the alternative splice site (Leff, Evans and Rosenfeld, 1987). A second gene encodes β-CGRP which differs only in one amino acid from α-CGRP (Amara *et al.*, 1985).

The tachykinin A gene represents another example of selective RNA splicing (Nawa *et al.*, 1983; Nawa, Kotani and Nakanishi, 1984; Krause *et al.*, 1987) (*Figure 4.2*). It is composed of seven exons with the third coding for substance P and the sixth for substance K (neurokinin A). The primary transcript can give rise to the

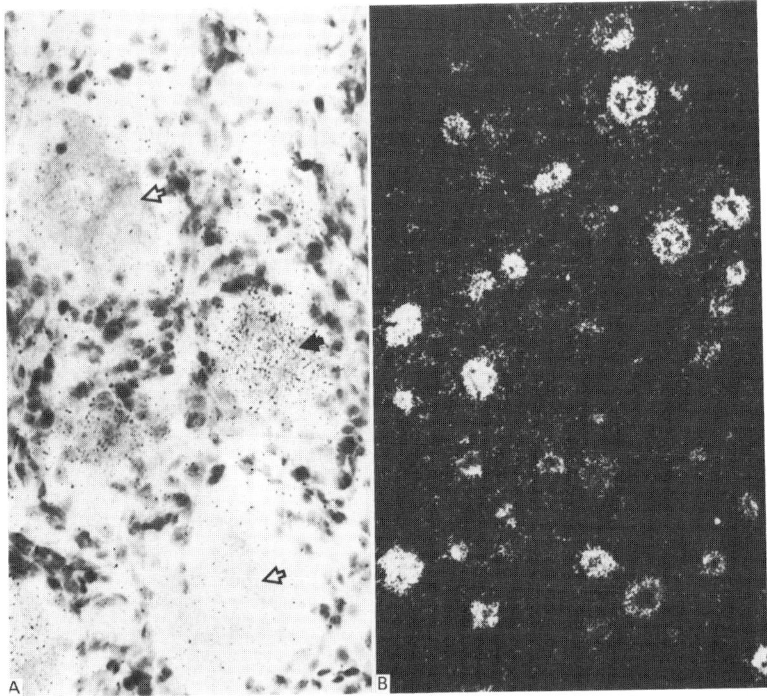

Figure 4.2 Cellular localization of preprotachykinin A mRNA in the bovine trigeminal ganglion. (*a*) Labelled and unlabelled (*arrows*) cell bodies can be seen. (*b*) Dark-field photomicrograph of a similar preparation as in (*a*)

precursors α-, β- and γ-preprotachykinin; α-preprotachykinin contains substance P, whereas β- preprotachykinin and γ-preprotachykinin contain both substance P and substance K. The difference results from the exclusion of the fourth exon in γ-preprotachykinin and of the sixth exon in α-preprotachykinin mRNA through alternative RNA splicing. There exists also tissue-specific processing of the tachykinin A gene, as evidenced by different ratios of α- and β-preprotachykinin. In nervous tissues, such as sensory ganglia, more α- than β-preprotachykinin is produced, whereas the reverse holds for endocrine tissues, such as the thyroid gland.

NEUROTRANSMITTER RECEPTORS

Nicotinic actylcholine receptors

The nicotinic acetylcholine receptor represents the best studied neurotransmitter receptor. Whereas a rigorous investigation of mammalian muscle and nervous system nicotinic receptors has only recently become possible through molecular biological techniques, the biochemical characterization of this receptor has made extensive use of the fact that it is exceptionally abundant in electroplaques from electric fish. It was also from this source that the nicotinic receptor was cloned and its complete amino acid sequence determined.

The nicotinic acetylcholine receptor from electroplaques is a pentameric protein of four subunits with a molar stoichiometry $\alpha_2\beta\gamma\delta$, and it appears that its mammalian muscle counterpart is organized in a similar manner (Numa *et al.*, 1983). Each of the four subunits forms a transmembrane protein encoded by a separate gene. Neuromuscular transmission represents the direct consequence of

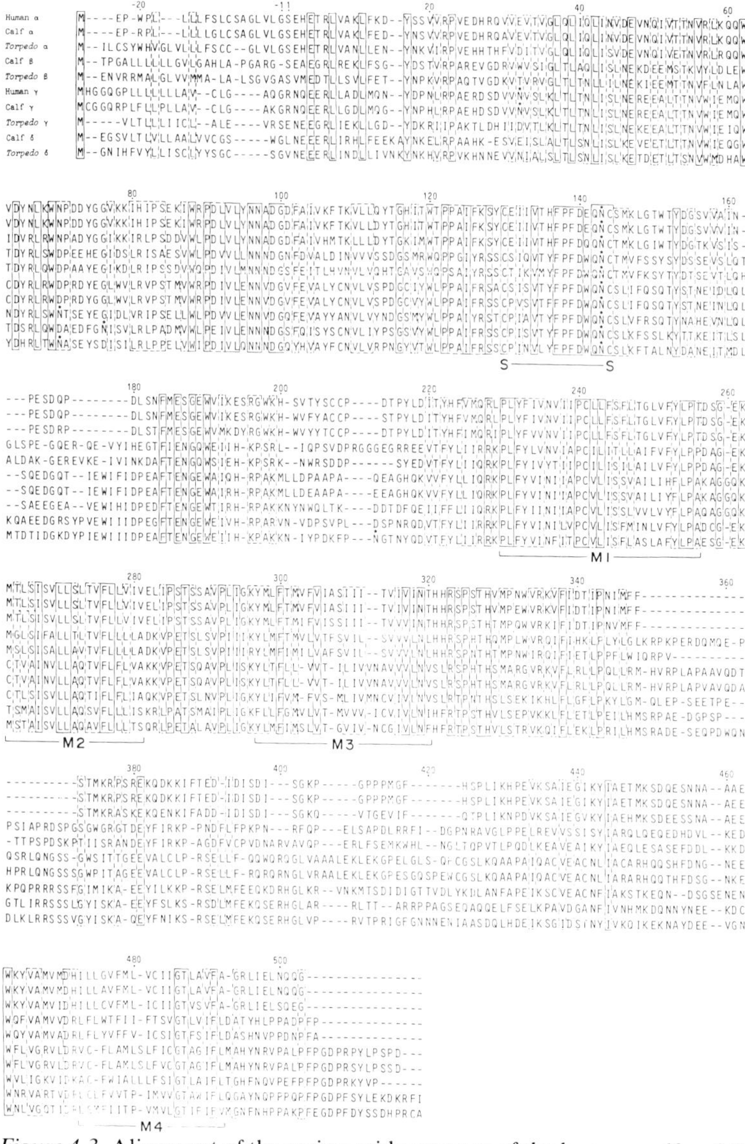

Figure 4.3 Alignment of the amino acid sequence of the human, calf and *Torpedo californica* acetylcholine receptor subunit precursors. Sets of identical residues are enclosed with solid lines and sets of conservative substitutions with dashed lines; gaps have been introduced to achieve maximum homology. Putative disulphide bridges and transmembrane segments are indicated; asterisks mark potential glycosylation sites. From Kubo *et al.* (1985), reprinted with permission from *European Journal of Biochemistry*, **149**, 5–13, and the authors

two molecules of acetylcholine binding to the extracellular portion of the α subunit, leading to a brief opening of the small cation-selective ion channel, and resulting in muscle depolarization and contraction. The α subunit carries the binding site for acetylcholine and various antagonists, whereas all subunits participate in the formation of the ion channel.

There exists substantial amino acid homology between the four subunits, suggesting the existence of an ancestral nicotinic acetylcholine receptor gene from which the different subunits have evolved through gene duplication. The α, β, γ and δ subunits from *Torpedo californica* are composed of 437, 469, 489 and 501 amino acids; they show similar hydrophilicity profiles and secondary structure (Numa *et al.,* 1983) (*Figure 4.3*). Each subunit contains four hydrophobic regions consisting of uncharged amino acids bounded by charged residues. These regions probably form four membrane-spanning regions (M1–M4). The amino-terminal segment preceding M1 is exposed on the extracellular surface of the membrane, whereas the region between M3 and M4, which constitutes approximately 25% of the mature subunit, is located on the cytoplasmic side of the membrane. It has been proposed that a fifth membrane-spanning region located between M3 and M4 may contribute to the ion channel (Finner-Moore and Stroud, 1984). This region could form an amphipathic helix partitioning charged and hydrophobic amino acids in such a way that the hydrophobic side of the helix could interact with the interior of the membrane.

Messenger RNA transcribed from full-length cDNA clones for the four subunits of the nicotinic receptor directs the synthesis of functional receptors in *Xenopus laevis* oocytes (Mishina *et al.,* 1984); this, in conjunction with site-directed mutagenesis, allows for the first time a detailed structure-activity analysis of a neurotransmitter receptor. Such studies have already demonstrated that the cysteine residues 192 and 193 of the α subunit are involved in agonist binding and signal transduction, whereas N-glycosylation of asparagine 141 and the cysteine residues 128 and 142 of all four subunits are required for a correct conformation of the extracellular region of the receptor (Mishina *et al.,* 1985). Since the determination of the amino acid sequence of the nicotinic receptor from fish electroplaques, acetylcholine receptor subunits have also been cloned and sequenced from skeletal muscle of chicken, calf, mouse and human. With one important exception they confirm the results obtained in electric fish, indicating that the structural features of the different subunits have been conserved through evolution.

In adult mammalian skeletal muscle acetylcholine receptors are mainly found at the neuromuscular junction, whereas in fetal and denervated adult muscle acetylcholine receptors are distributed over the whole muscle fibre. These two types of receptors differ in their conductance and gating properties and in their biochemical characteristics, suggesting that they might be different molecules. This long-standing issue has been resolved through the discovery of a fifth nicotinic receptor subunit in bovine muscle (Takai *et al.,* 1985). The ε subunit consists of 471 amino acids, and is most closely related to the γ subunit with which it is 53% homologous (*Figure 4.4*). It shares structural features common to all subunits; it contains four hydrophobic regions and an amphipathic segment, as well as cysteine residues in positions 128 and 142 and an asparagine residue in position 141. In calf muscle, mRNA for the γ subunit is found at fetal stages, whereas ϵ subunit mRNA is found postnatally (Mishina *et al.,* 1986). Thus, it appears that fetal and adult acetylcholine receptors are different molecules, with an $\alpha_2\beta\gamma\delta$ subunit composition

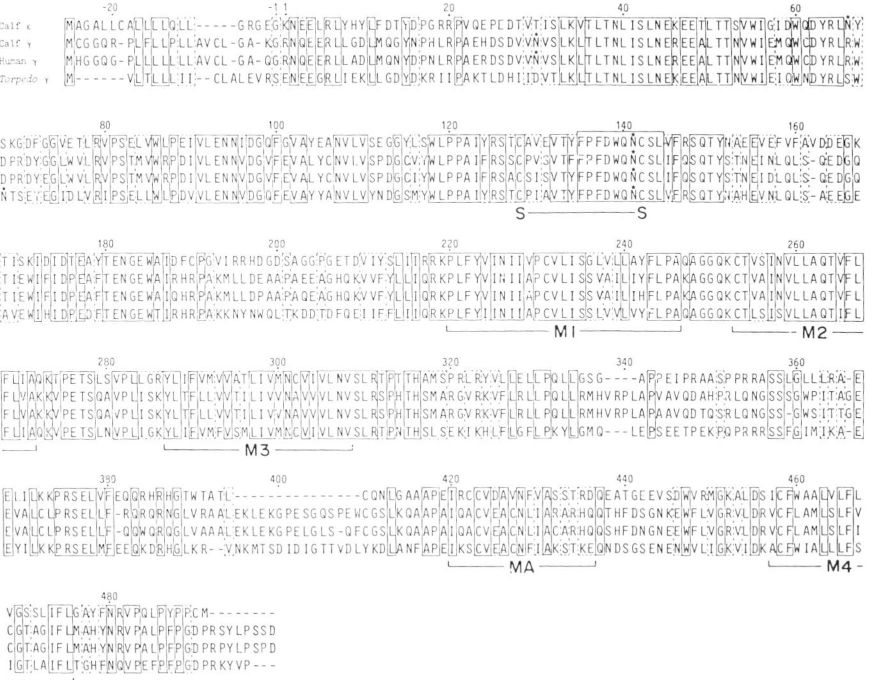

Figure 4.4 Alignment of the amino acid sequences of the calf muscle acetylcholine receptor ε subunit precursor and the calf, human and *Torpedo californica* γ subunit precursors. Sets of identical residues are enclosed with solid lines and sets of conservative residues with dashed lines; gaps have been introduced to achieve maximum homology. Putative disulphide bridges and transmembrane segments are indicated; asterisks mark potential glycosylation sites. From Takai *et al.* (1985), reprinted with permission from *Nature (London)*, **315**, 761–764, and the authors

in fetal muscle and an $\alpha_2\beta\epsilon\delta$ subunit composition in adult muscle. Most importantly, in *Xenopus laevis* oocytes $\alpha_2\beta\gamma\delta$ receptors exhibit electrophysiological characteristics similar to the fetal muscle receptor molecule, whereas $\alpha_2\beta\epsilon\delta$ receptors possess the properties of adult muscle receptors (Mishina *et al.*, 1986). It thus appears that during development innervation switches on the expression of the ε subunit gene, while repressing the γ subunit gene.

Nicotinic acetylcholine receptors in the vertebrate autonomic and central nervous systems are antigenically related to, but pharmacologically distinct from, the muscle receptor, suggesting a related but not identical amino acid sequence. This has been confirmed with the cloning and sequencing of a rat ganglionic-type acetylcholine receptor α subunit of 474 amino acids that is only expressed in medial habenula, substantia nigra pars compacta and ventral tegmental area (Boulter *et al.*, 1986). It is 47% homologous to the mouse muscle α subunit, possessing four hydrophobic domains, one amphipathic helix, conserved cysteine residues in positions 128, 142, 192 and 193, as well as an asparagine residue in position 141. It is 37 amino acids longer than the muscle type α subunit, the additional amino acids being located in the cytoplasmic domain between M3 and the amphipathic helix; in addition, there is not much sequence conservation in the region around the conserved cysteine residues, suggesting that this may account for some of the pharmacological differences between muscle and neuronal nicotinic receptors. Moreover, studies using low stringency hybridization have indicated the existence of several more α subunit related transcripts in the vertebrate central nervous system. One such gene has been cloned and sequenced (Goldman *et al.*, 1987). It contains four hydrophobic domains, an amphipathic helix prior to the fourth hydrophobic domain, an extracellular region with cysteine residues in positions 128, 142, 192 and 193, as well as an asparagine residue in position 141. It is 57% homologous to the ganglionic-type acetylcholine receptor α subunit and it is expressed in medial habenula, thalamus, hypothalamus and cerebral cortex. Circumstantial evidence suggests that this gene encodes an α subunit for a postsynaptic nicotinic receptor, whereas the ganglionic-type α subunit gene encodes an α subunit for a presynaptic nicotinic receptor.

GABA receptors

Gamma-aminobutyric acid (GABA) constitutes the major inhibitory neurotransmitter in the vertebrate brain. It mediates its hyperpolarizing action through binding to GABA receptors. As is the case of most well-documented neurotransmitters, several GABA receptor subtypes can be distinguished on pharmacological and electrophysiological grounds. $GABA_A$ sites are bicucilline-sensitive, linked to chloride channels and possess binding sites for benzodiazepines, barbiturates, β-carbolines and pictrotoxin. Benzodiazepines potentiate the response to GABA by increasing the frequency of chloride channel opening, whereas barbiturates do so by prolonging the channel lifetime. $GABA_B$ sites are baclofen-sensitive and bicucilline-insensitive. They are not linked to chloride channels but may activate a potassium conductance; they do not possess binding sites for benzodiazepines, barbiturates, β-carbolines or picrotoxin.

The $GABA_A$ receptor is a typical ligand-gated ion channel; it is made up of two distinct subunits in the stoichiometry $\alpha_2\beta_2$. The $GABA_A$ receptor from bovine brain has been cloned and both subunits sequenced (Schofield *et al.*, 1987). Expression studies in *Xenopus laevis* oocytes are consistent with the two subunits functioning as an authentic $GABA_A$ receptor. The 456 amino acid long α subunit and the 474 amino acid long β subunit are structurally related and show 35% amino acid homology. Each subunit contains four hydrophobic domains, suggesting the presence of four transmembrane segments (M1–M4). No sequence homology between the two subunits is found in the cytoplasmic region between M3 and M4 which varies in length and contains a cAMP-dependent phosphorylation site in the β subunit.

The domain organization of the $GABA_A$ receptor is remarkably similar to that of the nicotinic acetylcholine receptor, strongly suggesting that there exists a gene superfamily of ligand-gated ion channels. This probably includes excitatory amino acid receptors and the glycine receptor. The cloning and sequencing of a glycine receptor subunit has already provided evidence for substantial homologies with the nicotinic acetylcholine receptor (Grenningloh *et al.*, 1987). In limited regions in the amino-terminal hydrophilic region and in M1–M3 there exists amino acid homology between the $GABA_A$ receptor subunits and many of the bovine nicotinic acetylcholine receptor subunits. This suggests that ion channels share common features irrespective of their ion selectivity (sodium in the case of the nicotinic receptor and chloride in the case of the $GABA_A$ receptor). Interestingly, both subunits of the $GABA_A$ receptor contain two conserved cysteine residues in the putative ligand-binding region in a position analogous to that observed in the nicotinic acetylcholine receptor. The resulting disulphide bond would result in a structural loop that may represent a feature common to ligand-gated ion channel receptor proteins.

Muscarinic acetylcholine receptors

One fundamental difference between muscarinic and nicotinic cholinergic receptors is that the transduction between muscarinic receptor activation and membrane conductance changes is indirect, requiring linkage between the activated receptor and a GTP-binding protein. The muscarinic receptor belongs to a family of neurotransmitter and hormone receptors whose action is mediated through G-proteins. Biochemically, muscarinic receptors mediate inhibition of adenylate cyclase activity and hydrolysis of inositol phospholipids; electrophysiologically, muscarinic receptor activation leads to depolarizing and hyperpolarizing effects, at least some of which are mediated through potassium conductances. Pharmacologically, atropine and scopolamine antagonize these diverse actions. However, muscarinic receptors from various central nervous system regions exhibit different affinities for the antagonist pirenzepine, leading to the classification into M1 and M2 muscarinic receptor subtypes (Birdsall, Hulme and Burgen, 1980). The molecular cloning and sequencing of the porcine and rat M1 and M2 muscarinic cholinergic receptor genes puts this pharmacological classification on a firm structural basis (Kubo *et al.*, 1986a, 1986b; Peralta *et al.*, 1987; Bonner *et al.*, 1987) (*Figure 4.5*).

The muscarinic receptor from pig cerebral cortex consists of one subunit of 460 amino acids that shows no sequence homology with the nicotinic acetylcholine

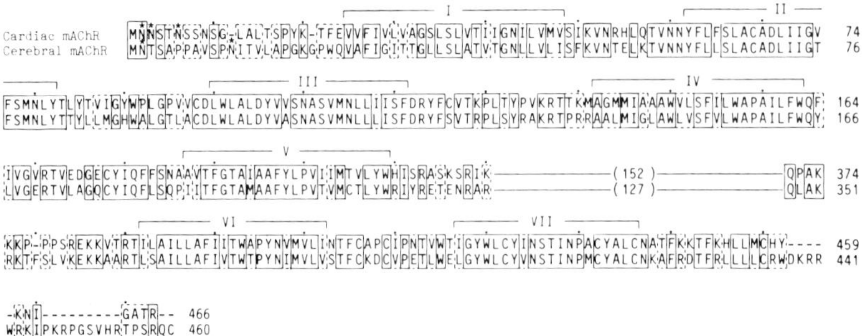

Figure 4.5 Alignment of the amino acid sequences of the porcine cardiac and cerebral muscarinic acetylcholine receptors. Sets of identical residues are enclosed with solid lines and sets of conservative residues with dashed lines; gaps have been introduced to achieve maximum homology. The non-homologous regions are shown as lines with the number of residues indicated in parentheses. The positions of the transmembrane segments I–VII are indicated; asterisks mark potential glycosylation sites. From Kubo *et al.* (1986), reprinted with permission from *FEBS Letters*, **209**, 367–371, and the authors

receptor (Kubo *et al.*, 1986a); its rat equivalent is encoded by a single intronless gene (Bonner *et al.*, 1987). The hydropathicity profile indicates the presence of seven hydrophobic regions (I–VII), one of which (VII) is less hydrophobic than the others, suggesting the presence of seven transmembrane regions. The only charged amino acids in these segments are two aspartic acid residues in segments II and III, and it appears likely that they are involved in ligand binding. The amino-terminus is exposed on the surface and contains two potential N-glycosylation sites. The loops between transmembrane segments are relatively short, with the exception of the cytoplasmic region between segments V and VI which consists of approximately 100 amino acids rich in serine, proline, glycine and acidic residues. The carboxy-terminal region of the receptor contains several serines and threonines which may represent potential phosphorylation sites.

A further two intronless muscarinic genes (called m3 and m4) that are distinct from the above receptor have been cloned and sequenced from rat cerebral cortex (Bonner *et al.*, 1987). Expression studies have indicated that their properties are those of the M1 receptor type. The m3 clone codes for a 589 amino acid protein and the m4 clone for a 478 amino acid protein. Both are highly homologous to the porcine M1 clone in the regions corresponding to the postulated transmembrane segments I–VII, with relatively little homology in the amino-terminal and carboxy-terminal regions (preceding segment I and following segment VII). Moreover, the large cytoplasmic loop between transmembrane regions V and VII is variable in size and shows little homology except for a large proportion of basic residues. *In situ* hybridization studies have indicated that clones corresponding to the rat equivalent of the porcine M1 gene, as well as the m3 and m4 genes, are differentially distributed throughout the cerebral cortex and hippocampus. This strongly suggests that the pharmacologically defined M1 receptor subtype is in fact a mixture of the M1, m3 and m4 receptor proteins.

The M2 muscarinic receptor subtype from pig heart is a protein of 466 amino acids that is encoded by a separate gene consisting of a single exon (Kubo *et al.*, 1986b; Peralta *et al.*, 1987). It shares the same general organization and extensive sequence homology with the cerebral M1 subtype. However, the two receptors have completely different amino acid sequences in the cytoplasmic region between transmembrane segments V and VI (a stretch of 152 amino acids for the M2 subtype and of 127 amino acids for the M1 subtype), strongly suggesting that this region may be responsible for the different pharmacological properties of the two subtypes. Expression studies in *Xenopus laevis* oocytes and mammalian tissue culture cells have confirmed that the cerebral muscarinic receptors represent the M1 subtype and the cardiac receptor the M2 subtype (Fukuda *et al.*, 1987; Peralta *et al.*, 1987; Bonner *et al.*, 1987).

The muscarinic cholinergic receptor protein from porcine tissues is similar in its general organization and amino acid sequence with rhodopsin and the β-adrenergic receptor. The degree of identity is 30% with the turkey erythrocyte β-adrenergic receptor and 23% with bovine opsin (Yarden *et al.*, 1986; Ovchinnikov, 1982). Rhodopsin and the β-adrenergic receptor are cell surface receptors that consist of seven transmembrane segments and are linked to GTP-binding proteins. An important generalization thus emerges, strongly suggesting that proteins with seven transmembrane segments represent a family of cell surface receptors linked to G-proteins. Moreover, the cone pigments share sequence homologies with rhodopsin (Nathans, Thomas and Hogness, 1986), and the mating receptors of yeast, the human *c-mas* oncogene, and mammalian tachykinin receptors comprise seven transmembrane segments (Burkholder and Hartwell, 1985; Nakayama, Miyajima and Arai, 1985; Young *et al.*, 1986; Masu *et al.*, 1987).

β-Adrenergic receptors

Adrenergic receptors which mediate the biological actions of adrenaline and noradrenaline represent the prototypes of a class of neurotransmitter receptors linked to second messengers. They can be classified into α-adrenergic receptors which respond to agonists in the order of potency adrenaline > noradrenaline ⩾ isoproterenol, and β-adrenergic receptors which respond in the order isoproterenol > adrenaline ⩾ noradrenaline. Radioligand binding studies and pharmacological experiments have led to a further subclassification into α_1 and α_2, as well as into β_1 and β_2 receptors. The different receptor types can also be distinguished by virtue of their second messengers; α_1-adrenergic receptors activate inositol phospholipid hydrolysis, α_2 receptors inhibit adenylate cyclase activity, whereas β_1- and β_2-adrenergic receptors stimulate adenylate cyclase activity. The last probably represent the most intensively studied receptors linked to adenylate cyclase through a GTP-binding protein. β-Adrenergic receptors from hamster lung, turkey erythrocytes and human brain and placenta have been cloned and their amino acid sequences deduced from the corresponding nucleotide sequences (Dixon *et al.*, 1986; Yarden *et al.*, 1986; Chung *et al.*, 1987; Kobilka *et al.*, 1987a). They show substantial amino acid homologies and their general organization is very similar. The human β_2-adrenergic receptor gene maps to the long arm of chromosome 5 and it contains no introns (Kobilka *et al.*, 1987a, 1987b).

The β-adrenergic receptor (β_2 subtype) from the hamster lung is a 418 amino acid protein with seven hydrophobic stretches, suggesting the presence of seven transmembrane segments (Dixon *et al.*, 1986) (*Figure 4.6*). As for the muscarinic cholinergic receptor, the amino-terminus is located extracellularly and contains two potential N-glycosylation sites. The carboxy-terminal end of the molecule is cytoplasmic and contains several serine and threonine residues that may represent sites for phosphorylation by protein kinases. The β-adrenergic receptor also contains two consensus cAMP-dependent protein kinase phosphorylation sites in its cytoplasmic region. The loops between transmembrane segments are relatively short, with the exception of the cytoplasmic region between transmembrane segments V and VI which consists of 54 amino acids. Site-directed mutagenesis has allowed the determination of which residues are involved in ligand binding and interactions of the receptor with the GTP-binding protein.

In the case of rhodopsin, which is structurally similar to the β_2 receptor, the stimulus, a photon of light, interacts with retinal which is covalently linked to a lysine residue in transmembrane segment VII (*see Figure 4.6*). Similarly, adrenergic agonists appear to intercalate among the hydrophobic transmembrane helices (Dixon *et al.*, 1987; Strader *et al.*, 1987a, 1987b). Most of the hydrophilic regions of the β-adrenergic receptor can be deleted without affecting ligand binding, suggesting that polar residues located in the transmembrane region comprise the ligand binding site. Replacement of the aspartic acid residue (position 113) in the third transmembrane segment by asparagine abolishes ligand binding; this suggests that binding of adrenergic ligands to the β receptor involves a hydrogen bond between the carboxylate group of aspartate 113 and the protonated amine group of the ligand. Residues located in the sixth hydrophilic region (amino

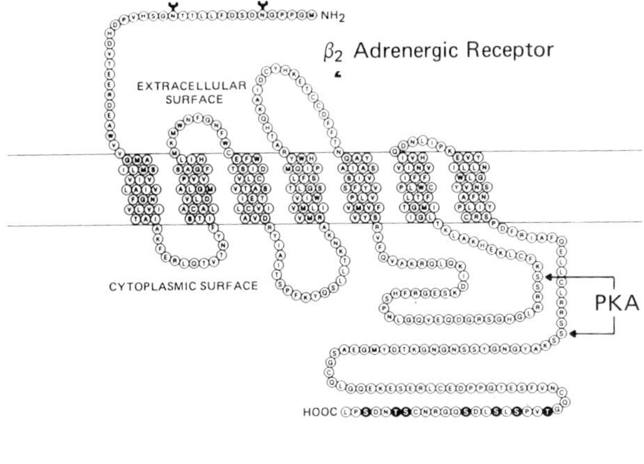

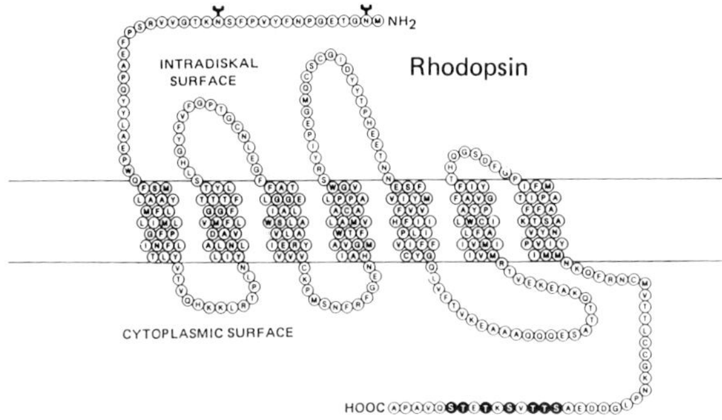

Figure 4.6 Comparison of the structure of the β_2-adrenergic receptor and rhodopsin. Black circles represent serine and threonine residues that constitute potential phosphorylation sites by β-adrenergic receptor kinase; PKA, potential phosphorylation sites by cAMP-dependent protein kinase; Y, potential N-linked glycosylation sites. From Lefkowitz *et al.* (1986), reprinted with permission from *Trends in Pharmacological Sciences,* **7,** 444–448 and the authors

acids 263–272 of the hamster receptor) are responsible for the coupling to the GTP-binding protein.

Prolonged exposure of cells to β-adrenergic agonists leads to a progressive reduction in the biological response (Siley and Lefkowitz, 1985). This phenomenon, known as desensitization, plays a widespread role in regulatory phenomena in general, and can also severely limit the clinical usefulness of many drugs. Homologous desensitization leads only to loss of sensitivity to the agonist with which the cells have been stimulated, whereas in heterologous desensitization stimulation by one agonist leads to a decreased responsiveness to a series of structurally unrelated substances working through distinct receptors. Both types of desensitization are well studied for the β-adrenergic receptor and the underlying mechanisms illustrate further similarities between the β-adrenergic receptor and the rhodopsin system.

Desensitization of the β-adrenergic receptor is the direct consequence of the phosphorylation of the receptor molecule, leading to a functional uncoupling of the agonist-occupied receptor from the GTP-binding protein and resulting in reduced adenylate cyclase stimulation. Homologous and heterologous desensitizations involve distinct protein kinases that appear to phosphorylate the β-adrenergic receptor at different sites. Heterologous desensitization is dependent on phorphorylation by cAMP-dependent protein kinase and probably by other protein kinases, such as protein kinase C (Benovic *et al.,* 1985). Homologous desensitization results from phosphorylation by a cAMP-independent protein kinase termed β-adrenergic receptor kinase (Benovic *et al.,* 1986). It possesses the remarkable property of phosphorylating the receptor only when it is occupied by an agonist. It appears likely that the phosphorylation takes place at the serine- and threonine-rich clusters located at the carboxy-terminus of the β-adrenergic receptor; these are distinct from the sites used by the cAMP-dependent protein kinase.

Light adaptation is a very similar phenomenon to receptor desensitization. It also involves phosphorylation of the rhodopsin molecule by a specific protein kinase called rhodopsin kinase, resulting in reduced coupling of photoactivated rhodopsin to cGMP phosphodiesterase (Kühn and Dreyer, 1972). Rhodopsin kinase acts on light bleached, but not unbleached, rhodopsin and phosphorylates the receptor molecule on its serine-rich and threonine-rich carboxy-terminus, leading to functional uncoupling of rhodopsin from transducin. Arrestin is a further protein involved in light adaptation; it binds specifically to phosphorylated rhodopsin and its role appears to be to accentuate uncoupling between rhodopsin and transducin (Kühn, 1978; Yamaki *et al.,* 1987). It remains to be shown whether a protein similar to arrestin acts on the phosphorylated β-adrenergic receptor. These studies thus underline the conserved nature of the structural components involved in the transduction of phenomena as different as a photon of light hitting the retina and a molecule of neurotransmitter or hormone binding to a cell.

NEUROPEPTIDE RECEPTORS

Tachykinin receptors

The first neuropeptide receptor gene has been cloned and sequenced in the form of the receptor for the tachykinin substance K (neurokinin A) (Masu *et al.,* 1987). Substance K belongs to the family of tachykinins that, in mammalian tissues, includes substance P and neuromedin K (neurokinin B). It is thought that each of these neuropeptides possesses a specific receptor to which it binds with high affinity, and that it can also bind with a reduced affinity to the receptors for the other two tachykinins. It is well established that the transmembrane events consequent to tachykinin receptor occupancy are mediated through inositol phospholipid hydrolysis.

The bovine substance K receptor consists of a protein of 384 amino acids, and hydropathicity analysis has indicated seven hydrophobic regions, suggesting the presence of seven transmembrane segments (Masu *et al.,* 1987). It shares substantial amino acid homologies with rhodopsin-type receptors, muscarinic receptors, the β-adrenergic receptor and the oncogene *c-mas*. The best homology is found in the hydrophobic regions, with only very little morphology in the amino-terminal and carboxy-terminal regions and in the cytoplasmic loop between transmembrane segments V and VI. As in the opsins and *c-mas* this stretch is very short in the substance K receptor. Like some other members of this family of receptors linked to GTP-binding proteins, the substance K receptor possesses two potential N-glycosylation sites in its amino-terminus, as well as several serine and threonine residues as potential phosphorylation sites in the carboxy-terminal region. Expression studies in *Xenopus laevis* oocytes have shown that, of the three mammalian tachykinins, this receptor possesses the highest affinity for substance K (half-maximal electrophysiological stimulation 6×10^{-8} M for substance K, 3.4×10^{-7} M for neuromedin K and 8.7×10^{-6} M for substance P).

CONCLUSIONS

The advent of modern molecular biological techniques has marked the beginning of a new era in the investigation of chemical neurotransmission. A truly molecular physiology and pharmacology of neurotransmission has now become a reality. It appears almost certain that the next decade will witness the elucidation of the primary structures of all major neurotransmitter receptors, and there can be little doubt that more than one surprise is in store. The central place of chemical neurotransmission in the functioning of the nervous system implies that neurology and psychiatry will greatly benefit from these advances through improved diagnosis and hopefully new therapies. A promising beginning has already been made in the case of manic-depressive illness and the catecholaminergic system. The elucidation of the physiological function of neuropeptides represents a major challenge in this field; it appears likely that the ever-increasing sophistication of recombinant DNA technology will greatly contribute towards an understanding of this outstanding question.

Acknowledgements

I am grateful to Drs R. J. Lefkowitz, J. Mallet, S. Numa, E. Ross and A. Ullrich for allowing me to reproduce some of their material and for providing manuscripts prior to publication. I am grateful to Mrs S. West for her help in the preparation of this manuscript.

References

AMARA, S. G., EVANS, R. M. and ROSENFELD, M. G. (1984) Calcitonin/calcitonin gene-related peptide transcription unit. Tissue-specific expression involves selective use of alternative polyadenylation sites. *Molecular and Cellular Biology,* **4,** 251–260

AMARA, S. G., ARRIZA, J. L., LEFF, S. E., SWANSON, L. W., EVANS, R. M. and ROSENFELD, M. G. (1985) Expression in brain of a messenger RNA encoding a novel neuropeptide homologous to calcitonin gene-related peptide. *Science,* **229,** 1094–1097

BANNON, M. J., ELLIOTT, P. J., ALPERT, J. E., GOEDERT, M., IVERSEN, S. D. and IVERSEN, L. L. (1983) Role of endogeneous substance P in stress-induced activation of mesocortical dopamine neurones. *Nature (London),* **306,** 791–792

BENOVIC, J. L., PIKE, L. J., CERIONE, R. A. *et al.* (1985) Phosphorylation of the mammalian beta-adrenergic receptor by cyclic AMP-dependent protein kinase. *Journal of Biological Chemistry,* **260,** 7094–7101

BENOVIC, J. L., STRASSER, R. H., CARON, M. G. and LEFKOWITZ, R. J. (1986) Beta-adrenergic receptor kinase: identification of a novel protein kinase that phosphorylates the agonist-occupied form of the receptor. *Proceedings of the National Academy of Sciences USA*, **83**, 2797–2801

BIRDSALL, N. J. M., HULME, E. C. and BURGEN, A. (1980) The character of muscarinic receptors in different regions of the rat brain. *Proceedings of the Royal Society, Series B*, **207**, 1–12

BLOOM, S. R., CHRISTOFIDES, N. D., DELAMARTER, J., BUELL, G., KAWASHIMA, E. and POLAK, J. M. (1983) Diarrhoea in vipoma patients associated with co-secretion of a second peptide (peptide histidine-isoleucine) explained by single coding gene. *Lancet*, **ii**, 1163–1165

BONNER, T. I., BUCKLEY, N. J., YOUNG, A. C. and BRANN, M. R. (1987) Identification of a family of muscarinic acetylcholine receptor genes. *Science*, **237**, 527–532

BOULTER, J., EVANS, K., GOLDMAN, D. *et al.* (1986) Isolation of a cDNA clone coding for a possible neural nicotinic acetylcholine receptor alpha subunit. *Nature (London)*, **319**, 368–374

BURKHOLDER, A. C. and HARTWELL, L. H. (1985) The yeast alpha factor receptor: Structural properties deduced from the sequence of the STE2 gene. *Nucleic Acids Research*, **13**, 8463–8475

CHUNG, F. Z., LENTES, K. U., GOCAYNE, J. *et al.* (1987) Cloning and sequence analysis of the human brain beta adrenergic receptor. *FEBS Letters*, **211**, 200–206

COMB, M., SEEBURG, P. H., ADELMAN, J., EIDEN, L. and HERBERT, E. (1982) Primary structure of the human met- and leu-enkephalin precursor and its mRNA. *Nature (London)*, **295**, 663–666

DALE, H. H. (1914) The action of certain esters and ethers of choline, and their relation to muscarine. *Journal of Pharmacology and Experimental Therapeutics*, **6**(14), 7–190

DARMON, M. C., GRIMA, B., CASH, C. D., MAITRE, M. and MALLET, J. (1986) Isolation of a rat pineal cDNA clone homologous to tyrosine and phenylalanine hydroxylases. *FEBS Letters*, **206**, 43–46

DESCHENES, R. J., LORENZ, L. J., HAUN, R. S., ROOS, B. A., COLLIER, K. J. and DIXON, J. E. (1984) Cloning and sequence analysis of cDNA encoding rat preprocholecystokinin. *Proceedings of the National Academy of Sciences USA*, **81**, 726–730

DETERA-WADLEIGH, S. D., BERRETTINI, W. H., GOLDIN, L. R., BOORMAN, D., ANDERSON, S. and GERSHON, E. S. (1987) Close linkage of *c* Harvey-*ras*-1 and the insulin gene to affective disorder is ruled out in three North American pedigrees. *Nature (London)*, **325**, 806–808

DIXON, R. A. F., KOBILKA, B. D., STRADER, D. J. *et al.* (1986) Cloning of the gene and cDNA for mammalian beta adrenergic receptor and homology with rhodopsin. *Nature (London)*, **321**, 75–79

DIXON, R. A. F., SIGAL, I. S., RANDS, E. *et al.* (1987) Ligand binding to the beta adrenergic receptor involves its rhodopsin-like core. *Nature (London)*, **326**, 73–77

EGELAND, J. A., GERHARD, D. S., PAULS, D. L. *et al.* (1987) Bipolar affective disorders linked to DNA markers on chromosome 11. *Nature (London)*, **325**, 783–787

EIPPER, B. A. and MAINS, R. E. (1980) Structure and biosynthesis of pro-adrenocorticotropin/endorphin and related peptides. *Endocrine Reviews*, **1**, 1–27

FINNER-MOORE, J. and STROUD, R. M. (1984) Amphipathic analysis and possible formation of the ion channel in an acetylcholine receptor. *Proceedings of the National Academy of Sciences USA*, **81**, 155–159

FUKUDA, K., KUBO, T., AKIBA, I., MAEDA, A., MISHINA, M. and NUMA, S. (1987) Molecular distinction between muscarinic acetylcholine receptor subtypes. *Nature (London)*, **327**, 623–625

GOEDERT, M., HUNTER, J. C. and NINKOVIC, M. (1984) Evidence for neurotensin as a non-adrenergic, non-cholinergic neurotransmitter in the guinea-pig ileum. *Nature (London)*, **311**, 59–62

GOLDMAN, D., DENERIS, E., LUYTEN, W., KOCHHAR, A., PATRICK, J. and HEINEMAN, S. (1987) Members of a nicotinic acetylcholine receptor gene family are expressed in different regions of the mammalian central nervous sytem. *Cell*, **48**, 965–973

GRENNINGLOH, G., RIENITZ, A., SCHMITT, B. *et al.* (1987) The strychnine-binding subunit of the glycine receptor shows homology with nicotinic acetylcholine receptors. *Nature (London)*, **328**, 215–220

GRIMA, B., LAMOUROUX, A., BLANOT, F., FAUCON BIGUET, N. and MALLET, J. (1985) Complete coding sequence of rat tyrosine hydroxylase mRNA. *Proceedings of the National Academy of Sciences USA*, **82**, 617–621

GRIMA, B., LAMOUROUX, A., BONI, C., JULIEN, J. F., JAVOY-AGID, F. and MALLET, J. (1987) A single human gene encodes multiple tyrosine hydroxylases predicted to differ in their functional characteristics. *Nature (London)*, **326**, 707–711

GUBLER, U., SEEBURG, P., HOFFMAN, B. J., GAGE, L. P. and UDENFRIEND, S. (1982) Molecular cloning establishes proenkephalin as precursor of enkephalin-containing peptides. *Nature (London)*, **295**, 206–208

GUBLER, U., CHUA, A. O., HOFFMAN, B. J., COLLIER, K. J. and ENG, J. (1984) Cloned cDNA to cholecystokinin mRNA predicts an identical preprocholecystokinin in pig brain and gut. *Proceedings of the National Academy of Sciences USA*, **81**, 4307–4310

HAMMER, R. E., BRINSTER, R. L., ROSENFELD, M. G., EVANS, R. M. and MAYO, K. E. (1985) Expression of human growth hormone-releasing factor in transgenic mice results in increased somatic growth. *Nature (London)*, **315**, 413–416

HODGKINSON, S., SHERRINGTON, R., GURLING, H. *et al.* (1987) Molecular genetic evidence for heterogeneity in manic depression. *Nature (London)*, **325**, 805–806

ITOH, N., OBATA, K., YANAIHARA, N. and OKAMOTO, H. (1983) Human preprovasocactive intestinal polypeptide contains a novel PHI-27-like peptide PHM-27. *Nature (London)*, **304**, 547–549

ITOH, N., SLEMMON, J. R., HAWKE, D. H. *et al.* (1986) Cloning of *Drosophila* choline acetyltransferase cDNA. *Proceedings of the National Academy of Sciences USA*, **83**, 4081–4085

JULIEN, J. F., LEGAY, F., DUMAS, S., TAPPAZ, M. and MALLET, J. (1987) Molecular cloning, expression and *in situ* hybridization of rat brain glutamic acid decarboxylase messenger RNA. *Neuroscience Letters*, **73**, 173–180

KAKIDANI, H., FURUTANI, Y., TAKAHASHI, H. *et al.* (1982) Cloning and sequence analysis of cDNA for porcine beta-neo-endorphin/dynorphin precursor. *Nature (London)*, **298**, 245–249

KAUFMAN, D. L., McGINNIS, J. F., KREIGER, N. R . and TOBIN, A. J. (1986) Brain glutamate decarboxylase cloned in lambda gt 11: Fusion protein produces gamma-aminobutyric acid. *Science*, **232**, 1138–1140

KOBILKA, B. K., DIXON, R. A. F., FRIELLE, T. *et al.* (1987a) cDNA for the human beta$_2$-adrenergic receptor: a protein with multiple membrane-spanning domains and encoded by a gene whose chromosomal location is shared with that of the receptor for platelet-derived growth factor. *Proceedings of the National Academy of Sciences USA*, **84**, 46–50

KOBILKA, B. K., FRIELLE, T., DOHLMAN, H. G. *et al.* (1987b) Delineation of the intronless nature of the genes for the human and hamster beta$_2$-adrenergic receptor and their putative promoter regions. *Journal of Biological Chemistry*, **262**, 7321–7327

KOTANI, H., HOSHIMARU, M., NAWA, H. and NAKANISHI, S. (1986) Structure and gene organization of bovine neuromedin K precursor. *Proceedings of the National Academy of Sciences USA*, **83**, 7074–7078

KRAUSE, J. E., CHIRGWIN, J. M., CARTER, M. S., XU, Z. S. and HERSHEY, A. D. (1987) Three rat preprotachykinin mRNAs encode the neuropeptides substance P and neurokinin A. *Proceedings of the National Academy of Sciences USA*, **84**, 881–885

KUBO, T., FUKUDA, K., MIKAMI, A. *et al.* (1986a) Cloning, sequencing and expression of complementary DNA encoding the muscarinic acetylcholine receptor. *Nature (London)*, **323**, 411–416

KUBO, T., MAEDA, A., SUGIMOTO, K. *et al.* (1986b) Primary structure of porcine cardiac muscarinic acetylcholine receptor deduced from the cDNA sequence. *FEBS Letters*, **209**, 367–371

KÜHN, H. (1978) Light-regulated binding of rhodopsin kinase and other proteins to cattle photoreceptor membranes. *Biochemistry*, **17**, 4389–4395

KÜHN, H. and DREYER, W. J. (1972) Light dependent phosphorylation of rhodopsin by ATP. *FEBS Letters*, **20**, 1–6

LAND, H., SCHÜTZ, G., SCHMALE, H. and RICHTER, D. (1982) Nucleotide sequence of cloned cDNA encoding bovine arginine vasopressin-neurophysin II precursor. *Nature (London)*, **295**, 299–303

LAND, H., GREZ, M., RUPPERT, S. *et al.* (1983) Deduced amino acid sequence from the bovine oxytocin-neurophysin I precursor cDNA. *Nature (London)*, **302**, 342–344

LECHAN, R. M., WU, P., JACKSON, I. M. D. *et al.* (1986) Thyrotropin-releasing hormone precursor. *Science*, **231**, 159–161

LEFF, S. E., EVANS, R. M. and ROSENFELD, M. G. (1987) Splice commitment dictates neuron-specific alternative processing in calcitonin/CGRP gene expression. *Cell*, **48**, 512–524

MASU, Y., NAKAYAMA, K., TAMAKI, H., HARADA, Y., KUNO, M. and NAKANISHI, S. (1987) cDNA cloning of bovine substance K receptor through oocyte expression system. *Nature (London)*,**329**, 836–838

MISHINA, M., KUROSAKI, T., TOBIMATSU, T. *et al.* (1984) Expression of functional acetylcholine receptor from cloned cDNAs. *Nature (London)*, **307**, 604–608

MISHINA, M., TOBIMATSU, T., IMOTO, K. *et al.* (1985) Location of functional regions of acetylcholine receptor alpha-subunit by site-directed mutagenesis. *Nature (London)*, **313**, 364–369

MISHINA, M., TAKAI, T., IMOTO, K. *et al.* (1986) Molecular distinction between fetal and adult forms of muscle acetylcholine receptor. *Nature (London)*, **321**, 406–411

MOSS, P. A. H., DAVIES, K. E., BONI, C., MALLET, J. and REEDERS, S. T. (1986) Linkage of tyrosine hydroxylase to four other markers on the short arm of chromosome 11. *Nucleic Acids Research*, **14**, 9927–9932

NAKANISHI, S., INOUE, A., KITA, T. *et al.* (1979) Nucleotide sequence of cloned cDNA for bovine corticotropin-beta-lipotropin precursor. *Nature (London)*, **278**, 423–427

NAKAYAMA, N., MIYAJIMA, A. and ARAI, K. (1985) Nucleotide sequences of STE2 and STE3, cell-type specific sterile genes from *Saccharomyces cerevisiae*. *EMBO Journal*, **4**, 2463–2468

NATHANS, J., THOMAS, D. and HOGNESS, D. S. (1986) Molecular genetics of human colour vision: The genes encoding blue, green and red pigment. *Science*, **232**, 193–202

NAWA, H., HIROSE, T., TAKASHIMA, H., INAYAMA, S. and NAKANISHI, S. (1983) Nucleotide sequences of cloned cDNAs for two types of bovine brain substance P precursor. *Nature (London)*, **306**, 32–36

NAWA, H., KOTANI, H. and NAKANISHI, S. (1984) Tissue-specific generation of two preprotachykinin mRNAs from one gene by alternative RNA splicing. *Nature (London)*, **312**, 729–734

NODA, M., FURUTANI, Y., TAKAHASHI, H. *et al.* (1982) Cloning and sequence analysis of cDNA for bovine adrenal preproenkephalin. *Nature (London)*, **295**, 202–206

NUMA, S., NODA, M., TAKAHASHI, H., TANABE, T., TOYOSATO, M., FURUTANI, Y. and KIKYOTANI, S. (1983) Molecular structure of the nicotinic acetylcholine receptor. *Cold Spring Harbor Symposia on Quantitative Biology*, **48**, 57–69

OVCHINNIKOV, Y. A. (1982) Rhodopsin and bacteriorhodopsin: Structure-function relationships. *FEBS Letters*, **148**, 179–191

PERALTA, E. G., WINSLOW, J. W., PETERSON, G. L. *et al.* (1987) Primary structure and biochemical properties of an M_2 muscarinic receptor. *Science*, **236**, 600–605

POWELL, J. F., BONI, C., LAMOUROUX, A., CRAIG, I. W. and MALLET, J. (1984) Assignment of the human tyrosine hydroxylase gene to chromosome 11. *FEBS Letters*, **175**, 37–40

ROSENFELD, M. G., MERMOD, J. J., AMARA, S. G. *et al.* (1983) Production of a novel neuropeptide encoded by the calcitonin gene via tissue-specific RNA processing. *Nature (London)*, **304**, 129–135

SCHOFIELD, P. R., DARLISON, M. G., FUJITA, N. *et al.* (1987) Sequence and functional expression of the $GABA_A$ receptor shows a ligand-gated receptor super-family. *Nature (London)*, **328**, 221–227

SCHUMACHER, M., CAMP, S., MAULET, Y. *et al.* (1986) Primary structure of *Torpedo californica* acetylcholinesterase deduced from its cDNA sequence. *Nature (London)*, **319**, 407–409

SILEY, D. R. and LEFKOWITZ, R. J. (1985) Molecular mechanisms of receptor desensitization using the beta-adrenergic receptor coupled adenylate cyclase as a model. *Nature (London)*, **317**, 124–129

STRADER, C. D., SIGAL, I. S., BLAKE, A. D. *et al.* (1987a) The carboxyl terminus of the hamster beta-adrenergic receptor expressed in mouse L cells is not required for receptor sequestration. *Cell*, **49**, 855–863

STRADER, C. D., SIGAL, I. S., REGISTER, R. B., CANCELORE, M. R., RANDS, E. and DIXON, R. A. F. (1987b) Identification of residues required for ligand binding to the beta-adrenergic receptor. *Proceedings of the National Academy of Sciences USA*, **84**, 4384–4388

TAKAI, T., NODA, M., MISHINA, M. *et al.* (1985) Cloning, sequencing and expression of a cDNA for a novel subunit of acetylcholine receptor from calf muscle. *Nature (London)*, **315**, 761–764

TAUSSIG, R. and SCHELLER, R. H. (1986) The *Aplysia* FMRF amide gene encodes sequences related to mammalian brain peptides. *DNA*, **5**, 453–461

VAN PRAAG, H. M. (1982) Depression. *Lancet*, **ii**, 1259–1264

YAMAKI, K., TAKAHASHI, Y., SAKURAGI, S. and MATSUBARA, K. (1987) Molecular cloning of the S-antigen from bovine retina. *Biochemical and Biophysical Research Communications*, **142**, 904–910

YARDEN, Y., RODRIGUEZ, H., WONG, S. K. F. *et al.* (1986) The avian beta-adrenergic receptor: primary structure and membrane topology. *Proceedings of the National Academy of Sciences USA*, **83**, 6795–6799

YOUNG, D., WAITCHES, G., BIRCHMEIER, C., FASANO, O. and WIGLER, M. (1986) Isolation and characterization of a new cellular oncogene encoding a protein with multiple potential transmembrane domains. *Cell*, **45**, 711–719

5
Proteins which regulate the development of the nervous system

David Pleasure

INTRODUCTION

Development of the nervous system is regulated by protein growth factors necessary for neuronal survival and neuroglial proliferation, and by cell surface and extracellular matrix proteins which modulate cellular adhesion and migration. Four such proteins are reviewed in this chapter: *nerve growth factor,* a small, positively charged molecule which is a survival and differentiation factor for neural crest-derived neurons of the peripheral nervous system (PNS) and for cholinergic neurons of the central nervous system (CNS); *platelet-derived growth factor,* a disulphide-linked dimeric protein homologous to the oncogenic protein *v-sis* that is a mitogen for CNS macroglia; *neural cell adhesion molecule,* a large intrinsic plasma membrane sialoglycoprotein expressed by neurons and neuroglia that, by homophilic interaction, regulates neuron–neuron and neuron–neuroglial interactions; and *laminin*, a very large, cruciate protein component of the basal lamina that provides a favourable substratum for neuron migration and neurite formation.

REGULATION OF CELL DIFFERENTIATION: NERVE GROWTH FACTOR

Nerve growth factor (NGF) was the first described and characterized of a group of extracellular proteins required for survival and differentiation of immature and regenerating neurons (Berg, 1984; Gurney *et al.*, 1986a, 1986b; Davies, Thoenen and Barde, 1986; Levi-Montalcini, 1987a, 1987b; Unsicker *et al.*, 1987; Lee, Ho and Gurney, 1987). The sequencing of NGF, completed before the advent of modern DNA and RNA methodologies, was facilitated by its extraordinarily high level of expression in mouse salivary glands (Hogue-Angeletti and Bradshaw, 1971; Levi-Montalcini, 1987a, 1987b).

NGF-dependent cells of the PNS are derived from neural crest, and include autonomic and immature sensory neurons and adrenal chromaffin cells (Acheson, Naujoks and Thoenen, 1984; Goedert *et al.*, 1984; Rich *et al.*, 1984; Unsicker, Skaper and Varon, 1985; Levi-Montalcini, 1987a, 1987b); placode-derived cranial nerve sensory neurons are not NGF-dependent (Lindsay and Rohrer, 1985; Davies and Lindsay, 1985). In the CNS, NGF affects the development of peptidergic and aminergic neurons in at least some species (Levi-Montalcini and Aloe, 1985), but its effects on cholinergic neuronal systems are best established (Hefti *et al.*, 1986). NGF enhances survival of CNS cholinergic neurons in tissue culture (Martinez *et al.*, 1985); *in vivo*, intracerebral or intraventricular NGF administration enhances cholinergic neuronal survival after axotomy and during ageing, and prevents age-dependent memory impairment attributable to dysfunction of CNS acetylcholine-dependent pathways (Fischer *et al.*, 1987; Kromer, 1987).

Structure and synthesis of nerve growth factor

Nerve growth factor is a dimer of two identical subunits linked by non-covalent bonds, and is readily prepared from mouse submandibular glands both in this form and as a complex with two other proteins, one a peptidase instrumental in NGF processing, the other of unknown function (Levi-Montalcini, 1987a, 1987b). The genes for mouse, human, chick and bovine NGF show a high degree of homology (Levi-Montalcini, 1987a, 1987b). The human gene for NGF is located on the short arm of chromosome 1, and encodes a polypeptide 307 amino acids long, which is subsequently processed to the 118 amino acid long, net positive charged mature NGF (Francke *et al.*, 1983). Variable splicing of the NGF mRNA in various cell-types has been reported (Edwards, Selby and Rutter, 1986).

Nerve growth factor is synthesized by tissues rich in sympathetic innervation (Bandtlow *et al.*, 1987), for example the iris and vas deferens, by certain sarcomas and teratocarcinomas (Dicou, Houlgatte and Brachet, 1986) and, in the nervous system, by Schwann cells and astroglia (Rush, 1984; Tarris, Weichsel and Fisher, 1986; Assouline *et al.*, 1987; Bandtlow *et al.*, 1987; Furukawa *et al.*, 1987). NGF synthesis is regulated largely at the level of transcription of its mRNA, and is downregulated by axonal contact (Shelton and Reichardt, 1984; Barth, Korsching and Thoenen, 1984; Korsching *et al.*, 1985; Heumann and Thoenen, 1986), but enhanced, in cultured fibroblasts, by catecholamines (Furukawa *et al.*, 1986). The perikarya of NGF-dependent neurons are supplied with NGF from two sources: local synthesis by neuroglia; and reverse axoplasmic transport of NGF taken up at nerve terminals (Hendry *et al.*, 1974; Taniuchi and Johnson, 1985; Taniuchi, Schweitzer and Johnson, 1986; Large *et al.*, 1986; Levi-Montalcini, 1987a, 1987b).

Mechanism of action of nerve growth factor

A specific receptor mediates binding to cells and internalization

The cellular action of NGF is initiated by binding to a specific transmembranous protein synthesized by NGF-responsive cells (Bernd and Greene, 1984). The NGF binding protein is encoded in man by a gene on chromosome 17 (Rettig *et al.*, 1986; Johnson *et al.*, 1986). This glycoprotein has three domains: an extracellular NGF binding site, a transmembranous hydrophobic segment, and a 155 amino acid cytoplasmic region (Johnson *et al.*, 1986). It is a component of two types of neuronal plasma membrane NGF receptor: the high affinity form, which mediates NGF internalization and biological response to NGF (Bernd and Greene, 1984;

Hosang and Shooter, 1987); and the low affinity form, which has as yet no known biological function (Bernd and Greene, 1984; Green and Greene, 1986; Green *et al.*, 1986). The biochemical basis for the distinction between the high-affinity and low-affinity neuronal NGF receptors is also unknown.

Plasma membrane NGF receptors are also expressed by neural crest progenitor cells (Bernd, 1986; Greiner, Lloyd and Guroff, 1986) and by Schwann cells, neurofibroma Schwann-like cells, and melanoma, but not by melanocytes (Zimmermann and Sutter, 1983; Grob *et al.*, 1985; Rohrer, 1985; Taniuchi, Clark and Johnson, 1986; Green and Greene, 1986; Pleasure *et al.*, 1986; Ross *et al.*, 1986; Yan and Johnson, 1987). The NGF receptors of Schwann cells are principally or solely of the low-affinity type (Green and Greene, 1986; Yasuda *et al.*, 1987), but it has been reported that exogenous NGF treatment augments Schwann cell surface expression of N-CAM (*see below*, p. 68). The NGF-binding protein component of the Schwann cell and melanoma NGF receptors is a product of the same gene as the neuronal NGF receptor (Ross *et al.*, 1986).

Nerve growth factor influences synthesis and phosphorylation of neuronal proteins

The late effects of NGF on the phenotype and survival of NGF-dependent cells are preceded by augmented synthesis of proteins necessary for neuronal differentiation, for example neurotransmitter synthesizing enzymes (Acheson, Naujoks and Thoenen, 1984), acetylcholine receptors (Patrick and Stallcup, 1977), sodium channels (Rudy, Kirschenbaum and Greene, 1982), microtubule-associated proteins (Greene *et al.*, 1983), neurofilament proteins (Lindenbaum, Carbonetto and Mushynski, 1987), and cell–surface adhesion molecules (McGuire, Greene and Furano, 1978; *see below*, p. 70). These events are preceded by even shorter latency effects of NGF on the levels of phosphorylation of neuronal enzymes and cytoskeletal components (Halegous and Patrick, 1980; Seeley *et al.*, 1984; Landreth and Rieser, 1985; Togari and Guroff, 1985; Hama *et al.*, 1986; Lindenbaum, Carbonetto and Mushynski, 1987; Rowland *et al.*, 1987; Matsuda and Guroff, 1987).

REGULATION OF CELL PROLIFERATION: PLATELET-DERIVED GROWTH FACTOR

Platelet-derived growth factor (PDGF) is a potent mitogen and chemotactic agent for mesenchyme-derived cells in non-neural tissues (Grotendorst *et al.*, 1981, 1982; Seppa *et al.*, 1982; Deuel, Tong and Huang, 1985; Cochran, 1985; Westermark and Heldin, 1986). Platelet-derived growth factor can be thought of as a hormone that circulates in the bloodstream and exerts effects on target cells far from its site of synthesis, but under most circumstances functions more as a *paracrine growth factor,* secreted by PDGF-unresponsive cells (for example, endothelial cells) in the neighbourhood of responsive cells (for example, arterial smooth muscle cells) or as an *autocrine growth factor*, synthesized by cells (for example, fibroblasts) that are themselves thereby rendered competent to proliferate (Westermark and Heldin, 1986; Paulsson *et al.*, 1987a; 1987b). PDGF is a '*competence factor*' that primes its target cells to pass through G_0-G_1 and to mitose only if provided with '*progression factors*' such as epidermal growth factor and somatomedin C (Pledger *et al.*, 1977; Smith and Stiles, 1982; Wharton *et al.*, 1982; Olashaw and Pledger, 1983). The signal provided by PDGF is transduced to the interior of its target cells by phosphorylation of its plasma membrane receptor (Sibley *et al.*, 1987).

Soon after PDGF was initially characterized, it was recognized to be a mitogen for CNS neuroglia (Heldin, Wasteson and Westermark, 1977). More recent studies suggest that PDGF plays an important role in normal CNS development (Asch *et al.*, 1986; Dubois-Dalcq, 1987), in the response of the CNS to injury (Takamiya *et al.*, 1986), and as an autocrine factor for gliomas (Lens, Altena and Nusse, 1986).

Structure and synthesis of platelet-derived growth factor

Platelet-derived growth factor is composed of two peptides linked by disulphide bonds; the A peptide is encoded by a gene on human chromosome 7, and the B peptide gene is on chromosome 22. The two genes were probably derived from duplication of a common ancestral gene, but have diverged during evolution, and the amino acid sequences of the fully processed A and B peptides are only 60% homologous (Betsholtz *et al.*, 1986; Westermark and Heldin, 1986).

Among the cells that synthesize PDGF are megakaryocytes, endothelial cells, macrophages, smooth muscle cells, fibroblasts, and placental cytotrophoblasts (Leslie, Musson and Henson, 1984; Barrett *et al.*, 1984; Shimokado *et al.*, 1985; Collins *et al.*, 1985; Paulsson *et al.*, 1987). The PDGF that is stored in human platelet granules is a heterodimer of peptides A and B, but porcine platelet PDGF is a B-B homodimer, and cells other than megakaryocytes in various species synthesize A-A or B-B homodimers as well (Ross, Raines and Bowen-Pope, 1986; Westermark and Heldin, 1986; Collins, Bonthron and Ortein, 1987; Paulsson *et al.*, 1987; Tong *et al.*, 1987). Further PDGF heterogeneity is made possible by an N-glycosylation site on the A but not the B peptide, variable proteolytic clipping of the A and B chains, and alternative splicing of A chain mRNA (Johnsson *et al.*, 1984; Niman, Houghten and Bowen-Pope, 1984; Betsholtz *et al.*, 1986; Westermark and Heldin, 1986; Collins, Bonthron and Orkin, 1987; Tong *et al.*, 1987).

The simian sarcoma virus-derived transforming *v-sis* peptide (Hannink and Donoghue, 1986; Yeh, Pierce and Deuel, 1987), expressed by sarcomas, glioblastomas, hepatomas and T-cell leukaemia cells (Owen, Pantazis and Antoniades, 1984; Westermark and Heldin, 1986; Lens, Altena and Nusse, 1986; Ross, Raines and Bowen-Pope, 1986), is 94% homologous with PDGF B chains, and hence PDGF B peptide is also referred to as *c-sis* (Johnsson *et al.*, 1984). Homology of the mRNAs for *v-sis* and *c-sis* is considerably lower than that of the fully processed peptides, since the hydrophobic signal sequence of *v-sis* is transcribed from a fused viral nucleotide sequence rather than from the B peptide gene (Martinet *et al.*, 1986; Hannick and Donohue, 1986; Westermark and Heldin, 1986).

Synthesis of PDGF by smooth muscle cells falls with maturation, rises in macrophages with activation, is affected by the surface upon which cells are grown, and is induced in cultured fibroblasts by exogenous PDGF in a positive feedback loop (Barrett *et al.*, 1984; Shimokado *et al.*, 1985; Martinet *et al.*, 1986; Westermark and Heldin, 1986; Ross, Raines and Bowen-Pope, 1986; Paulsson *et al.*, 1987). Because PDGF is a cationic, hydrophobic protein which adheres avidly to extracellular matrix and many other surfaces, its turnover after release from cells is very rapid (Smith *et al.*, 1982; Singh, Chaikin and Stiles, 1982; Bowen-Pope *et al.*, 1984; Ross, Raines and Bowen-Pope, 1986). Much of the PDGF in plasma circulates in a complex with a PDGF-binding protein, which may protect the mitogen from degradation but may also inhibit its binding to target cells (Huang, Huang and Deuel, 1983; Raines *et al.*, 1984; Ross, Raines and Bowen-Pope, 1986).

Mechanism of action of platelet-derived growth factor

Platelet-derived growth factor-responsive cells express specific cell–surface PDGF receptors with intrinsic tyrosine protein kinase activity

Platelet-derived growth factor binds to specific receptors expressed on the surfaces of target cells (Bowen-Pope and Ross, 1982; Huang *et al.*, 1982; Heldin, Wasteson and Westermark, 1982; Cochran, 1985; Hart *et al.*, 1987; Daniel *et al.*, 1987). The PDGF receptor, encoded in humans by a gene located on chromosome 5, has an aminoterminal extracellular PDGF binding domain, a hydrophobic transmembranous sequence, and a carboxyterminal intracellular domain which includes a protein kinase region. When activated by PDGF binding to the extracellular domain, this intrinsic protein kinase catalyses autophosphorylation of receptor tyrosine residues and also the phosphorylation of tyrosines in several lower molecular weight peptides (Pike *et al.*, 1983; Heldin, Ek and Ronnstrand, 1983; Frackelton, Tremble and Williams, 1984; Cochran, 1985; Daniel *et al.*, 1985, 1987; Yarden *et al.*, 1986).

PDGF–PDGF receptor complexes are rapidly internalized at 37 °C but not at 5 °C (Rosenfeld, Bowen-Pope and Ross, 1984; Hart *et al.*, 1987; Daniel *et al.*, 1987). PDGF–PDGF receptor binding is inhibited by the heterocyclic anionic compound, suramin, and by anti-PDGF antibodies, and these reagents block both the biological response to PDGF and cellular transformation by simian sarcoma virus (Williams *et al.*, 1984; Leal *et al.*, 1985; Johnsson *et al.*, 1985; Westermark and Heldin, 1986; Yarden *et al.*, 1986; Coffey *et al.*, 1987).

Effects of platelet-derived growth factor on cell metabolism

Platelet-derived growth factor renders its target cells competent to progress to mitosis if subsequently provided with PDGF-poor plasma or with epidermal growth factor and somatomedin C (Pledger *et al.*, 1977; Smith and Stiles, 1981; Olashaw and Pledger, 1983). This competence, which is only temporary, is a consequence not of persistent association of PDGF with the target cells but of a series of immediate (latency 1–10 min) and rapid (latency 30 min to 3 h) alterations in cell metabolism that follow autophosphorylation of the PDGF receptor. The immediate events include increased turnover of phosphatidylinositol, intracellular alkalinization, increased flux through the hexose monophosphate pathway, diminished plasma membrane epidermal growth factor binding, and reorganization of actin microfilaments (Heldin, Wasteson and Westermark, 1982; Collins, Sinnett-Smith and Rozengurt, 1983; Bockus and Stiles, 1984; Habenicht *et al.*, 1985; Davis and Czech, 1985; Reed *et al.*, 1987). Rapid events include induction of *c-myc* and *c-fos* oncogenes and of plasma membrane receptors for low density lipoprotein, enhanced plasma membrane binding of somatomedin C, and stimulation of chemotaxis (Witte *et al.*, 1982; Kelly *et al.*, 1983; Muller *et al.*, 1984; Cochran, 1985; Womer *et al.*, 1987; Paulsson *et al.*, 1987; Bierman *et al.*, 1987).

Effects of platelet-derived growth factor on the nervous system

Platelet-derived growth factor participates in the pathogenesis of gliomas. Intracranial injection of simian sarcoma virus elicits astrocytomas in marmosets (Deinhardt, 1980). Gliomas induced by ethylnitrosourea express high levels of *c-sis* transcripts, whereas premalignant cells in animals treated with this carcinogen do not, suggesting that an autocrine mechanism involving *c-sis* may be important in transition of glial cells in these animals from a premalignant to a malignant phenotype (Lens, Altena and Nusse, 1986). Human gliomas also express a PDGF-like protein and PDGF receptors (Nister *et al.*, 1986).

Platelet-derived growth factor also influences the proliferation and phenotype of normal CNS neuroglia and of neuroglia reacting to a brain injury. Addition of PDGF to the medium stimulates tyrosine protein kinase activity and mitosis of human glial cells cultured in the presence of serum (Heldin, Wasteson and Westermark, 1977; Ek *et al.*, 1982), and induces synthesis of the adhesive protein, thrombospondin (Frazier, 1987) by glia maintained in the absence of serum or other progression factors (Asch *et al.*, 1986). In rats, a species in which a progenitor cell for oligodendroglia and type 2 astroglia can be identified by its surface binding of the antiganglioside antibody, A2B5 (Raff, Abney and Miller, 1984; Noble and Murray, 1984; Temple and Raff, 1985), PDGF appears to be a mitogen for these progenitor cells (Dubois-Dalcq, 1987). PDGF has also been reported to induce the mitosis of surface galactocerebroside-positive differentiated rat oligodendroglia (Besnard *et al.*, 1987). In rats subjected to a brain wound, treatment with a PDGF antagonist partially prevents astrogliosis (Takamiya *et al.*, 1986).

REGULATION OF CELL–CELL ADHESION: NEURAL CELL ADHESION MOLECULE

Appropriate cell–cell adhesion could be specified during development of the nervous system and other tissues either by a sufficiently large number of surface components to ensure discrete recognition by each interacting cell pair (Meyer and Sperry, 1973) or by a few cell surface adhesion molecules that are rendered specific by developmental modulation of their properties (Edelman, 1983). Probably both mechanisms occur. For example, interactions between carbohydrate-binding cell surface lectins on one neuron and cell surface glycoconjugates on another appear to contribute to the organization of neurons within dorsal root ganglia (Regan *et al.*, 1986). Interactions between molecules of neural cell adhesion molecule (N-CAM) on neighbouring axons contribute to axonal fasciculation, and those between N-CAM molecules on neurons and neuroglia to adhesion of migrating neurons to immature neuroglia (Edelman, 1983; Edelman and Thiery, 1985; Martini and Schachner, 1986; Rathjen *et al.*, 1987; Chang, Rathjen and Raper, 1987).

The adhesion mediated by N-CAM is *homotypic* and *divalent cation-independent*, and in these ways can be discriminated from heterotypic mechanisms (for example, those mediated by neuron–glial cell adhesion molecule, Ng-CAM) and adhesion that is calcium-dependent (for example, mediated by uvomorulin or by liver cell adhesion molecule, L-CAM) (Grabel *et al.*, 1983; Ogou, Yoshida-Mor and Takeichi, 1983; Shirayoshi, Okada and Takeichi, 1983; Edelman, 1983, 1984; Cunningham *et al.*, 1984; Faissner *et al.*, 1984; Rathjen and Schachner, 1984; Rathjen *et al.*, 1987). Further, N-CAM is expressed very early in embryonic development, in multiple germ layers, and on neuroglia as well as on neurons, whereas Ng-CAM appears later, is restricted to the nervous system, and in the nervous system to neurons (Grumet *et al.*, 1984; Thiery *et al.*, 1985; Keilhauer, Faissner and Schachner, 1985; Bhat and Silberberg, 1985; Hoffman *et al.*, 1986; Daniloff *et al.*, 1986; Rieger *et al.*, 1986; Covault and Sanes, 1986; Chuong, Crossin and Edelman, 1987; Richardson *et al.*, 1987). Also, unlike adhesive molecules like uvomorulin which participate specifically in the assembly of tight junctions and desmosomes, and hence can be classified as junction adhesion molecules (JAMs) (Cowin and Garrod, 1983; Boller, Vestweber and Kemmler, 1985; Obrink, 1986), N-CAM is more generally distributed on neural cell surfaces.

Structure and synthesis of neural cell adhesion molecule

The human N-CAM gene is on the distal long arm of chromosome 11, and is 90% homologous with the mouse gene (Nguyen *et al.*, 1986). Different N-CAM polypeptides are translated from N-CAM mRNA species that arise by tissue-specific alternate splicing of transcripts from the single N-CAM gene (Murray *et al.*, 1986a, 1986b; Covault *et al.*, 1986); these peptides differ in the lengths of their intracellular, carboxyterminal segment (Frelinger and Rutishauser, 1986), but not in their extracellular domains (Cunningham *et al.*, 1987).

The primary sequence of N-CAM includes five intrachain tandem repeats near the aminoterminal end, centred on five pairs of cysteine residues 50–56 residues apart. Since there appear to be no free sulphydryl groups or interchain disulphide bonds in N-CAM, these cysteine residues are likely to form intrachain bonds, allowing for a linear array of five loops (Cunningham *et al.*, 1987). This structure is similar to the sequence of repeats in the variable and constant immunoglobulin domains, and to repeat sequences in myelin-associated glycoprotein (MAG), expressed by oligodendroglia and Schwann cells, lymphocyte T4 protein, and *thy*-1, which is present, like N-CAM, on the surfaces of neurons and astroglia (Pruss, 1979; Fields, Currie and Dutton, 1982; Williams and Gagnon, 1982; van Rus *et al.*, 1985; Hemperly *et al.*, 1986; Arquint *et al.*, 1987; Lai *et al.*, 1987; Parekh *et al.*, 1987; Cunningham *et al.*, 1987).

Analysis of N-CAM epitopes, plotted using a panel of monoclonal antibodies, has defined three extracellular domains. One, closest to the plasma membrane, is a region bearing polysialylated carbohydrate chains that crossreact with antibodies to the carbohydrate portion of MAG (Kruse *et al.*, 1984; Noronha *et al.*, 1986). The other two, situated more distally from the plane of the plasma membrane, are a binding site for heparin and the homophilic N-CAM binding site (Frelinger and Rutishauser, 1986).

Developing brain contains an enzyme, poly-α-2,8-sialosyl sialyltransferase, which catalyses the elongation of the polysialyl carbohydrate side chain of N-CAM (McCoy, Vimr and Troy, 1985) to a linear component at least 12 sialyl residues in length (Finne and Makela, 1985). CNS enzymes responsible for degradation of these unusual polysialylated carbohydrate chains have not yet been characterized.

Mechanism of action of neural cell adhesion molecule

The adhesivity of N-CAM is developmentally regulated by two mechanisms: modulation of its level of surface expression; and shortening of its polysialylated side chains. N-CAM is expressed continuously in CNS, transiently in neural crest and placodes of the developing peripheral nervous system, and also transiently during development of heart, kidney, limb buds, and other tissues (Edelman, 1983; Maier *et al.*, 1986). N-CAM covers all plasma membrane surfaces of immature neurons, and is particularly abundant on axonal growth cones, but its expression is considerably diminished on the surface of adult neurons (van den Pol, di Porzio and Rutishauser, 1986; Rieger *et al.*, 1986; Martini and Schachner, 1986). Similarly, N-CAM is uniformly expressed on the surface of myotubes, lost from all but the junctional membrane after innervation is established, and reappears over the entire muscle fibre surface after denervation (Daniloff *et al.*, 1986; Covault and Sanes, 1986; *see also* Walsh and Modi, Chapter 6 of this volume). Expression of N-CAM on the surfaces of Schwann cells and PC12 phaeochromocytoma cells is induced by nerve growth factor (Prentice *et al.*, 1987; Seilheimer and Schachner, 1987).

The embryonic form of N-CAM is microheterogeneous in sialic acid; this heterogeneity may contribute to the variable adhesivity necessary for specificity during development (Edelman, 1983). During maturation, N-CAM is progressively converted from the embryonic high sialic acid form to an adult form lower in sialic acid content (Chuong and Edelman, 1984; Richardson *et al.*, 1987). N-CAM adhesivity is augmented rather than inhibited by desialylation (Edelman, 1983, 1984). Transition from the embryonic to the adult form of N-CAM is retarded in the staggerer mutant mouse (Edelman and Chuong, 1982).

The precision of neural cell interactions is diminished by inhibition of the homotypic N-CAM interaction system by antibody administration (Fraser *et al.*, 1984; Silver and Rutishauser, 1984); such antibodies also inhibit cellular interactions during development of other tissues in which N-CAM is transiently expressed (Maier *et al.*, 1986). Antibodies have also been useful in localizing the domains of N-CAM necessary for homotypic interaction; those recognizing epitopes in the sialic acid-rich domain of N-CAM do not block adhesion, whereas both the heparan sulphate proteoglycan analogue heparin and antibodies against epitopes in the heparan sulphate-binding domain do (Watanabe, Frelinger and Rutishauser, 1986; Cole, Loewy and Glaser, 1986; Cole *et al.*, 1986).

REGULATION OF CELL–SUBSTRATUM ADHESION: LAMININ

Laminin, a very large, cruciate cell-binding protein (Engel *et al.*, 1981; Palm, McCarthy and Furcht, 1985; Horwitz *et al.*, 1985) with binding sites for both type IV collagen and heparan sulphate proteoglycan (Charons *et al.*, 1986; Aumailley *et al.*, 1987), is one of a group of extracellular matrix proteins that participate in the regulation of neuronal migration and neurite formation (Yamada, 1983; Cole and Glaser, 1984; Grumet *et al.*, 1985; Kruse *et al.*, 1984; Bronner-Fraser, 1986; Hantaz-Abrose, Vigny and Koenig, 1987).

Structure of laminin

Laminin is a very large glycoprotein (M_r 800 000–900 000) that is composed of three peptides, an A chain (M_r 400 000), a β_1 chain (M_r 225 000), and a β_2 chain (M_r

205 000) (Engel *et al.*, 1981; Edgar, Timp and Thoenen, 1984; Palm, McCarthy and Furcht, 1985; Engvall *et al.*, 1986; Aumailley *et al.*, 1987; Graf *et al.*, 1987; Sasaki *et al.*, 1987). The β_1 gene has been cloned, and contains a cysteine-rich segment homologous to epidermal growth factor (Sasaki *et al.*, 1987) that is one of the principal sites mediating cell attachment and cell migration (Graf *et al.*, 1987). Laminin is cross-shaped, with one long arm and two short arms (Engel *et al.*, 1981; Palm, McCarthy and Furcht, 1985). The heparan sulphate proteoglycan binding domain is located at the globular end of the long arm, a domain participating in neurite outgrowth is more proximal on the long arm, at least one of the regions for binding to cells (Acheson *et al.*, 1986) is in the cross-arm region, and type IV collagen domains are on the short arms (Edgar, Timpl and Thoenen, 1984; Charons *et al.*, 1986; Engvall *et al.*, 1986).

Mechanism of action of laminin

Receptors for laminin

Several monoclonal antibodies have been raised that recognize cell surface epitopes necessary for laminin binding (Horwitz *et al.*, 1985; Hall, Neugebauer and Reichardt, 1987; Cohen *et al.*, 1987). *In vivo* administration of such antibodies induces abnormalities in neural tube development and neural crest migration (Bronner-Fraser, 1986), and in tissue culture they block laminin-dependent neurite development (Cohen *et al.*, 1987). A cDNA for a laminin receptor protein expressed by human carcinoma cells has been partially sequenced (Weyer *et al.*, 1986). Laminin also adheres to sulphogalactosylceramide (sulphatide), and this glycolipid may contribute to cell membrane laminin binding (Roberts *et al.*, 1985).

Biological effects of laminin

Temporal and spatial laminin gradients, as well as age-dependent changes in neuronal affinity for laminin, help to dictate neuronal migration and neurite formation during development (Acheson *et al.*, 1986; Rogers *et al.*, 1986; Hammarback and Letourneau, 1986; Hall, Neugebauer and Riechardt, 1987; Gundersen, 1987). Contact with laminin also favours neuronal survival and differentiation (Adler, Jerdan and Hewitt, 1985; Acheson *et al.*, 1986). Laminin binds to other extracellular matrix components (*see above*) and exerts its effects on neurons in conjunction with them (Hantaz-Ambrose, Vigny and Koenig, 1987); for example, antilaminin antibodies do not block the stimulatory effect of basal lamina complex on neurite formation, but this stimulation is blocked by an antibody against an epitope expressed by the laminin–heparan sulphate proteoglycan complex (Chiu, Matthew and Patterson, 1986).

References

ACHESON, A. L., NAUJOKS, K. and THOENEN, H. (1984) Nerve growth factor-mediated enzyme induction in primary cultures of bovine adrenal chromaffin cells: specificity and level of regulation. *Journal of Neuroscience*, **4**, 1771–1780

ACHESON, A., EDGAR, D., TIMPL, R. and THOENEN, H. (1986) Laminin increases both levels and activity of tyrosine hydroxylase in calf adrenal chromaffin cells. *Journal of Cell Biology,* **102,** 151–159

ADLER, R., JERDAN, L. and HEWITT, A. T. (1985) Responses of cultured neural retinal cells to substratum-bound laminin and other extracellular matrix molecules. *Developmental Biology,* **112,** 100–114

ARQUINT, M., RODER, J., CHIA, L. S. *et al.* (1987) Molecular cloning and primary structure of myelin-associated glycoprotein. *Proceedings of the National Academy of Sciences USA,* **84,** 600–604

ASCH, A. S., LEUNG, L. L. K., SHAPIRO, J. and NACHMAN, R. L. (1986) Human brain glial cells synthesize thrombospondin. *Proceedings of the National Academy of Sciences USA,* **83,** 2904–2908

ASSOULINE, A. G., BOSCH, P., LIM, R., KIM, I. S., KENSEN, R. and PANTAZIS, N. J. (1987) Rat astrocytes and Schwann cells in culture synthesize nerve growth factor-like neurite-promoting factors. *Developmental Brain Research,* **31,** 103–118

AUMAILLEY, M., NURCOMBE, V., EDGAR, D., PAULSSON, M. and TIMPL, R. (1987) The cellular interactions of laminin fragments. *Journal of Biological Chemistry,* **262,** 11532–11538

BANDTLOW, C. E., HEUMANN, R., SCHWAB, M. E. and THOENEN, H. (1987) Cellular localization of nerve growth factor synthesis by *in situ* hybridization. *EMBO Journal,* **6,** 891–899

BARRETT, T. B., GAJDUSEK, C. M., SCHWARTZ, S. M., McDOUGALL, J. K. and BENDITT, E. P. (1984) Expression of the *sis* gene by endothelial cells in culture and *in vivo. Proceedings of the National Academy of Sciences USA,* **81,** 6772–6774

BARTH, E. M., KORSCHING, S. and THOENEN, H. (1984) Regulation of nerve growth factor synthesis and release in organ cultures of rat iris. *Journal of Cell Biology,* **99,** 839–843

BERG, D. K. (1984) New neuronal growth factors. *Annual Reviews of Neuroscience,* **7,** 149–170

BERND, P. (1986) Characterization of nerve growth factor binding to cultured neural crest cells: evidence of an early developmental form of the NGF receptor. *Developmental Biology,* **115,** 415–424

BERND, P. and GREENE, L. A. (1984) Association of ^{125}I-nerve growth factor with PC12 pheochromocytoma cells. *Journal of Biological Chemistry,* **259,** 15509–15516

BESNARD, F., PERRAUD, F., SENSENBRENNER, M. and LABOURDETTE, G. (1987) Platelet-derived growth factor is a mitogen for glial but not for neuronal rat brain cells *in vitro. Neuroscience Letters,* **73,** 287–292

BETSHOLTZ, C., JOHNSSON, A., HELDIN, C. H. *et al.* (1986) cDNA sequence and chromosomal localization of human platelet-derived growth factor A-chain and its expression in tumour cell lines. *Nature (London),* **320,** 695–699

BHAT, S. and SILBERBERG, D. H. (1985) Rat oligodendrocytes have cell adhesion molecules. *Developmental Brain Research,* **19,** 139–145

BIERMAN, A. J., TERTOOLEN, L. G. L., DE LAAT, S. W. and MOOLENAAR, W. H. (1987) The Na^+/H^+ exchanger is constitutively activated in P19 embryonal carcinoma cells, but not in a differentiated derivative. *Journal of Biological Chemistry,* **262,** 9621–9628

BOCKUS, B. J. and STILES, C. D. (1984) Regulation of cytoskeletal architecture by platelet-derived growth factor, insulin, and epidermal growth factor. *Experimental Cell Research,* **153,** 186–197

BOLLER, K., VESTWEBER, D. and KEMLER, R. (1985) Cell-adhesion molecule uvomorulin is localized in the intermediate junctions of adult intestinal epithelial cells. *Journal of Cell Biology,* **100,** 327–332

BOWEN-POPE, D. F. and ROSS, R. (1982) Platelet-derived growth factor. II. Specific binding to cultured cells. *Journal of Biological Chemistry,* **257,** 5161–5171

BOWEN-POPE, D. F., MALPASS, T. W., FOSTER, D. M. and ROSS, R. (1984) Platelet-derived growth factor *in vivo*: levels, activity, and rate of clearance. *Blood,* **64,** 458–469

BRONNER-FRASER, M. (1986) An antibody to a receptor for fibronectin and laminin perturbs cranial neural crest development *in vivo. Developmental Biology,* **117,** 528–536

CHANG, S., RATHJEN, F. G. and RAPER, J. A. (1987) Extension of neurites on axons is impaired by antibodies againsts specific neural crest surface glycoproteins. *Journal of Cell Biology,* **104,** 355–362

CHARONS, A. S., TSILIBAR, E. C., SAKU, T. and FURTHMAYR, H. (1986) Inhibition of laminin self-assembly and interactions with type IV collagen by antibodies to the terminal domain of the long arm. *Journal of Cell Biology,* **103,** 1681–1697

CHIU, A. Y., MATTHEW, W. D. and PATTERSON, P. H. (1986) A monoclonal antibody that blocks the activity of a neurite regeneration-promoting factor: studies on the binding site and its localization *in vivo. Journal of Cell Biology,* **103,** 1383–1398

CHUONG, C. M., and EDELMAN, G. M. (1984) Alterations in neural cell adhesion molecules during development of different regions of the nervous system. *Journal of Neuroscience,* **4,** 2354–2368

CHUONG, C. M., CROSSIN, K. L. and EDELMAN, G. M. (1987) Sequential expression and differential function of multiple adhesion molecules during the formation of cerebellar cortical layers. *Journal of Cell Biology,* **104,** 331–342

COCHRAN, B. H. (1985) The molecular action of platelet-derived growth factor. *Advances in Cancer Research,* **45,** 183–216

COFFEY, R. J. JR, LEOF, E. B., SHIPLEY, G. D. and MOSES, H. L. (1987) Suramin inhibition of growth factor receptor binding and mitogenicity in AKR-2B cells. *Journal of Cellular Physiology,* **132,** 143–148

COHEN, J., BURNE, J. F., McKINLAY, C. and WINTER, J. (1987) The role of laminin and the laminin/fibronectin receptor complex in the outgrowth of retinal ganglion cell axons. *Developmental Biology,* **122,** 407–418

COLE, G. J. and GLASER, L. (1984) Cell–substratum adhesion in embryonic chick central nervous system is mediated by a 170000-mol-wt neural-specific peptide. *Journal of Cell Biology,* **99,** 1605–1612

COLE, G. J., LOEWY, A. and GLASER, L. (1986) Neuronal cell-cell adhesion depends on interactions of N-CAM with heparin-like molecules. *Nature (London),* **320,** 445–449

COLE, G. J., LOEWY, A., CROSS, N. V., AKESON, R. and GLASER, L. (1986) Topographic localization of the heparin-binding domain of the neural cell adhesion molecule N-CAM. *Journal of Cell Biology,* **103,** 1739–1744

COLLINS, M. K. L., SINNETT-SMITH, J. W. and ROZENGURT, E. (1983) Platelet-derived growth factor treatment decreases the affinity of the epidermal growth factor receptors of Swiss 3T3 cells. *Journal of Biological Chemistry,* **258,** 11689–11693

COLLINS, T., BONTHRON, D. T. and ORKIN, S. H. (1987) Alternative RNA splicing affects function of encoded platelet-derived growth factor A chain. *Nature (London),* **328,** 621–625

COLLINS, T., GINSBURG, D., BOSS, J. M., ORKIN, S. H. and POBER, J. (1985) Cultured human endothelial cells express platelet-derived growth factor chain 2: cDNA cloning and structural analysis. *Nature (London),* **316,** 748–750

COVAULT, J. and SANES, J. R. (1986) Distribution of N-CAM in synaptic and extrasynaptic portions of developing and adult skeletal muscle. *Journal of Cell Biology,* **102,** 716–730

COVAULT, J., MERLIE, J. P., GORIDIS, C. and SANES, J. R. (1986) Molecular forms of N-CAM and its RNA in developing and denervated skeletal muscle. *Journal of Cell Biology,* **102,** 731–739

COWIN, P. and GARROD, D. R. (1983) Antibodies to epithelial desmosomes show wide tissue and species cross-reactivity. *Nature (London),* **302,** 148–150

CUNNINGHAM, B. A., LEUTZINGER, Y., GALLIN, W. J., SORKIN, B. C. and EDELMAN, G. M. (1984) Linear organization of the liver cell adhesion molecule L-CAM. *Proceedings of the National Academy of Sciences USA,* **81,** 5787–5791

CUNNINGHAM, B. A., HEMPERLY, J. J., MURRAY, B. A., PREDIGER, E. A., BRACKENBURY, R. and EDELMAN, G. M. (1987) Neural cell adhesion molecule: structure, immunoglobulin-like domains, cell surface modulation, and alternative RNA splicing. *Science,* **236,** 799–806

DANIEL, T. O., TREMBLE, P. M., FRACKELTON, A. R. and WILLIAMS, L. T. (1985) Purification of the platelet-derived growth factor receptor by using an antiphosphotyrosine antibody. *Proceedings of the National Academy of Sciences USA,* **82,** 2684–2687

DANIEL, T. O., MILFAY, D. F., ESCOBEDO, J. and WILLIAMS, L. T. (1987) Biosynthetic and glycosylation studies of cell surface platelet-derived growth factor receptors. *Journal of Biological Chemistry,* **262,** 9778–9784

DANILOFF, J. K., LEVI, G., GRUMET, M., RIEGER, F. and EDELMAN, G. M. (1986) Altered expression of neuronal cell adhesion molecules induced by nerve injury and repair. *Journal of Cell Biology,* **103,** 929–945

DAVIES, A. M. and LINDSAY, R. M. (1985) The cranial sensory ganglia in culture: differences in the response of placode-derived and neural crest-derived neurons to nerve growth factor. *Developmental Biology,* **111,** 62–72

DAVIES, A. M., THOENEN, H. and BARDE, Y. A. (1986) Different factors from the central nervous system and periphery regulate the survival of sensory neurones. *Nature (London),* **319,** 497–499

DAVIS, R. J. and CZECH, M. P. (1985) Platelet-derived growth factor mimics phorbol diester action on epidermal growth factor receptor phosphorylation at threonine-654. *Proceedings of the National Academy of Sciences USA,* **82,** 4080–4084

DEINHARDT, F. (1980) Biology of primate retroviruses. In: *Viral Oncology,* edited by G. Klein, pp. 357–398. New York: Raven Press

DEUEL, T. F., TONG, B. D. and HUANG, J. S. (1985) Platelet-derived growth factor: structure, function, and roles in normal and transformed cells. *Current Topics in Cellular Regulation,* **26,** 51–64

DICOU, E., HOULGATTE, R. and BRACHET, P. (1986) Synthesis and secretion of β-nerve growth factor by mouse teratocarcinoma cell lines. *Experimental Cell Research,* **167,** 287–294

DUBOIS-DALCQ, M. (1987) Characterization of a slowly proliferative cell along the oligodendrocyte differentiation pathway. *EMBO Journal,* **6,** 2587–2595

EDELMAN, G. M. (1983) Cell adhesion molecules. *Science,* **219,** 450–457

EDELMAN, G. M. (1984) Cell adhesion and morphogenesis: the regulator hypothesis. *Proceedings of the National Academy of Sciences USA,* **81,** 1460–1464

EDELMAN, G. M. and CHUONG, C. M. (1982) Embryonic to adult conversion of neural cell adhesion molecules in normal and staggerer mice. *Proceedings of the National Academy of Sciences USA,* **79,** 7036–7040

EDGAR, D., TIMPL, R. and THOENEN, H. (1984) The heparin binding domain of laminin is responsible for its effects on neurite outgrowth and neuronal survival. *EMBO Journal,* **3,** 1463–1468

EDWARDS, R. H., SELBY, M. J. and RUTTER, W. J. (1986) Differential RNA splicing predicts two distinct nerve growth factor precursors. *Nature (London),* **319,** 784–787

EK, B., WESTERMARK, B., WASTESON, A. and HELDIN, C. H. (1982) Stimulation of tyrosine-specific phosphorylation by platelet-derived growth factor. *Nature (London),* **295,** 419–420

ENGEL, J., ODERMATT, E., ENGEL, A., MADRI, J. A., FURTHMAYR, H., ROHDE, H. and TIMPL, R. (1981) Shapes, domain organizations and flexibility of laminin and fibronectin, two multifunctional proteins of the extracellular matrix. *Journal of Molecular Biology,* **150,** 97–120

ENGVALL, E., DAVIS, G. E., DICKERSON, K., RUOSLAHTI, E., VARON, S. and MANTHORPE, M. (1986) Mapping of domains in human laminin using monoclonal antibodies: localization of the neurite-promoting site. *Journal of Cell Biology,* **103,** 2457–2465

FAISSNER, A., KRUSE, J., GORIDIS, C., BOCK, E. and SCHACHNER, M. (1984) The neural cell adhesion molecule L1 is distinct from the N-CAM related group of surface antigens BSP-2 and D2. *EMBO Journal,* **3,** 733–737

FIELDS, K. L., CURRIE, D. N. and DUTTON, G. R. (1982) Development of *thy*-1 antigen on cerebellar neurons in culture. *Journal of Neuroscience,* **2,** 663–673

FINNE, J. and MAKELA, P. H. (1985) Cleavage of the polysialosyl units of brain glycoproteins by a bacteriophage endosialidase. *Journal of Biological Chemistry,* **260,** 1265–1270

FISCHER, W., WICTORIN, K., BJORKLUND, A., WILLIAMS, L. R., VARON, S. and GAGE, F. H. (1987) Amelioration of cholinergic neuron atrophy and spatial memory impairment in aged rats by nerve growth factor. *Nature (London),* **329,** 65–68

FRACKELTON, A. R. JR, TREMBLE, P. M. and WILLIAMS, L. T. (1984) Evidence for the platelet-derived growth factor-stimulated tyrosine phosphorylation of the platelet-derived growth factor receptor *in vivo*. *Journal of Biological Chemistry,* **259,** 7909–7915

FRANCKE, U., DeMARTINVILLE, B., COUSSENS, L. and ULLRICH, A. (1983) The human gene for the β subunit of nerve growth factor is located on the proximal short arm of chromosome 1. *Science,* **222,** 1248–1250

FRASER, S. E., MURRAY, B. A., CHUONG, C. M. and EDELMAN, G. M. (1984) Alteration of the retinotectal map in *Xenopus* by antibodies to neural cell adhesion molecules. *Proceedings of the National Academy of Sciences USA,* **81,** 4222–4226

FRAZIER, W. A. (1987) Thrombospondin: a nodular adhesive glycoprotein of platelets and nucleated cells. *Journal of Cell Biology,* **105,** 625–632

FRELINGER, A. L. III and RUTISHAUSER, U. (1986) Topography of N-CAM structural and functional determinants. II. Placement of monoclonal antibody epitopes. *Journal of Cell Biology,* **103,** 1729–1737

FURUKAWA, Y., FURUKAWA, S., SATOYOSHI, E. and HAYASHI, K. (1986) Catecholamines induce an increase in nerve growth factor content in the medium of mouse L-M cells. *Journal of Biological Chemistry,* **261,** 6039–6047

FURUKAWA, S., FURUKAWA, Y., SATOYOSHI, E. and HAYASHI, K. (1987) Synthesis/secretion of nerve growth factor is associated with cell growth in cultured mouse astroglial cells. *Biochemical and Biophysical Research Communications,* **142,** 395–402

GOEDERT, M., OTTEN, U., HUNT, S. P., BOND, A., CHAPMAN, D., SCHLUMPF, M. and LICHTENSTEIGER, W. (1984) Biochemical and anatomical effects of antibodies against nerve growth factor on developing rat sensory ganglia. *Proceedings of the National Academy of Sciences USA,* **81,** 1580–1584

GRABEL, L. B., SINGER, M. S., MARTIN, G. R. and ROSEN, S. D. (1983) Teratocarcinoma stem cell adhesion: the role of divalent cations and a cell surface lectin. *Journal of Cell Biology,* **96,** 1532–1537

GRAF, J., IWAMOTO, Y., SASAKI, M., MARTIN, G. R., KLEINMAN, H. K., ROBEY, F. A. and YAMADA, Y. (1987) Identification of an amino acid sequence in laminin mediating cell attachment, chemotaxis, and receptor binding. *Cell,* **48,** 989–996

GREEN, S. H. and GREENE, L. A. (1986) A single M_r = 103 000 ^{125}I-β-nerve growth factor-affinity-labeled species represents both the low and high affinity forms of the nerve growth factor receptor. *Journal of Biological Chemistry,* **261,** 15316–15326

GREEN, S. H., RYDEL, R. E., CONNOLLY, J. L. and GREENE, L. A. (1986) PC12 cell mutants that possess low- but not high-affinity nerve growth factor receptors neither respond to nor internalize nerve growth factor. *Journal of Cell Biology,* **102,** 830–843

GREENE, L. A., DREXLER, S. A., CONNOLLY, J. L., RUKENSTEIN, A. and GREEN, S. H. (1986) Selective inhibition of responses to nerve growth factor and of microtubule-associated protein phosphorylation by activators of adenylate cyclase. *Journal of Cell Biology,* **103,** 1967–1978

GREENE, L. A., LIEM, R. K. H. and SHELAWSKI, M. L. (1983) Regulation of a high molecular weight microtubule-associated protein in PC12 cells by nerve growth factor. *Journal of Cell Biology,* **96,** 76–83

GREINER, C. A. M., LLOYD, A. T. and GUROFF, G. (1986) Ontogeny of the nerve growth factor receptor in primary cultures of neural crest cells. *Developmental Brain Research,* **26,** 145–150

GROB, P. M., ROSS, A. H., KOPROWSKI, H. and BOTHWELL, M. (1985) Characterization of the human melanoma nerve growth factor receptor. *Journal of Biological Chemistry,* **260,** 8044–8049

GROTENDORST, G., SEPPA, H. E. J., KLEINMAN, H. K. and MARTIN, G. (1981) Attachment of smooth muscle cells to collagen and their migration toward platelet-derived growth factor. *Proceedings of the National Academy of Sciences USA,* **78,** 3669–3672

GROTENDORST, G. R., CHANG, T., SEPPA, H. E. J., KLEINMAN, H. K. and MARTIN, G. R. (1982) Platelet-derived growth factor is a chemoattractant for vascular smooth muscle cells. *Journal of Cellular Physiology,* **113,** 261–266

GRUMET, M., HOFFMAN, S., CHUONG, C. M. and EDELMAN, G. M. (1984) Polypeptide components and binding functions of neuron–glia adhesion molecules. *Proceedings of the National Academy of Sciences USA,* **81,** 7989–7993

GRUMET, M., HOFFMAN, S., CROSSIN, K. L. and EDELMAN, G. M. (1985) Cytotactin, an extracellular matrix protein of neural and non-neural tissues that mediates glia–neuron interaction. *Proceedings of the National Academy of Sciences USA,* **82,** 8075–8079

GUNDERSEN, R. W. (1987) Response of sensory neurites and growth cones to patterned substrata of laminin and fibronectin *in vitro. Developmental Biology,* **121,** 423–431

GURNEY, M. E., HEINRICH, S. A., LEE, M. R. and YIN, H. S. (1986a) Molecular cloning and expression of neuroleukin, a neuronotrophic factor for spinal and sensory neurons. *Science,* **234,** 566–574

GURNEY, M. E., APATOFF, B. R., SPEAR, G. T., BAUMEL, M. J., ANTEL, J. P., BANA, M. B. and REIDER, A. T. (1986b) Neuroleukin: a lymphokine product of lectin-stimulated T-cells. *Science,* **234,** 574–581

HABENICHT, A. J. R., GLOMSET, J. A., GOERIG, M., GRONWALD, R., GRULICK, J., LOTH, U. and SCHETTLER, G. (1985) Cell cycle-dependent changes in arachidonic acid and glycerol metabolism in Swiss 3T3 cells stimulated by platelet-derived growth factor. *Journal of Biological Chemistry,* **260,** 1370–1373

HALL, D. E., NEUGEBAUER, K. M. and REICHARDT, L. F. (1987) Embryonic neural retinal cell response to extracellular matrix proteins: developmental changes and effects of the cell substratum attachment antibody (CSAT). *Journal of Cell Biology,* **104,** 623–634

HALEGOUS, S. and PATRICK, J. (1980) Nerve growth factor mediates phosphorylation of specific proteins. *Cell,* **22,** 571–581

HAMA, T., HUANG, K. P. and GUROFF, G. (1986) Protein kinase C as a component of a nerve growth factor-sensitive phosphorylation system in PC12 cells. *Proceedings of the National Academy of Sciences USA,* **83,** 2353–2357

HAMMARBACK, J. A. and LETOURNEAU, P. C. (1986) Neurite extension across regions of low cell-substratum adhesivity: implications for the guidepost hypothesis of axonal pathfinding. *Developmental Biology,* **117,** 655–662

HANNINK, M. and DONOGHUE, D. J. (1986) Biosynthesis of the *v-sis* gene product: signal sequence, cleavage, glycosylation, and proteolytic processing. *Molecular and Cellular Biology,* **6,** 1343–1348

HANTAZ-AMBROSE, D., VIGNY, M. and KOENIG, J. (1987) Heparan sulfate proteoglycan and laminin mediate two different types of neurite outgrowth. *Journal of Neuroscience,* **7,** 2293–2304

HART, C. E., SEIFERT, R. A., ROSS, R. and BOWEN-POPE, D. F. (1987) Synthesis, phosphorylation, and degradation of multiple forms of the platelet-derived growth factor receptor studied using a monoclonal antibody. *Journal of Biological Chemistry,* **262,** 10780–10785

HEFTI, F., HARTIKKA, J., SALVATIERRA, A., WEINER, W. J. and MASH, D. C. (1986) Localization of nerve growth factor receptors in cholinergic neurons of the human basal forebrain. *Neuroscience Letters,* **69,** 37–41

HELDIN, C. H., EK, B. and RONNSTRAND, L. (1983) Characterization of the receptor for platelet-derived growth factor on human fibroblasts. *Journal of Biological Chemistry,* **258,** 10054–10061

HELDIN, C. H., WASTESON, A. and WESTERMARK, B. (1977) Partial purification and characterization of platelet factors stimulating the multiplication of normal human glial cells. *Experimental Cell Research,* **109,** 429–437

HELDIN, C. H., WASTESON, A. and WESTERMARK, B. (1982) Interaction of platelet-derived growth factor with its fibroblast receptor. *Journal of Biological Chemistry,* **257,** 4216–4221

HEMPERLY, J. J., MURRAY, B. A., EDELMAN, G. M. and CUNNINGHAM, B. A. (1986) Sequence of a cDNA clone encoding the polysialic acid-rich and cytoplasmic domains of the neural cell adhesion molecule N-CAM. *Proceedings of the National Academy of Sciences USA*, **83**, 3037–3041

HENDRY, I. A., STOCKEL, K., THOENEN, H. and IVERSEN, L. L. (1974) The retrograde axonal transport of nerve growth factor. *Brain Research*, **68**, 103–121

HEUMANN, R. and THOENEN, H. (1986) Comparison between the time course of changes in nerve growth factor protein levels and those of its messenger RNA in the cultured rat iris. *Journal of Biological Chemistry*, **261**, 9246–9249

HEUMANN, R., KORSCHING, S., SCOTT, J. and THOENEN, H. (1984) Relationship between levels of nerve growth factor (NGF) and its messenger RNA in sympathetic ganglia and peripheral target tissues. *EMBO Journal*, **3**, 3183–3189

HOFFMAN, S., FRIEDLAND, D. R., CHUONG, C. M., GRUMET, M. and EDELMAN, G. M. (1986) Differential contributions of Ng-CAM and N-CAM to cell adhesion in different neural regions. *Journal of Cell Biology*, **103**, 145–158

HOGUE-ANGELETTI, R. and BRADSHAW, R. A. (1971) Nerve growth factor from mouse submaxillary glands: amino acid sequence. *Proceedings of the National Academy of Sciences USA*, **68**, 2417–2420

HORWITZ, A., DUGGAN, K., GREGGS, R., DECKER, C. and BUCK, C. (1985) The cell substrate attachment (CSAT) antigen has properties of a receptor for laminin and fibronectin. *Journal of Cell Biology*, **101**, 2134–2144

HOSANG, M. and SHOOTER, E. M. (1987) The internalization of nerve growth factor by high-affinity receptors on pheochromocytoma PC12 cells. *EMBO Journal*, **6**, 1197–1202

HUANG, J. S., HUANG, S. S., KENNEDY, B. and DEUEL, T. F. (1982) Platelet-derived growth factor. Specific binding to target cells. *Journal of Biological Chemistry*, **257**, 8130–8136

HUANG, J. S., HUANG, S. S. and DEUEL, T. F. (1983) Human platelet-derived growth factor: radioimmunoassay and discovery of a specific plasma-binding protein. *Journal of Cell Biology*, **97**, 383–388

JOHNSON, D., LANAGAN, A., BUCK, C. R. *et al.* (1986) Expression and structure of the human NGF receptor. *Cell*, **47**, 545–554

JOHNSSON, A., HELDIN, C. H., WASTESON, A. *et al.* (1984) The *c-sis* gene encodes a precursor of the B chain of platelet-derived growth factor. *EMBO Journal*, **3**, 921–928

JOHNSSON, A., BETSHOLTZ, C., VON DER HELM, K., HELDIN, C. H. and WESTERMARK, B. (1985) Platelet-derived growth factor agonist activity of a secreted form of the *v-sis* oncogene product. *Proceedings of the National Academy of Sciences USA*, **82**, 1721–1725

KEILHAUER, G., FAISSNER, A. and SCHACHNER, M. (1985) Differential inhibition of neurone–neurone, neurone–astrocyte and astrocyte–astrocyte adhesion by L1, L2 and N-CAM antibodies. *Nature (London)*, **316**, 728–730

KELLY, K., COCHRAN, B. H., STILES, C. D. and LEDER, P. (1983) Cell-specific regulation of the *c-myc* gene by lymphocyte mitogens and platelet-derived growth factor. *Cell*, **35**, 603–610

KORSCHING, S., AUBURGER, G., HEUMANN, R., SCOTT, J. and THOENEN, H. (1985) Levels of nerve growth factor and its mRNA in the central nervous system of the rat correlate with cholinergic innervation. *EMBO Journal*, **4**, 1389–1393

KROMER, L. F. (1987) Nerve growth factor treatment after brain injury prevents neuronal death. *Science*, **235**, 214–216

KRUSE, J., MAILHAMMER, R., WENECKE, H., FAISSNER, A., SOMMER, I., GORIDIS, C. and SCHACHNER, M. (1984) Neural cell adhesion molecules and myelin-associated glycoprotein share a common carbohydrate moiety recognized by monoclonal antibodies L2 and HNK-1. *Nature (London)*, **311**, 153–155

LAI, C., BROW, M. A., NAVE, K. A. *et al.* (1987) Two forms of 1B236/myelin-associated glycoprotein, a cell adhesion molecule for postnatal neural development are produced by alternative splicing. *Proceedings of the National Academy of Sciences USA*, **84**, 4337–4341

LANDRETH, G. E. and RIESER, G. D. (1985) Nerve growth factor- and epidermal growth factor-stimulated phosphorylation of a PC12 cytoskeletally associated protein *in situ*. *Journal of Cell Biology*, **100**, 677–683

LARGE, T. H., BODARY, S. C., CLEGG, D. O., WESKAMP, G., OTTEN, U. and REICHARDT, L. F. (1986) Nerve growth factor gene expression in the developing rat brain. *Science*, **234**, 352–355

LEAL, F., WILLIAMS, L. T., ROBBINS, K. C. and AARONSON, S. A. (1985) Evidence that the *v-sis* gene product transforms by interaction with the receptor for platelet-derived growth factor. *Science*, **230**, 327–330

LEE, M. R., HO, D. D. and GURNEY, M. E. (1987) Functional interaction and partial homology between human immunodeficiency virus and neuroleukin. *Science*, **237**, 1047–1051

LENS, P. F., ALTENA, B. and NUSSE, R. (1986) Expression of *c-sis* and platelet-derived growth factor in *in vitro*-transformed glioma cells from rat brain tissue transplacentally treated with ethylnitrosourea. *Molecular and Cellular Biology*, **6**, 3537–3540

LESLIE, C. C., MUSSON, R. A. and HENSON, P. M. (1984) Production of growth factor activity for fibroblasts by human monocyte-derived macrophages. *Journal of Leukocyte Biology*, **36**, 143–159

LEVI-MONTALCINI, R. (1987a) The nerve growth factor: thirty-five years later. *EMBO Journal*, **6**, 1145–1154

LEVI-MONTALCINI, R. (1987b) The nerve growth factor 35 years later. *Science*, **237**, 1154–1162

LEVI-MONTALCINI, R. and ALOE, L. (1985) Differentiating effects of murine nerve growth factor in the peripheral and central nervous systems of *Xenopus laevis* tadpoles. *Proceedings of the National Academy of Sciences USA*, **82**, 7111–7115

LINDENBAUM, M. H., CARBONETTO, S. and MUSHYNSKI, W. E. (1987) Nerve growth factor enhances the synthesis, phosphorylation, and metabolic stability of neurofilament proteins in PC12 cells. *Journal of Biological Chemistry*, **262**, 605–610

LINDSAY, R. M. and ROHRER, H. (1985) Placodal sensory neurons in culture: nodose ganglion neurons are unresponsive to NGF, lack NGF receptors but are supported by a liver-derived neurotrophic factor. *Developmental Biology*, **112**, 30–48

MAIER, C. E., WATANABE, M., SINGER, M., McQUARRIE, I. G., SUNSHINE, J. and RUTISHAUSER, U. (1986) Expression and function of neural cell adhesion molecule during limb regeneration. *Proceedings of the National Academy of Sciences USA*, **83**, 8395–8399

MARTINET, Y., BITTERMAN, P. B., MORNEX, J. F., GROTENDORST, G. R., MARTIN, G. R. and CRYSTAL, R. G. (1986) Activated human monocytes express the *c-sis* proto-oncogene and release a mediator showing PDGF-like activity. *Nature (London)*, **319**, 158–160

MARTINEZ, H. J., DREYFUS, C. F., JONAKAIT, G. M. and BLACK, I. B. (1985) Nerve growth factor promotes cholinergic development in brain striatal cultures. *Proceedings of the National Academy of Sciences USA*, **82**, 7777–7781

MARTINI, R. and SCHACHNER, M. (1986) Immunoelectron microscopic localization of neural cell adhesion molecules (L1, N-CAM, and MAG) and their shared carbohydrate epitope and myelin basic protein in developing sciatic nerve. *Journal of Cell Biology*, **103**, 2439–2448

MATSUDA, Y. and GUROFF, G. (1987) Purification and mechanism of action of a nerve growth factor-sensitive S6 kinase from PC12 cells. *Journal of Biological Chemistry*, **262**, 2832–2844

McCOY, R. D., VIMR, E. R. and TROY, F. A. (1985) CMP-NeuNAc:poly-α-2,8-sialosyl sialyltransferase and the biosynthesis of polysialosyl units in neural cell adhesion molecules. *Journal of Biological Chemistry*, **260**, 12695–12699

McGUIRE, J. C., GREENE, L. A. and FURANO, A. (1978) NGF stimulates incorporation of fucose or glucosamine into an external glycoprotein in cultured rat PC12 pheochromocytoma cells. *Cell*, **15**, 357–365

MEYER, R. L. and SPERRY, R. W. (1973) Tests for neuroplasticity in the anuran retinotectal system. *Experimental Neurology*, **40**, 525–530

MOBLEY, W. C., RUTKOWSKI, J. L., TENNEKOON, G. I., BUCHANAN, K. and JOHNSTON, M. V. (1985) Choline acetyltransferase activity in striatum of neonatal rats increased by nerve growth factor. *Science*, **239**, 284–287

MURRAY, B. A., HEMPERLY, J. J., PREDIGER, E. A., EDELMAN, G. M. and CUNNINGHAM, B. A. (1986a) Alternatively spliced mRNAs code for different polypeptide chains of the chick neural cell adhesion molecule (N-CAM). *Journal of Cell Biology*, **102**, 189–193

MURRAY, B. A., OWENS, G. C., PREDIGER, E. A., CROSSIN, K. L., CUNNINGHAM, B. A. and EDELMAN, G. M. (1986b) Cell surface modulation of the neural cell adhesion molecule resulting from alternative mRNA splicing in a tissue-specific developmental sequence. *Journal of Cell Biology*, **103**, 1431–1439

MULLER, R., BRAVO, R., BURCKHARDT, J. and CURRAN, T. (1984) Induction of *c-fos* gene and protein by growth factors precedes activation of *c-myc*. *Nature (London)*, **312**, 716–720

NGUYEN, C., MATTEI, M. G., MATTEI, J. F., SANTONI, M. J., GORIDIS, C. and JORDAN, B. R. (1986) Localization of the human NCAM gene to band q23 of chromosome 11: the third gene coding for a cell interaction molecule mapped to the distal portion of the long arm of chromosome 11. *Journal of Cell Biology*, **102**, 711–715

NIMAN, H. L., HOUGHTEN, R. A. and BOWEN-POPE, D. F. (1984) Detection of high molecular weight forms of platelet-derived growth factor by sequence-specific antisera. *Science*, **226**, 701–703

NISTER, M., HELDIN, C. H. and WESTERMARK, B. (1986) Clonal variation in the production of a platelet-derived growth factor-like protein and expression of corresponding receptors in a human malignant glioma. *Cancer Research*, **46**, 332–340

NOBLE, M. and MURRAY, K. (1984) Purified astrocytes promote the *in vitro* division of a biopotential glial progenitor cell. *EMBO Journal*, **3**, 2243–2247

NORONHA, A. B., ILYAS, A., ANTONICEK, H., SCHACHNER, M. and QUARLES, R. H. (1986) Molecular specificity of L2 monoclonal antibodies that bind to carbohydrate determinants of neural cell adhesion molecules and their resemblance to other monoclonal antibodies recognizing the myelin-associated glycoprotein. *Brain Research,* **385,** 237–244

OBRINK, B. (1986) Epithelial cell adhesion molecules. *Experimental Cell Research,* **163,** 1–21

OGOU, S. I., YOSHIDA-NOR, C. and TAKEICHI, M. (1983) Calcium-dependent cell-cell adhesion molecules common to hepatocytes and teratocarcinoma stem cells. *Journal of Cell Biology,* **97,** 944–948

OLASHAW, N. E. and PLEDGER, W. J. (1983) Association of platelet-derived growth factor-induced protein with nuclear material. *Nature (London),* **306,** 272–274

OWEN, A. J., PANTAZIS, P. and ANTONIADES, H. N. (1984) Simian sarcoma virus-transformed cells secrete a mitogen identical to platelet-derived growth factor. *Science,* **225,** 54–56

PALM, S. L., McCARTHY, J. B. and FURCHT, L. T. (1985) Alternative model for the internal structure of laminin. *Biochemistry,* **24,** 7753–7760

PAREKH, T. B., TSE, A. G. D., DWEK, R. A., WILLIAMS, A. F. and RADEMACHER, T. W. (1987) Tissue-specific N-glycosylation, site-specific oligosaccharide patterns and lentil lectin recognition of rat *thy*-1. *EMBO Journal,* **6,** 1233–1244

PATRICK, J. and STALLCUP, W. (1977) α-Bungarotoxin binding and cholinergic receptor function on a rat sympathetic nerve line. *Journal of Biological Chemistry,* **252,** 8629–8633

PAULSSON, Y., BYWATER, M., HELDIN, C. H. and WESTERMARK, B. (1987a) Effects of epidermal growth factor and platelet-derived growth factor on *c-fos* and *c-myc* mRNA levels in normal human fibroblasts. *Experimental Cell Research,* **171,** 186–194

PAULSSON, Y., HAMMACHER, A., HELDIN, C. H. and WESTERMARK, B. (1987b) Possible positive autocrine feedback in the prereplicative phase of human fibroblasts. *Nature (London),* **328,** 715–717

PIKE, L. J., BOWEN-POPE, D. F., ROSS, R. and KREBS, E. G. (1983) Characterization of platelet-derived growth factor-stimulated phosphorylation in cell membranes. *Journal of Biological Chemistry,* **258,** 9383–9390

PLEASURE, D., KREIDER, B., SOBUE, G., ROSS, A., KOPROWSKI, H., SONNENFELD, K. and RUBENSTEIN, A. (1986) Schwann-like cells cultured from human dermal neurofibromas: immunohistological identification and response to Schwann cell mitogens. *Annals of the New York Academy of Science,* **486,** 227–240

PLEDGER, W. J., STILES, C. D., ANTONIADES, H. N. and SCHER, C. D. (1977) Induction of DNA synthesis in BALB/c 3T3 cells by serum components: reevaluation of the commitment process. *Proceedings of the National Academy of Sciences USA,* **74,** 4481–4485

PRENTICE, H. M., MOORE, S. E., DICKSON, J. G., DOHERTY, P. and WALSH, F. S. (1987) Nerve growth factor-induced changes in neural cell adhesion molecule (N-CAM) in PC12 cells. *EMBO Journal,* **6,** 1859–1863

PRUSS, R. M. (1979) *Thy*-1 antigen on astrocytes in long-term cultures of rat central nervous system. *Nature (London),* **280,** 688–690

RAFF, M. C., ABNEY, E. R. and MILLER, R. H. (1984) Two glial lineages diverge prenatally in rat optic nerve. *Developmental Biology,* **106,** 53–60

RAINES, E. W., BOWEN-POPE, D. F. and ROSS, R. (1984) Plasma binding proteins for platelet derived growth factor that inhibit its binding to cell-surface receptors. *Proceedings of the National Academy of Sciences USA,* **81,** 3424–3428

RATHJEN, F. G. and SCHACHNER, M. (1984) Immunocytological and biochemical characterization of a new neuronal cell surface component (L1 antigen) which is involved in cell adhesion. *EMBO Journal,* **3,** 1–10

RATHJEN, F. G., WOLFF, J. M., FRANK, R., BONHOEFFER, F., RUTISHAUSER, U. and SCHOEFFSKI, A. (1987) Membrane glycoproteins involved in neurite fasciculation. *Journal of Cell Biology,* **104,** 343–353

REED, B. Y., KING, M. T., GITOMER, W. L. and VEECH, R. L. (1987) Early metabolic effects of platelet-derived growth factor and transforming growth factor-β in rat liver *in vivo. Journal of Biological Chemistry,* **262,** 8712–8715

REGAN, L. J., DODD, J., BARONDES, S. H. and JESSELL, T. M. (1986) Selective expression of endogenous lactose-binding lectins and lactoseries glycoconjugates in subsets of rat sensory neurons. *Proceedings of the National Academy of Sciences USA,* **83,** 2248–2252

RENNERT, P. D. and HEINRICH, G. (1986) Nerve growth factor mRNA in brain: localization by *in situ* hybridization. *Biochemical and Biophysical Research Communications,* **138,** 813–818

RETTIG, W. J., THOMSON, T. M., SPENGLER, B. A., BIEDLER, J. L. and OLD, L. J. (1986) Assignment of human nerve growth factor receptor gene to chromosome 17 and regulation of receptor expression in somatic cell hybrids. *Somatic Cell and Molecular Genetics,* **12,** 441–447

RICH, K. M., YIP, H. K., OSBORNE, P. A., SCHMIDT, R. E. and JOHNSON, E. M. JR (1984) Role of nerve growth factor in the adult dorsal root ganglia neuron and its response to injury. *Journal of Comparative Neurology,* **230,** 110–118

RICHARDSON, G. P., CROSSIN, J. L., CHUONG, C. M. and EDELMAN, G. M. (1987) Expression of cell adhesion molecules during embryonic induction. III. Development of the otic placode. *Developmental Biology,* **119,** 217–230

RIEGER, F., DANILOFF, J. K., PINCON-RAYMOND, M., CROSSIN, K. L., GRUMET, M. and EDELMAN, G. M. (1986) Neuronal cell adhesion molecules and cytotactin are colocalized at the node of Ranvier. *Journal of Cell Biology,* **103,** 379–391

ROBERTS, D. D., RAO, C. N., MAGNANI, J. L., SPITALNIK, S.L., LIOTTA, L. A. and GINSBURG, V. (1985) Laminin binds specifically to sulfated glycolipids. *Proceedings of the National Academy of Sciences USA,* **82,** 1306–1310

ROGERS, S. L., EDSON, K. J., LETOURNEAU, P. C. and McLOON, S. C. (1986) Distribution of laminin in the developing peripheral nervous system of the chick. *Developmental Biology,* **113,** 429–435

ROHRER, H. (1985) Non-neuronal cells from chick sympathetic and dorsal root sensory ganglia express catecholamine uptake and receptors for nerve growth factor during development. *Developmental Biology,* **111,** 95–107

ROSENFELD, M. E., BOWEN-POPE, D. F. and ROSS, R. (1984) Platelet-derived growth factor: morphologic and biochemical studies of binding, internalization, and degradation. *Journal of Cellular Physiology,* **121,** 263–274

ROSS, R., RAINES, E. W. and BOWEN-POPE, D. F. (1986) The biology of platelet-derived growth factor. *Cell,* **46,** 155–169

ROSS, A., HERLYN, M., MAUL, G. G. *et al.* (1986) The nerve growth factor receptor in normal and transformed neural crest cells. *Annals of the New York Academy of Science,* **486,** 115–131

ROWLAND, E. A., MULLER, T., GOLDSTEIN, M. and GREENE, L. A. (1987) Cell-free detection and characterization of a novel nerve growth factor-activated protein kinase in PC12 cells. *Journal of Biological Chemistry,* **262,** 7504–7513

RUDY, B., KIRSCHENBAUM, B. and GREENE, L. A. (1982) Nerve growth factor-induced increase in saxitoxin binding to rat PC12 pheochromocytoma cells. *Journal of Neuroscience,* **2,** 1405–1411

RUSH, R. A. (1984) Immunohistochemical localization of endogenous nerve growth factor. *Nature (London),* **312,** 364–367

SASAKI, M., KATO, S., KOHNO, K., MARTIN, G. and YAMADA, Y. (1987) Sequence of the cDNA encoding the laminin B1 chain reveals a multidomain protein containing cysteine-rich repeats. *Proceedings of the National Academy of Sciences USA,* **84,** 935–939

SEELEY, P. J., RUKENSTEIN, A., CONNOLLY, J. L. and GREENE, L. A. (1984) Differential inhibition of nerve growth factor and epidermal growth factor effects on the PC12 pheochromocytoma line. *Journal of Cell Biology,* **98,** 417–426

SEILHEIMER, B. and SCHACHNER, M. (1987) Regulation of neural cell adhesion molecule expression on cultured mouse Schwann cells by nerve growth factor. *EMBO Journal,* **6,** 1611–1616

SEPPA, H., GROTENDORST, G., SEPPA, S., SCHIFFMANN, E. and MARTIN, G. R. (1982) Platelet-derived growth factor is chemotactic for fibroblasts. *Journal of Cell Biology,* **92,** 584–588

SHELTON, D. L. and REICHARDT, L. F. (1984) Expression of the β-nerve growth factor gene correlates with the density of sympathetic innervation in effector organs. *Proceedings of the National Academy of Sciences USA,* **83,** 7951–7955

SHIMOKADO, K., RAINES, E. W., MADTES, D. K., BARRETT, T. B., BENDITT, E. P. and ROSS, R. (1985) A significant part of macrophage-derived growth factor consists of at least two forms of PDGF. *Cell,* **43,** 277–286

SHIRAYOSHI, Y., OKADA, T. S. and TACKEICHI, M. (1983) The calcium-dependent cell–cell adhesion system regulates inner cell mass formation and cell surface polarization in early mouse development. *Cell,* **35,** 631–638

SIBLEY, D. R., BENOVIC, J. L., CARON, M. G. and LEFKOWITZ, R. J. (1987) Regulation of transmembrane signaling by receptor phosphorylation. *Cell,* **48,** 813–922

SILVER, J. and RUTISHAUSER, U. (1984) Guidance of optic axons *in vivo* by a preformed adhesive pathway on neuroepithelial endfeet. *Developmental Biology,* **106,** 485–499

SINGH, J. P., CHAIKIN, M. A. and STILES, C. D. (1982) Phylogenetic analysis of platelet-derived growth factor by radio-receptor assay. *Journal of Cell Biology,* **95,** 667–671

SMITH, J. C. and TILES, C. D. (1981) Cytoplasmic transfer of the mitogenic response to platelet-derived growth factor. *Proceedings of the National Academy of Sciences USA,* **78,** 4363–4367

SMITH, J. C., SINGH, J. P., LILLQUIST, J. S., GOON, D. S. and STILES, C. D. (1982) Growth factors adherent to cell substrate are mitogenically active *in situ. Nature (London),* **296,** 154–156

TAKAMIYA, Y., KOHSAKA, S., TOYA, S., OTANI, M., MIKOSHIBA, K. and TSUKADA, Y. (1986) Possible association of platelet-derived growth factor (PDGF) with the appearance of reactive astrocytes following brain injury *in situ. Brain Research,* **383,** 305–309

TANIUCHI, M. and JOHNSON, E. M. JR (1985) Characterization of the binding properties and retrograde axonal transport of a monoclonal antibody directed against the rat nerve growth factor receptor. *Journal of Cell Biology,* **101,** 1180–1186

TANIUCHI, M., CLARK, H. B. and JOHNSON, E. M. JR (1986) Induction of nerve growth factor receptor in Schwann cells after axotomy. *Proceedings of the National Academy of Sciences USA,* **83,** 4094–4098

TANIUCHI, M., SCHWEITZER, J. B. and JOHNSON, E. M. JR (1986) Nerve growth factor receptor molecules in rat brain. *Proceedings of the National Academy of Sciences USA,* **83,** 1950–1954

TARRIS, R. H., WEICHSEL, M. E. JR and FISHER, D. A. (1986) Synthesis and secretion of a nerve growth-stimulating factor by neonatal mouse astrocyte cells *in vitro. Pediatric Research,* **20,** 367–372

TEMPLE, S. and RAFF, M. C. (1985) Differentiation of a bipotential glial progenitor cell in single cell microculture. *Nature (London),* **313,** 223–225

THIERY, J. P., DELOUVEE, A., GRUMET, M. and EDELMAN, G. M. (1985) Initial appearance and regional distribution of the neuron-glia cell adhesion molecule in the chick embryo. *Journal of Cell Biology,* **100,** 442–456

TOGARI, A. and GUROFF, F. (1985) Partial purification and characterization of a nerve growth factor-sensitive kinase and its substrate from PC12 cells. *Journal of Biological Chemistry,* **260,** 3804–3811

TONG, B. D., AUER, D. E., JAYE, M. *et al.* (1987) cDNA clones reveal differences between human glial and endothelial cell platelet-derived growth factor A-chains. *Nature (London),* **328,** 619–621

UNSCKER, K., REICHERT-PREIBSCH, H., SCHMIDT, R., PETTMANN, B., LABOURDETTE, G. and SENSENBRENNER, M. (1987) Astroglial and fibroblast growth factors have neurotrophic functions for cultured peripheral and central nervous system neurons. *Proceedings of the National Academy of Sciences USA,* **84,** 5459–5463

UNSCKER, K., SKAPER, S. D. and VARON, S. (1985) Developmental changes in the responses of rat chromaffin cells to neuronotrophic and neurite-promoting factors. *Developmental Biology,* **111,** 425–433

VAN DEN POL, A. N., DI PORZIO, U. and RUTISHAUSER, U. (1986) Growth cone localization of neural cell adhesion molecule on central nervous system neurons *in vitro. Journal of Cell Biology,* **102,** 2281–2294

VAN RUS, J., GIGUERE, V., HURST, J., VAN AGTHOVEN, Г., VAN KESSEL, G., GOYERT, S. and GROSVELD, F. (1985) Chromosomal localization of the human *thy*-1 gene. *Proceedings of the National Academy of Sciences USA,* **82,** 5832–5835

WATANABE, M., FRELINGER, A. L. III and RUTISHAUSER, U. (1986) Topography of N-CAM structural and functional determinants. I. Classification of monoclonal antibody epitopes. *Journal of Cell Biology,* **103,** 1721–1727

WESTERMARK, B. and HELDIN, C. H. (1986) Platelet-derived growth factor as a mediator of normal and neoplastic cell proliferation. *Medical Oncology and Tumor Pharmacotherapy,* **3,** 177–183

WEYER, U. M., LIOTTA, L. A., JAYE, M. *et al.* (1986) Altered levels of laminin receptor mRNA in various human carcinoma cells that have different abilities to bind laminin. *Proceedings of the National Academy of Sciences USA,* **83,** 7137–7141

WHARTON, W., LEOF, E., PLEDGER, W. J. and O'KEEFE, E. J. (1982) Modulation of the epidermal growth factor receptor by platelet-derived growth factor and choleragen: effects on mitogenesis. *Proceedings of the National Academy of Sciences USA,* **79,** 5567–5571

WILLIAMS, A. F. and GAGNON, J. (1982) Neuronal cell *thy*-1 glycoprotein: homology with immunoglobulin. *Science,* **216,** 696–703

WILLIAMS, L. T., TREMBLE, P. M., LAVIN, M. F. and SUNDAY, M. E. (1984) Platelet-derived growth factor receptors form a high affinity state in membrane preparations. *Journal of Biological Chemistry,* **259,** 5287–5294

WITTE, L. D., CORNICELLI, J. A., MILLER, R. W. and GOODMAN, D. S. (1982) Effects of platelet-derived and endothelial cell-derived growth factors on the low density lipoprotein receptor pathway in cultured human fibroblasts. *Journal of Biological Chemistry,* **257,** 5392–5401

WOMER, R. B., FRICK, K., MITCHELL, C. D., ROSS, A. H., BISHAYEE, S. and SCHER, C. D. (1987) PDGF induces c-myc mRNA expression in MG-63 human osteosarcoma cells but does not stimulate cell replication. *Journal of Cellular Physiology,* **132,** 65–72

YARDEN, Y., ESCOBEDO, J. A., KUANG, W. J. *et al.* (1986) Structure of the receptor for platelet-derived growth factor helps define a family of closely related growth factor receptors. *Nature (London),* **323,** 226

YAMADA, K. M. (1983) Cell surface interactions with extracellular materials. *Annual Reviews of Biochemistry,* **52,** 761–799

YAN, Q, and JOHNSON, E. M. JR (1987) A quantitative study of the developmental expression of nerve growth factor (NGF) receptor in rats. *Developmental Biology,* **121,** 139–148

YASUDA, T., SOBUE, G., MOKUNO, K., KREIDER, B. and PLEASURE, D. (1987) Cultured rat Schwann cells express low affinity receptors for nerve growth factor. *Brain Research,* in press

YEH, H. J., PIERCE, G. F. and DEUEL, T. F. (1987) Ultrastructural localization of a platelet-derived growth factor/*v-sis*-related protein(s) in cytoplasm and nucleus of simian sarcoma virus-transformed cells. *Proceedings of the National Academy of Sciences USA,* **84,** 2317–2321

ZIMMERMANN, A. and SUTTER, A. (1983) Beta-nerve growth factor (β-NGF) receptors on glial cells. Cell–cell interaction between neurones and Schwann cells in cultures of chick sensory ganglia. *EMBO Journal,* **2,** 879–885

6
Gene expression in skeletal muscle

Frank S. Walsh and Girish Modi

INTRODUCTION

One of the fundamental problems of modern cell biology is how tissue-specific gene expression is controlled and how it results in the differentiated adult phenotype. Skeletal muscle is an ideal model system for a molecular analysis of the differentiated state as its formation and development are associated with major morphological and biochemical changes. In addition, a variety of model systems using tissue culture or animal lesions have been described and are amenable to molecular studies. A major expectation of the detailed analysis of gene expression in muscle is that it may provide insights into the molecular processes that may be abnormally regulated in skeletal muscle disease. If specific defects cannot be identified, at least the stage at which abnormal regulation becomes apparent should be more clear.

Skeletal muscle develops from the embryonic mesoderm, and a schematic diagram of the major stages in muscle development is shown in *Figure 6.1*. The precursor cells giving rise to the myogenic lineage are also thought to be the source

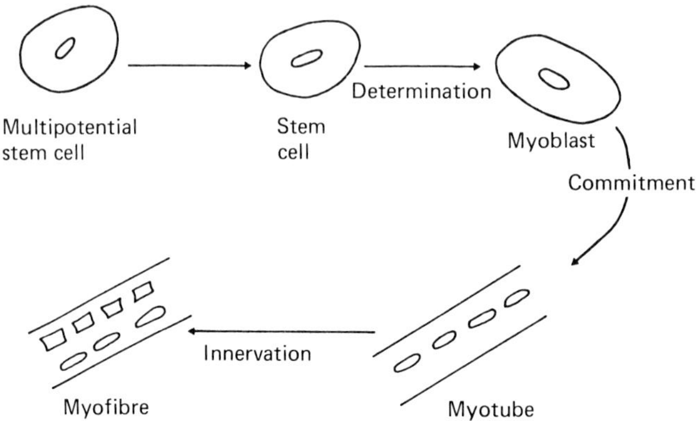

Figure 6.1 Schematic diagram of the origin and development of skeletal muscle cells from a stem cell through to an innervated myofibre

of fibroblast and chondrocyte lineages. These mesodermal cells are arranged as somites around the central cavity of the early embryo and generate the cell lineages that give rise to the limb buds. The molecular correlates of determination to a particular lineage are not known and much effort has been put into defining model systems to study this process. One *in vitro* model system that appears to be particularly promising is the conversion of C3H10T½ 'fibroblasts' to myoblasts and other lineages (Taylor and Jones, 1979; Konieczny and Emerson, 1984; Lassar, Paterson and Weintraub, 1986). Treatment of 10T½ fibroblasts with 5-azacytidine, an agent that demethylates DNA, results in transformation of these undifferentiated cells into clonally stable populations of myoblasts, adipocytes and chondrocytes, with the majority of clones choosing the myoblast pathway. Thus 10T½ cells appear to be a multipotential mesodermal stem cell that offer the opportunity of analysing the mechanisms of determination to a specific lineage. Transfection studies have shown that DNA from parental 10T½ cells do not cause the transformation to myogenic cells, while transfection with genomic DNA from 5-azacytidine-induced myoblasts or the mouse C2 muscle cell line does (Lassar, Paterson and Weintraub, 1986). As the conversion occurs at low frequency it can be concluded that 5-azacytidine is modifying a single locus but the nature of this locus, what changes in gene expression occur, and whether there are gene rearrangements remain undefined. The main advantage of 10T½ cells is that they represent an *in vitro* system where large numbers of clonally homogeneous cells can be grown in order to analyse muscle determination.

Myoblasts are the earliest recognizable cells of the muscle lineage, but the number of cell lineages in muscle is unclear. Muscle fibres are classically divided into fast, slow and mixed, usually based on a histochemical classification and the ultimate fibre type that a myofibre acquires has been suggested to be based purely on the characteristics of the innervating motor neuron. The assumption from these studies has been that there is only one type of muscle stem cell and that the ultimate diversity of phenotype found is neurally induced although there have been no rigorous tests of this hypothesis.

One of the problems has been the lack of availability of specific reagents that would allow fine distinctions to be made between different fibre types. The generation of monoclonal antibodies (McAb) to myosin heavy chain (MHC) isoforms has allowed a detailed analysis to be made of myogenic cell lineages in the chick embryo (Miller and Stockdale, 1986a, 1986b), and it is likely that the results of these studies will also alter current views of muscle stem cell development in mammals.

In the avian embryo myogenesis occurs in two phases. The first starts at about day 3 in ovo, is innervation-independent and produces primary myotubes. The second is innervation-dependent, starts at day 8 and produces secondary myotubes. Using specific McAbs to MHC it has been shown that early chick embryo myoblasts placed in tissue culture can form clonally inheritable and stable populations of fast, slow or mixed myofibres. Myoblasts from late embryos only produced fast myofibres (Miller and Stockdale, 1986a, 1986b). These results indicate that myofibre diversity can occur in the absence of neural influence and suggest that it is the myoblast itself rather than the exogenous influences to which it is subjected, that ultimately determines myofibre type. These studies have not yet been extended to mammals, nor has the place of the muscle satellite cell in this scheme been determined.

The myoblasts eventually line up end to end and fuse to form multinucleated myotubes (*see Figure 6.1*) that synthesize and assemble myofibrillar proteins. *In vivo* they then grow in length as further myoblast fusion occurs to form the fetal muscle fibres. The assembly of myofilaments results in the formation of myofibrils with highly organized sarcomeres. The myofibrils first appear near the centre of the cell, and further myofilaments are added to the outside of the existing ones, eventually causing the nucleus to be displaced to the periphery of the fibre. The sarcoplasmic reticulum forms later and is followed by ingrowth of the transverse tubule system.

These developmental changes, some of which occur after neural innervation, result in the formation and characteristic appearance of mature or adult skeletal muscle. The number of muscle fibres does not increase once differentiation is complete. Further growth in length follows an increase in sarcomere length and number, while increases in fibre diameter result from synthesis of myofibrils. The processes continue well into postnatal life, with myoblasts probably fusing with established muscle fibres to increase the number of nuclei in each. Some myoblasts persist into adult life as satellite cells which lie close to muscle fibres beneath their basement membranes (Bischoff, 1970). These satellite cells remain dormant and divide rarely, their role being to provide a source of myoblasts for muscle regeneration following injury or disease (Bischoff, 1970). However, whether satellite cells are a separate cell lineage and at which stage of their developmental programme they are 'frozen' is unknown. There are probably cell surface differences between myofibres and satellite cells since satellite cells are in contact with the myofibres but they only spontaneously fuse when 'activated'.

MOLECULAR CORRELATES OF MUSCLE DIFFERENTIATION

The morphological differentiation of muscle from the fusion of cells to give a syncytia of multinucleated myotubes is accompanied by changes in expression of specific proteins. These changes are specifically directed to mediate the functioning of the differentiated state. These events occur just prior to fusion – that is, once commitment to differentiation has been established. In addition to a quantitative change in protein concentration, myogenesis is more specifically associated with a qualitative change characterized by a transition in isoform expression of specific proteins. The studies of Devlin and Emerson (1978) have indicated that changes in gene expression are coordinately regulated for at least some of the proteins of the contractile apparatus. However, this concept now appears to be an oversimplification as more sensitive Northern blotting analyses show temporal differences in the accumulation of various mRNA species that were previously thought to be coordinately regulated.

THE MYOFIBRILLAR PROTEINS

The myofibrillar proteins are the most abundant proteins in muscle and isoforms have been found for all of the proteins that have been examined. They also show characteristic changes in their expression during myogenesis. The best characterized myofibrillar proteins, namely actin, myosin and tropomyosin, are present as distinct isoforms in different types of muscle tissue as well as in a wide variety of non-muscle tissues. In skeletal muscle development the complexity of myofibrillar protein expression is indicated by the presence of developmental stage specific isoforms (such as fetal, newborn, adult); developmental transitions in isoform expression (non-muscle to muscle type) take place concomitant with commitment to differentiation and developmental coexpression of isoforms (Buckingham, 1985).

Myosin is the major protein of the A filament and is a hexameric molecule consisting of four light chains (regulatory light chains and alkali light chains) and two heavy chains. In the sarcomere the heavy chains extend along the length of the tail largely in an α-helical conformation, where they are associated as a coiled coil. The two heads of the heavy chain result from a less complex secondary structure and contain one alkali light chain and one regulatory light chain. Thus at least three different polypeptide chains comprise a myosin molecule. These molecules demonstrate different ATPase activity in fast and slow fibres in adult skeletal muscle, fetal skeletal muscle, cardiac muscle and smooth muscle. In non-muscle tissues at least two different myosins have been found (brain and platelet), the functions of which are unknown. During skeletal muscle development a transition from non-muscle to muscle type myosin occurs. The mRNAs encoding the different myosins are distinct and have therefore facilitated use of cDNA probes in their analysis. Proliferating myoblasts predominantly express the non-muscle RNAs with low levels of muscle type mRNAs (Medford *et al.*, 1984). The change in expression of the different populations of myosins accompanying myogenic differentiation has been shown to be regulated in part by an alteration in the chromatin conformation of the genes encoding muscle and non-muscle myosins.

Not surprisingly, the genetic control of myosin synthesis appears complex in that there are at least seven genes for the heavy chains in mammals and almost an equal number for the light chains. Analysis of cDNA and genomic recombinant clones have indicated that the myosin heavy chains are expressed by separate genes, and it has been found that the α and β cardiac MHC are arranged in tandem on the genome as are the skeletal embryonic, neonatal and adult MHC isoforms. Diversity in the myosin light chains is generated largely by alternative RNA splicing mechanisms (Buckingham, 1985). In addition, thyroid hormones have been shown to regulate individual members of the MHC gene family in a highly specific manner depending on which muscle is analysed. Thus, in addition to neural influences, the MHC genes are under the control of hormones.

Actin is the major constituent of the I filament. At least six different isotypes have been isolated (Vandekerckove and Weber, 1979). The cytoplasmic β and γ actins are coexpressed in all non-muscle cells whilst α actin is expressed as distinct isoforms in cardiac and skeletal muscle. The striking feature of the actins is the strong conservation in their amino acid sequences. Sequence analysis of each of the mRNAs encoding the different actins has indicated that they are sufficiently different to be encoded by separate genes which are present at single copy per haploid genome. The amino terminus of the actin molecule which is involved in the major actin–myosin interactions is highly conserved. During skeletal muscle development there is an initial transition from non-muscle to muscle phenotype with proliferating myoblasts expressing the β and γ actins and myotubes predominantly α actin. In developing muscle, both types of α actin (cardiac and skeletal) are expressed, but in the neonate the skeletal isoform predominates. No difference in actin expression is noted between fast and slow skeletal muscle. The conservation in the structure of actins, as well as the lack of distinct skeletal muscle isoforms would indicate that the role of actin in the contractile system is similar in all muscle types. In the I filament of skeletal muscle actin interacts with various other myofibrillar proteins which exist in different isoforms in different muscle types, and it is therefore probable that the lack of actin heterogeneity serves to accommodate the maintenance of these interactions.

The tropomyosins and troponins constitute the regulatory proteins of the contractile mechanism. The tropomyosins exist as distinct muscle and non-muscle isoforms which are generated by distinct and separate genes, as well as by alternative splicing mechanisms (MacLeod *et al.*, 1985). In addition to the differential expression of these isoforms they exhibit a distinct developmental profile during myogenesis. Proliferating myoblasts express a large number of isoforms (characterized hitherto as muscle and non-muscle isoforms), while myotubes selectively express adult type tropomyosins and developed skeletal muscle expresses restricted muscle isoforms designated α and β.

CONTROL OF MUSCLE GENE EXPRESSION

The differential expression of muscle proteins during myogenesis, manifesting as a transition in phenotype, is regulated predominantly at the level of mRNA transcription, as is evident from the increased abundance of mRNAs for the muscle specific isoforms of these proteins. Characterization of the structure and organization of some of these proteins has indicated that this transcriptional change is mediated through the activation of distinct and separate genes encoding individual isoforms, as well as the differential activation of multiple transcripts from a single gene through a variety of alternative splicing mechanisms.

In eukaryotic organisms the generation of multiple isoforms by alternative RNA splicing has been described for various genes including those encoding α amylase, rat calcitonin, and *Drosophila* alcohol dehydrogenase. The calcitonin gene is of interest because alternative splicing results in the formation of different transcripts that are expressed in different tissues where they mediate different functions (Leff, Evans and Rosenfeld, 1987). Skeletal muscle genes are often regulated through alternative splicing and three general mechanisms of splicing have been identified. *Figure 6.2* shows a schematic diagram of the three mechanisms that predominate. The first (*Figure 6.2a*) is the use of multiple promoter regions, and this is a mechanism used for the myosin alkali light chains of fast muscle ($MLC1_F$, $MLC3_F$). These proteins have common COOH-terminal regions but different NH_2-terminal regions in both chick (Nabeshima *et al.*, 1984) and mouse muscle (Robert *et al.*, 1984) suggesting that the observed protein sequence diversity is generated by use of alternative promoters.

The second general mechanism is the generation of alternative spliced isoforms by use of alternative polyadenylation sites. An example of this pathway can be found by analysing the expression of tropomyosin in human fibroblasts (MacLeod *et al.*, 1985). Human fibroblasts express a number of tropomyosin genes, one of which is a 1.1 kb mRNA encoding a 284 amino acid muscle type tropomyosin. A comparison of the sequence of this transcript with human skeletal muscle β-tropomyosin shows regions of identity and non-identity. The boundaries of these are similar to those found for other non-muscle tropomyosins and it is possible to predict the model shown in *Figure 6.2b*. This study is also important as it shows that the same gene can produce functionally distinct products in different cell types and the splicing of the myofibrillar genes is clearly not restricted to muscle cells. The third mechanism involves the production of alternatively spliced transcripts where the promoter and polyadenylation sites are common. An example is the splicing of troponin T (*Figure 6.2c*).

The physiological basis for the selective advantage of encoding different protein isoforms in a single gene rather than in different genes is not clear. The muscle-specific isoforms of the myofibrillar proteins are assembled into the sarcomere, which is differentially modulated in a developmental and tissue specific manner. This requires the conservation of domains involved in intermolecular assembly, and moreover a capacity to generate diversity in the domains involved in modulation of function. It is therefore probable that maintaining the constant domain of a protein in a single copy, capable of combining with multiple variable domains with specialized characteristics, offers selective advantage despite the higher risk for lethal mutations apparently associated with single copy genes (Breitbart *et al.*, 1985). This represents one possible explanation for alternative RNA splicing as a mechanism for the regulation of gene expression – that is, it offers the flexibility of generating diversity in a reversible fashion, while maintaining the constant regions involved, in conserved intermolecular interactions.

The molecular mechanisms controlling alternative splicing are not well defined. In the examples given in *Figure 6.2a* and *b* it is possible that *cis*-acting regulatory sequences are present that cause progressive splicing in a 5′–3′ direction. The exclusion of a given exon from a mature mRNA presumably occurs as a result of flanking splice junction consensus sequences being overlooked. However, this cannot be the case for troponin T, where different mRNAs are spliced from identical primary transcripts in a regulated manner (Breitbart *et al.*, 1985). The pattern of splicing of this gene must involve *trans*-acting factors. The involvement of *trans*-acting factors in the generation of isoform diversity by alternative RNA splicing has also been shown for the rat calcitonin-CGRP gene (Leff, Evans and Rosenfeld, 1987). The biochemical nature of such *trans*-acting factors is not known at present. Various small RNA species capable of hybridizing to splice consensus sequences of mRNA have been implicated, but the universality of spliced consensus sequences in the introns argues against specific recognition of particular exons by this means. It has also been proposed that these RNA species bind to exon sequences, lending specificity to the splicing, and that changes in the secondary or tertiary structure of the differentially spliced exons might be involved in the modulation of the splicing pathway (Breitbart *et al.*, 1985). The formation of secondary structures of the gene transcript is presumed to allow the juxtaposition of particular exons and the exclusion of others. Allowable splices would be favoured by local conformations of the RNA having the least free energy, either alone or further stabilized by additional intramolecular secondary structure (*cis*), or by interaction with other factors (*trans*). Moreover, splices executed in this way at one point in the RNA molecule may generate conformational changes in the intermediate that affect splicing elsewhere.

An important aspect of myogenesis is the fact that differentiation is associated with coordinately activated/repressed proteins to mediate the needs of the differentiated state. The regulation of muscle gene expression therefore has to take into consideration the coordinate nature of gene activation since to a large extent the genes involved are located on different chromosomes. Therefore regulation by single *cis*-acting regulatory sequences is unlikely, and a complex version where a *trans*-acting signal regulates a set of dispersed genes, each containing a related *cis*-acting regulatory sequence, is more likely.

Although evidence for such mechanisms has been accumulating over the past few years there appears to be no universal sequence homology in muscle promotors. The troponin I gene contains in its 5′ upstream regions sequences homologous with that of muscle α actin genes (in the chick and rat) and myosin light chain (chicken) and cardiac myosin heavy chain gene (rat). This sequence is a 17 base pair (bp) repeat located upstream of the transcription start site (Konieczny, Baldwin and Emerson, 1986). The sequence aligns into three regions: a G-C rich region and an A-T rich region that flank a central core region of more variable sequence: GCCCGA cacc CAAATAT. The finding of homology between cardiac and skeletal genes supports the validity of these sequences as it is apparent that cardiac and skeletal genes tend to be coexpressed during muscle development. It has therefore been proposed that these homologous sequences are *cis*-acting transcriptional control elements involved in the coordinate control of muscle genes, analogous to consensus short repeat sequences involved in the regulation of heat shock genes of *Drosophila* and the steroid-induced genes. In addition, Buckingham

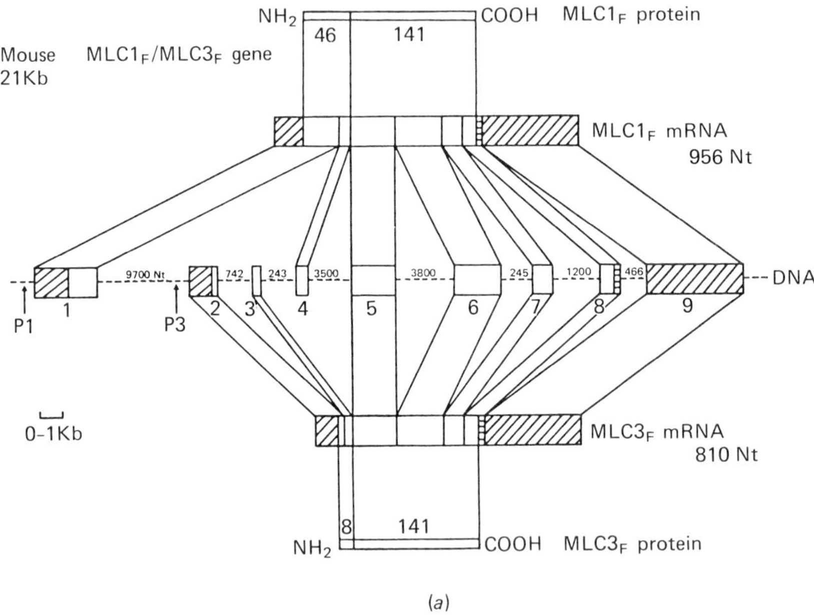

Figure 6.2 Different mechanisms of alternative splicing of myofibrillar proteins in skeletal muscle. (*a*) Myosin alkali light chain (MLC): two promoters and alternate splicing. The figure is based on Robert *et al.*, 1984, and shows the intron–exon relationship of the MLC gene and the pattern of alternative splicing. Dashed boxes show non-coding regions. Nt indicates nucleotides. Arrows indicate the positions of the promoter region for $MLC1_F$ (P1) and $MLC3_F$ (P3). (*b*) Human tropomyosin (TM): alternative polyadenylation and splicing patterns. The model assumes the existence within the functional gene of at least two exons encoding the amino acid 188 to 213 region, one containing the fibroblast (hscp36) sequence, and one containing the skeletal (SK) muscle sequence. There must also be similar duplication of an exon encoding the amino acid 258 to 284 region, and the 3′ untranslated region as well since this is specific to each mature mRNA (based on MacLeod *et al.*, 1985). (*c*) Rat α and β troponin T: alternative splicing using common promoter and polyadenylation sites (based on Medford *et al.*, 1984). Boxes represent exons and lines represent introns

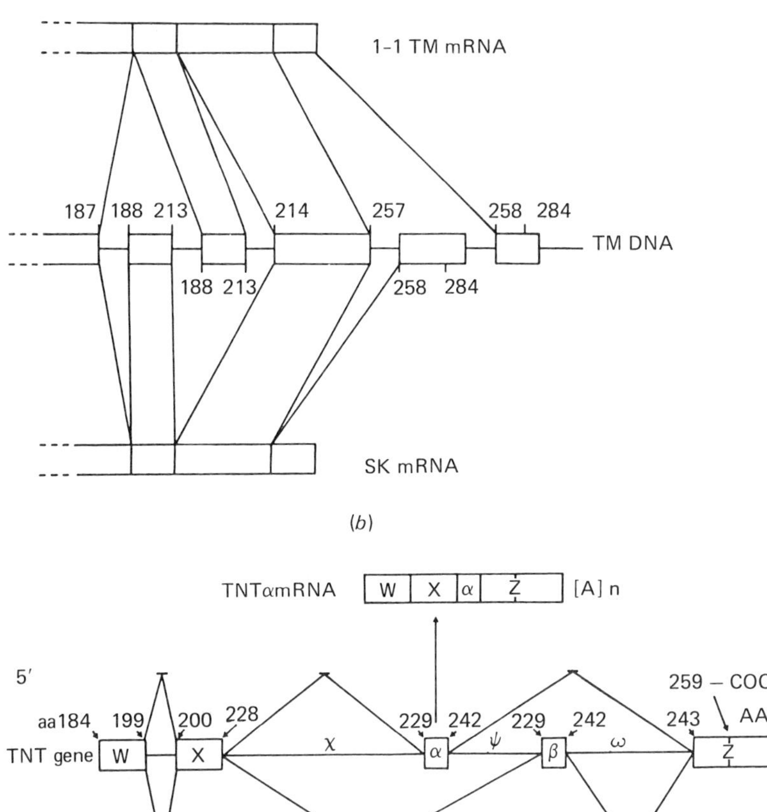

(*b*)

(*c*)

et al. (1986) have shown sequence homology in the promotor region of α cardiac actin and $MLCI_F$ genes with an adenovirus E1A-type enhancer core element. While this sequence on its own does not control myogenesis it may make the enhancer susceptible to the muscle environment. It seems likely that further detailed analysis of the promoter regions of muscle genes should shed light on the precise mechanisms that control gene expression.

THE CELL SURFACE OF SKELETAL MUSCLE CELLS

The cell surface of skeletal muscle is of importance in regulating cellular behaviour at crucial stages in its differentiation and development. Cell–cell interactions in muscle occur between like cells of similar developmental stages – for example, myoblast–myoblast – different stages – for example, myoblast–myotube – or between dissimilar cells – for example myoblast – fibroblast, myotube–neuron. It is likely that the observed high degree of specificity found in cell–cell interactions is reflected in specific patterns of expression of glycoconjugates at the cell surface. The major expectation of the analysis of cell surface glycoconjugates of muscle is that it will provide a detailed mechanistic picture of specific interactions such as myoblast fusion and the interaction of myotubes with neurons to form synapses.

Much effort has been devoted to identifying specific muscle surface proteins that are developmentally regulated. Vectorial labelling studies of the cell surface of muscle cells in culture identified a number of specific markers, but immunochemical methods have been more powerful. McAb studies have demonstrated a number of specific phenotypic markers of muscle and also a number of developmentally regulated components. Lee and Kaufman (1981) identified a number of surface antigens some of which are downregulated with fusion. They also identified an antigen reaction with McAb H36, which was expressed by myoblasts but not myotubes or fibroblasts. A similar antigenic determinant has been identified in human muscle by the McAb 24.1D5. Again this antigen is expressed by myoblasts but not myotubes or fibroblasts. The function of these two antigens is unknown at present, but as cDNA clones encoding the 24.1D5 antigen have recently been obtained and polyclonal antibodies prepared to bacterial fusion proteins, then this question can now be addressed using appropriate model systems (Walsh, unpublished observations).

One additional McAb that has been extensively used in analysing the muscle cell surface is the McAb 5.1H11 (Walsh and Ritter, 1981). This antibody reacts with myoblasts and myotubes but not fibroblasts. As such, it has been valuable for the identification of myoblasts in cell cultures from normal and diseased human skeletal muscle. It has also been used in fluorescence activated cell sorting analyses of human muscle and has allowed the isolation of pure populations of human myoblasts. This is particularly important in diseases of human muscle where there are often large number of fibroblasts contaminating cell cultures, and these cells inhibit the differentiation of myoblasts. All these immunochemical markers have been useful in describing the dynamic changes in the muscle membrane during development, but they have not yet been of value in functional studies. It is reasonable to expect that detailed analysis of one family of molecules, namely the neural cell adhesion molecules (N-CAMs) will change this.

N-CAM was originally described in brain on the basis of inhibition of a functional cell adhesion assay, and is therefore likely to be crucially important in mediating cell–cell interactions. N-CAM is expressed as a number of specific isoforms in brain of 180 kilodaltons (kd), 140 kd, and 120 kd, that can be differentially glycosylated and phosphorylated. Recently it has been shown that N-CAM is a single copy gene and that the different isoforms arise via alternative splicing.

In skeletal muscle a detailed picture of N-CAM expression has emerged (Walsh and Moore, 1985; Moore and Walsh, 1986). N-CAM is expressed on myoblasts and myotubes in cell cultures and *in vivo* is expressed by embryonic and neonatal myofibres. In adult muscle N-CAM is localized to the neuromuscular junction, but its localization is not coincident with that of acetylcholine receptors. A number of experimental or pathological lesions can reinduce N-CAM expression in muscle. Paralysis, denervation and nerve crush cause a rapid re-expression of N-CAM in myofibres. It seems likely that innervation is not the sole determining factor in regulating N-CAM levels, as experimental thyroidectomy also activates N-CAM expression and treatment of hypothyroid rats with thyroxine resulted in downregulation of N-CAM.

Western blot analysis of N-CAM isoforms has allowed changes in expression to be studied further. Myoblasts express a 145 kD transmembrane isoform that can be phosphorylated. This isoform is downregulated with fusion and is not expressed by myotubes or myofibres. Myotubes and embryonic and denervated myofibres in contrast express isoforms of 125 and 155 kD. These two isoforms are novel in that they are linked to the plasma membrane via a phosphatidylinositol linkage. This means that these isoforms can be removed from the myotube membrane by phosphatidylinositol specific phospholipase C to generate an N-CAM free myotube membrane. It is not known whether this is a mechanism for local clearing of N-CAM at the cell surface during cell–cell interactions, or for rapid release of cell–cell bonds. *Figure 6.3* is a schematic diagram of the various N-CAM isoforms in muscle.

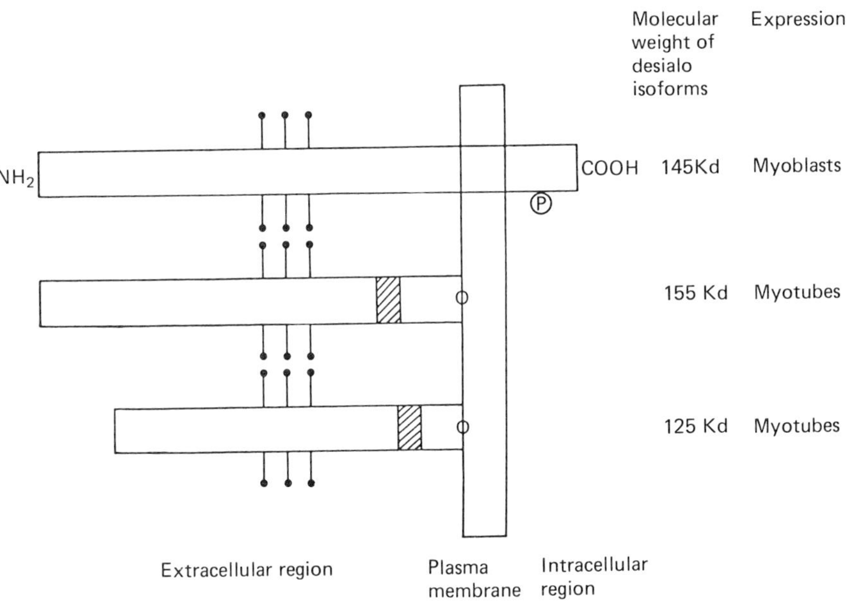

Figure 6.3 Schematic diagram of the likely structure of specific N-CAM isoforms in muscle membrane during myogenesis. ● Represents sites of carbohydrate attachment; ○ represents the phosphatidylinositol linkage; Ⓟ represents sites of phosphorylation and □ the muscle specific exon. The sizes given are for the desialo isoforms. The explanation for the size difference between the 125 kD and 155 kD isoforms is not known at present

Additional complexity in N-CAM structure was found from an analysis of cDNAs. A number of cDNA clones for muscle N-CAM have been sequenced and these encode the transmembrane and phosphatidylinositol-linked isoforms (Dickson *et al.*, 1987). Comparison of their sequences showed that the non-transmembrane clones are homologous to clones encoding 120 kD N-CAM from brain. However, in the extracellular domain there is an extra exon of 37 amino acids that encodes a muscle-specific sequence. Northern analysis has not identified this sequence in any other N-CAM isoform. The appearance of the lipid linked N-CAM isoforms occurs after myoblast fusion via an alternative splicing mechanism, so the acquisition of the differentiated muscle phenotype is accompanied by the expression of this splicing mechanism. It will be of interest to identify the role of this muscle specific exon and to define the nature of the *trans*-acting factor that controls its expression. The precision with which it is expressed and its location suggest that it may be functionally important, perhaps in mediating nerve–muscle interactions or in enhancing the stability of these N-CAM isoforms. Thus a detailed cellular and molecular analysis of N-CAM has demonstrated versatility and diversity of gene expression at the cell surface. Further use of these approaches is likely to be highly productive and informative, especially as full length cDNA clones are now available. The use of these in conjunction with transfection systems using sense, anti-sense or mutagenized constructs should ultimately define the role of this class of recognition molecule.

References

BISCHOFF, R. (1970) The myogenic stem cell in development of skeletal muscle. In: *Regeneration of Striated Muscle and Myogenesis*, edited by A. Mauro, S. A. Schafig and A. T. Milhorat, pp. 218–231. Amsterdam: Excerpta Medica

BREITBART, R. E., NGUYGEN, H. T., MEDFORD, R. M., DESTREE, A. T., MAHDAVI, V. and NADAL-GINARD, B. (1985) Intricate combinational patterns of exon splicing generate multiple regulated troponin T isoforms from a single gene. *Cell*, **41**, 67–82

BUCKINGHAM, M. E. (1985) Actin and myosin multigene families: their expression during the formation of skeletal muscle. In: *Essays in Biochemistry*, Vol. 20, edited by P. N. Campbell and R. D. Marshall, pp. 77–109. New York: Academic Press

BUCKINGHAM, M., ALONSO, S., BARTON, P. *et al.* (1986) Actin and myosin multigenic families: their expression during the formation and maturation of skeletal muscle. *American Journal of Medical Genetics*, **25**, 623–634

DEVLIN, R. and EMERSON, C. (1978) Coordinate regulation of contractile protein synthesis during myoblast differentiation. *Cell*, **13**, 599–611

DICKSON, G., GOWER, H. J., BARTON, C. H. *et al.* (1987) Human muscle neural cell adhesion molecule (N-CAM): identification of a muscle specific sequence in the extracellular domain. *Cell*, **50**, 1119–1130

KONIECZNY, S. F. and EMERSON, C. P. (1984) 5-Azacytidine induction of stable mesodermal stem cell lineages from 10T½ cells: evidence for regulatory genes controlling determination. *Cell*, **38**, 791–800

KONIECZNY, S. F., BALDWIN, A. S. and EMERSON, C. P. (1985) Myogenic determination of differentiation of 10T½ cell lineages: evidence for a simple genetic regulatory system. In: *Molecular Biology of Muscle Development, UCLA Symposium on Molecular and Cellular Biology*, new series Vol. 29, edited by C. Emerson, D. A. Fischman, B. Nadal-Ginard and M. A. Q. Siddiqui, pp. 21–34. New York: A. R. Liss Inc.

LASSAR, A. B., PATERSON, B. M. and WEINTRAUB, H. (1986) Transfection of a DNA locus that mediates the conversion of 10T½ fibroblasts to myoblasts. *Cell*, **47**, 64–656

LEE, H. U. and KAUFMAN, S. J. (1981) Use of monoclonal antibodies in the analysis of myoblast development. *Developmental Biology*, **81**, 81–95

LEFF, S. E., EVANS, R. M. and ROSENFELD, M. G. (1987) Splice commitment dictates neuron-specific alternative RNA processing in calcitonin-CGRP gene expression. *Cell*, **48**, 517–524

MacLEOD, A. R., HOULKER, C., REINACH, F. S. *et al.* (1985) Expression of a muscle-type tropomyosin in human non-muscle cells by an alternative RNA splicing mechanism. *Proceedings of the National Academy of Sciences USA*, **82**, 7835–7839

MEDFORD, R. M., NGUYEN, H. T., DESTREE, A. T., SUMMERS, E. and NADAL-GINARD, B. (1984) A novel mechanism of alternative splicing for the developmentally regulated generation of troponin T isoforms from a single gene. *Cell*, **38**, 409–421

MILLER, J. B. and STOCKDALE, F. E. (1986a) Developmental regulation of the multiple myogenic cell lineages of the avian embryo. *Journal of Cell Biology*, **103**, 2197–2208

MILLER, J. B. and STOCKDALE, F. E. (1986b) Developmental origins of skeletal muscle fibers: clonal analysis of myogenic cell lineages based on fast and slow myosin heavy chain expression. *Proceedings of the National Academy of Sciences USA*, **83**, 3860–3864

MOORE, S. E. and WALSH, F. S. (1985) Specific regulation of N-CAM/D2-CAM cell adhesion molecule during skeletal muscle development. *EMBO Journal,* **4,** 623–630

MOORE, S. E. and WALSH, F. S. (1986) Nerve dependent regulation of N-CAM expression in skeletal muscle. *Neuroscience,* **18,** 499–505

NABESHIMA, Y., FUJII-KURIYAMA, Y., MURAMATSU, M. and OGATA, K. (1984) Alternative transcription and two modes of splicing results in two myosin light chains from one gene. *Nature (London),* **308,** 333–338

ROBERT, B., DAUBAS, P., AKIMENKO, M. A. *et al.* (1984) A single locus in the mouse encodes both myosin light chain 1 and 3, a second locus corresponds to a related pseudogene. *Cell,* **39,** 129–140

TAYLOR, S. M. and JONES, P. A. (1979) Multiple new phenotypes induced in 10T½ and 3T3 cells treated with 5-azacytidine. *Cell,* **17,** 771–779

VANDEKERCKOVE, J. and WEBER, K. (1979) The complete amino acid sequence of actins from bovine aorta, bovine heart, bovine fast skeletal muscle and rabbit slow skeletal muscle. *Differentiation,* **14,** 123–133

WALSH, F. S. and RITTER, M. A. (1981) Surface antigen differentiation during human myogenesis in culture. *Nature (London),* **289,** 60–64

WALSH, F. S. and MOORE, S. E. (1985) Expression of cell adhesion molecule N-CAM in diseases of adult human skeletal muscle. *Neuroscience Letters,* **59,** 73–78

7
Host and viral genetic factors which influence viral neurotropism

Kenneth L. Tyler

INTRODUCTION

The capacity of viruses selectively to infect and injure specific regions of the central nervous system (CNS), and even specific populations of cells within these regions ('tropism'), depends on a series of complex interactions between viral genes and proteins and a number of host factors. In this chapter recent insights into the molecular and genetic basis for viral neurotropism are reviewed.

HOST FACTORS INFLUENCING SUSCEPTIBILITY TO VIRAL INFECTION

Studies using inbred strains of mice with well-defined genetic backgrounds have led to a number of important discoveries concerning the many mechanisms by which genes in the host can influence susceptibility to viral infection. The list of host genes which affect susceptibility to neurotropic viruses is extensive, and the following examples are meant to illustrate several mechanisms by which these genes appear to act rather than to serve as an exhaustive catalogue (*see* Brinton and Nathanson (1981); Brinton, Blank and Nathanson (1984); Rosenstreich, Weinblatt and O'Brien (1982) for review). In the case of some neurotropic viral diseases, such as that caused by lymphocytic choriomeningitis virus (LCM), and the chronic paralysis and demyelinating disease induced by Theiler's virus, immune-mediated tissue injury is a cardinal feature of the infection. In both these cases, differences in susceptibility to clinically apparent disease have been mapped to genes within the major histocompatibility complex (H2 in the mouse) (Buchmeier *et al.*, 1980; Clatch *et al.*, 1985). A discussion of the genetic factors that influence this type of immune-mediated tissue injury is beyond the scope of this chapter, and interested readers are referred elsewhere for more information (Notkins and Oldstone, 1985; 1986).

There are several examples where differential susceptibility to infection with viruses in different strains of inbred mice correlates with variations in the magnitude or nature of the immune response to the inciting virus. Interestingly, the host genes involved are often not linked to immune response (*Ir*) genes within the mouse major histocompatibility complex (H2). For example, A/WySnJ mice are extremely susceptible to intraperitoneal inoculations of street rabies virus, whereas SJL/J and CBA/J strain mice are relatively resistant. These differences correlate with the magnitude of the antirabies virus antibody titre, with the resistant mice having higher titres then the susceptible ones. High serum antibody titres, in turn, seem to result in decreased viral replication in the CNS (Lodmell and Ewalt, 1985). The magnitude of the serum neutralizing antibody response in inbred mice immunized with a vaccine strain of rabies virus also seems to be genetically determined. For example, C3H/J mice are 'high responders' and C57BL/6 mice are 'low responders' in terms of serum antibody levels following vaccination. Studies of the F1 hybrid cross between these two strains indicates that the 'high response' is dominant, and that it is not linked to H2 genes (Templeton *et al.,* 1986).

Another mechanism of genetically controlled resistance to viral infection is seen in A2G strain mice infected with influenza virus. Although influenza is not, strictly speaking, usually considered a neurotropic virus, the general principles involved may turn out to have broader applicability to paramyxoviruses in general and possibly other unrelated viral groups. A2G mice have a dominant allele ('Mx+') located on chromosome 16 which confers resistance to influenza virus infection. This effect appears to be mediated via α and β (not γ) interferon. Following infection with influenza virus, but not other unrelated viruses, Mx+ mice produce an interferon-induced 75 kD protein that appears to be essential for the specific anti-influenza viral resistant state. This protein alone, even in the absence of other interferon-induced proteins, is sufficient to mediate the antiviral state in these mice, although its mechanism of action remains obscure (Staehli *et al.,* 1986).

An obvious method by which host genes may affect susceptibility to viral infection is by encoding viral receptors. Mice lacking or carrying an altered form of the receptor gene would presumably be resistant to receptor-mediated infection by the homologous virus. This mechanism may very well explain differences in the susceptibility of some strains of mice to infection with mouse hepatitis virus (MHV), a coronavirus. BALB/c mice are susceptible to MHV infection, whereas SJL/J mice are resistant. The resistant locus ('Mhv-1') has been mapped to chromosome 7 (Knobler *et al.,* 1984). Recent studies have indicated that the resistant BALB/c mice lack receptor for MHV on intestinal epithelial cells and hepatocytes (Boyle, Weismiller and Holmes, 1987). It appears that the Mhv-1 locus actually encodes the gene for the MHV receptor (J. Boyle, personal communication).

A novel mechanism of receptor-related genetically controlled resistance to viral infection is exemplified by resistance of some mouse strains to infection with ecotropic murine retroviruses. Some of these viruses can produce paralytic disease in mice associated with spongiform neuropathological changes in the CNS. Retroviruses are RNA viruses which transcribe their genomic RNA into DNA with the aid of the enzyme RNA-dependent DNA polymerase ('reverse transcriptase'). The viral DNA can then become integrated into the genomic DNA of the host target cells. At this stage in its life cycle the retrovirus is often referred to as a 'provirus'. In some cases the integrated proviral DNA does not contain a full complement of all the retroviral genes required for encoding all the viral proteins ('defective provirus'). A defective provirus is unable to activate from its integrated state to produce infectious virion particles. Mice carrying the Fv-4 allele in their genome inherit a defective endogenous ecotropic retrovirus. The envelope (*env*) gene of the provirus, which is the viral cell attachment protein, is transcribed in host cells and viral envelope protein is produced. This envelope protein then binds to the receptor on the host cell surface used by ecotropic retroviruses, and prevents infection (due to receptor blockade) by exogenous ecotropic retroviruses (Yuen *et al.*, 1986).

Host genes are also related to susceptibility of mice to infection with the still unidentified agent responsible for scrapie. Dr Stanley Prusiner and his colleagues have suggested that scrapie is caused by an infectious protein ('prion') devoid of detectable nucleic acid. The major component of scrapie prions is a 27–30 kD protein designated PrP 27–30. This protein is encoded by a cellular gene, located on mouse chromosome 2, that is found in both scrapie-infected and uninfected animals. mRNA transcribed from this gene is also found in both healthy and infected animals, although the ultimate protein product appears to differ in terms of several biological properties in the two instances. One of the genes that influences susceptibility to scrapie infection in the mouse ('Prn-1') is closely linked to the gene encoding PrP 27–30 (Carlson *et al.*, 1986). It remains to be established whether other genes known to influence scrapie susceptibility or incubation time (for example, the scrapie incubation or *Sinc* gene) are identical to Prn-1, allelic, or unrelated but also linked to the PrP 27–30 gene (*see* Hope and Kimberlin, 1987, for review).

Host genes also appear to play a role in determining susceptibility to latent as well as acute viral infections. Following inoculation of herpes simplex virus type 1 (HSV-1) into the lip of an adult mouse, virus replicates locally and then spreads retrogradely through the peripheral branches of the trigeminal nerve to reach the trigeminal ganglion. During the acute phase of viral infection (which lasts for about 2 weeks) HSV-1 can be isolated from tissue homogenates of the trigeminal ganglion. Following this time period virus is no longer detectable in trigeminal tissue homogenates. However, virus can be recovered if the trigeminal ganglion is explanted, minced into small pieces, and co-cultivated with an indicator cell line (for example, CV-1 cells). When a large number of inbred strains of mice are tested for their susceptibility to latent HSV-1 infection, some strains (for example, C57BL, BALB/cByJ) are resistant to the development of latent infections whereas others are extremely susceptible (for example, PL/J, LP/J, CBA/J). Resistance appears to be a dominant trait, which is not linked to H2 genes (Kastrukoff, Lau and Puterman, 1986). Interestingly, resistance to development of latent HSV-1 infection, and resistance to acute infection with HSV-1, do not always cosegregate in inbred strains of mice (Lopez, 1975; Kastrukoff, Lau and Puterman, 1986). At least two independently segregating, non-H2-linked, genes govern resistance to acute HSV-1 infection in the mouse (Lopez, 1980).

The mechanism(s) responsible for the resistance of mice to either acute or latent infections with HSV remain unknown. Interferon may play a role in some types of resistance. Evidence supporting this view comes from studies in which administration of anti-interferon (α/β) antibodies converted resistant strains of mice to susceptibility to HSV-1 infection (Gresser *et al.*, 1976). The specific pathway and cellular targets through which endogenous interferon acts to induce HSV-1 resistance are not known. Macrophages would appear to be a likely candidate cell, as macrophages from resistant mouse strains do not appear to support HSV-1 replication to nearly the same degree as cells from susceptible strains. Anti-interferon (α/β) antibody treatment of the macrophages from resistant mice renders them susceptible to HSV-1 infection. Support for the idea that it is the macrophages that are the target cell for endogenous interferon action comes from studies with C3H/HeJ mice. These mice carry an autosomal gene defect (Lps^{d}) on chromosome 4 that makes them resistant to the effects of endotoxin (lipopolysaccharide). Resistance to endotoxin appears to be due to abnormalities in macrophage function. It has been suggested that one abnormality in macrophage function in these mice is the deficient production of endogenous interferon, and that this accounts for the heightened susceptibility of these mice to infection by a wide variety of pathogens including viruses (Vogel and Fertsch, 1987). So in this case, susceptibility to viral infection appears to be linked to a gene that regulates specific aspects of macrophage function, a similar mechanism could theoretically occur to account for some cases of HSV-1 susceptibility as well. It is important to recognize, however that other arms of the immune response including T-lymphocytes, macrophages, and natural-killer cells have all been shown to play a role in resistance to HSV infections (*see* Rosenstreich, Weinblatt and O'Brien, 1982 for review).

The examples reviewed above illustrate the fact that genetically controlled resistance to viral infections may operate through a wide variety of mechanisms. Interestingly, in virtually every case identified to date the genes mediating resistance and susceptibility appear to be unique for different viruses, and seem to segregate independently. This suggests that the mechanisms by which host cell genes influence susceptibility to viral infections may very well be as diverse as viruses themselves.

VIRAL FACTORS

Site of entry and route of spread

It has been known for many years that neurotropic viruses depend primarily on spread through the bloodstream or via nerves to reach the CNS from their diverse sites of entry into the host (*see* Johnson, 1982, for review). Viruses which are capable of spreading via nerves include rabies, herpes simplex, herpes simiae, polioviruses and reovirus type 3. Since nerves provide specific pathways of connectivity between sites in the periphery and discrete, defined regions of the CNS, it is not surprising that for virtually all neurally spreading viruses that the ultimate distribution within the CNS can be profoundly influenced by their initial site of entry into the host (*see*, for example, Sabin and Olitsky, 1937, 1938; Howe and Bodian, 1942; and, more recently, Martin and Dolivo, 1983; Anderson and Field, 1983). The effect of site of entry on CNS tropism has been less clearly defined for blood-borne viruses. Since the pathway of spread of a virus may affect its CNS tropism, it is appropriate to review briefly the current state of our understanding of the role played by viral genes in determining specific patterns of viral spread within an infected host.

Recent studies with reovirus reassortants and intertypic recombinants of HSV have provided some insights into the molecular and genetic basis for viral spread. Although reoviruses are not important human pathogens, they have proved to be valuable experimental models for identifying the role of individual viral genes and the proteins they encode at specific stages in viral pathogenesis (*see* Tyler and Fields, 1985; Sharpe and Fields, 1983, 1985, for review). Reoviruses have a genome consisting of ten discrete segments of double-stranded RNA ('genes'), each segment encoding a unique mRNA, that is translated into at least one protein product. If susceptible cells are simultaneously coinfected with reoviruses of two different serotypes, intertypic reassortants (in other words, viruses containing gene segments derived from each of the two parental serotypes) can be generated. By comparing the behaviour of these reassortant viruses to that of the two parental viruses, the role of specific viral genes and proteins in determining specific biological properties of the virus can often be determined. This approach has been used to identify the importance of the S1 viral gene in determining the capacity of reoviruses to utilize either neural or haematogenous routes to spread to the CNS in infected animals (Tyler, McPhee and Fields, 1986).

After inoculation into either the hindlimb or forelimb footpad of neonatal mice reovirus type 3 spreads to the CNS via nerves, whereas type 1 spreads via the bloodstream. Studies with intertypic (T1 × T3) reassortants, which contain various combinations of genes derived from reoviruses types 1 and 3, indicate that viruses containing an S1 gene segment derived from reovirus type 3 spread to the CNS via nerves, whereas viruses with an S1 gene segment derived from reovirus type 1 spread haematogenously (Tyler, McPhee and Fields, 1986). The S1 gene encodes a viral outer capsid protein, σ 1, that also serves as the viral cell attachment protein. Data derived from kinetic and pharmacological experiments strongly suggest that the neural spread of reovirus type 3 is via the microtubule-associated pathway of fast axonal transport, although the mechanism by which the σ 1 protein mediates this transport remains unknown.

Information is also rapidly accumulating about genes that appear to be important in determining both the capacity of HSV to spread to the CNS from peripheral sites, and its virulence after this type of inoculation. For example, by studying the spread of herpes virus from the cornea to the CNS in mice, Oakes, Gray and Lausch (1986) established that HSV-1 DNA sequences that lie between 0.31 and 0.44 map units (mu) are important in determining the capacity of the virus to spread to the CNS. When HSV-1 (strain 17) is inoculated onto the cornea of mice, it replicates in corneal epithelial cells and then spreads rapidly to the trigeminal ganglia and brain. By contrast, although HSV-2 (strain 186) replicates in the cornea, it reaches the trigeminal ganglia only in small quantities and is not detectable in the brain following corneal inoculation. Both strains of virus are equally lethal after direct intracerebral inoculation into mice. In order to determine which HSV-1 genes were responsible for these differences, an intertypic recombinant virus was generated which contained a complete HSV-2 genome except for the portion between 0.31 and 0.44 mu which was derived from HSV-1. This recombinant virus spread to the CNS in a fashion indistinguishable from HSV-1 indicating that the transferred HSV-1 DNA sequences are important in directing the capacity of herpes virus to spread to the CNS. The region in question contains genes coding for a large number of viral proteins including the nucleocapsid protein p40, glycoprotein gB, the DNA-binding protein ICP8, the viral DNA polymerase, and an origin of DNA replication. It is unclear which of these proteins play the critical role in viral spread.

A sequence in the herpesvirus genome between mu 0.762 and 0.787 appears to be important in determining the virulence of HSV-1 isolates for mice after intraperitoneal inoculation (Becker *et al.*, 1986). An overlapping portion of the genome (0.71–0.83 mu) has been shown to be a determinant of HSV-1 virulence after intracerebral inoculation into mice (Thompson *et al.*, 1985).

Cell receptors and viral cell attachment proteins

A series of classic studies performed by J. J. Holland and L. C. McLaren in the late 1950s and early 1960s using poliovirus helped to establish the principle that one of the major determinants of the tropism of viruses was the availability of receptors on target cells. In the subsequent quarter century there has been tremendous growth in understanding of the nature of viral cell attachment proteins, and in the identification of putative host cell receptors for virus. In this section selected examples designed to highlight current knowledge in this area as it pertains to neurotropic viruses are reviewed.

All three serotypes of poliovirus compete for attachment to a common receptor which differs from the receptors utilized by other enteroviruses (such as coxsackie, echovirus) and picornaviruses (such as rhinovirus, Theiler's virus, mengovirus). The poliovirus receptor is an integral membrane protein, present in several thousand copies per cell (Krah and Crowell, 1985). The gene encoding the receptor is located on human chromosome 19 (Miller *et al.*, 1974). It has recently been demonstrated that previously poliovirus-resistant cells, such as mouse L-cell fibroblasts, can be made susceptible to poliovirus infection if they are transfected with specific DNA fragments derived from a poliovirus susceptible human (HeLa) cell line (Mendelsohn *et al.*, 1986). The transfected DNA presumably contains the gene coding for the poliovirus receptor, and further characterization of the transfected DNA may lead the way toward isolation and characterization of the mammalian poliovirus receptor.

The application of the techniques of X-ray crystallography, nucleotide sequence analysis of viral genomic RNA, and the use of monoclonal antibodies has resulted in an extraordinary synergistic growth of knowledge concerning structure–function relationships for poliovirus and other related picornaviruses. This is well illustrated in the case of polioviruses.

(1) The complete nucleotide sequence of the entire genome of all three serotypes of poliovirus has now been established (*see*, for example, Stanway *et al.*, 1984).
(2) The three-dimensional crystal structure of poliovirus has recently been solved at extremely high resolution (<3 Å) (Hogle, Chow and Filman, 1985). This has allowed the accurate determination of the three-dimensional position of virtually all the amino acids composing the viral structural proteins.
(3) Neutralizing monoclonal antibodies (N-MAbs) have been generated against the major antigenic epitopes on the poliovirus surface, and variant viruses have been generated which resist neutralization by these N-MAbs (Diamond *et al.*, 1985).

These variants typically contain only single amino acid substitutions in individual viral proteins. Transposing information gained from studies with MAbs and MAb-resistant variant viruses to the three-dimensional crystal structure of the virus has allowed the identification of specific regions on the poliovirus surface that serve as the major antigenic domains for neutralizing antibody production in the host, as well as the identification of a candidate region for the viral cell attachment protein.

The poliovirus crystal structure shows that the viral outer surface is studded with 'peaks' composed of the outer capsid protein VP1. These peaks are surrounded by broad 'valleys', the floors of which are composed of portions of proteins VP1 and VP3 (Hogle, Chow and Filman, 1985). Structural features analogous to the poliovirus valley have been seen in the three-dimensional crystal structures of rhinovirus ('canyon') (Rossman *et al.*, 1985) and mengovirus ('pit') (Luo *et al.*, 1987). It has been suggested that it is these regions of these related viruses which serve as the viral attachment site to host cell receptors.

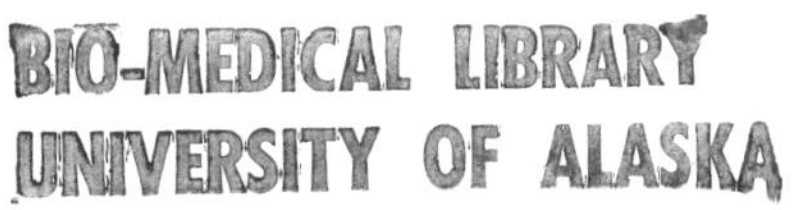

Studies with rabies virus have provided additional insights concerning the nature of viral cell attachment proteins and their host-cell receptors. Following intramuscular inoculation into experimental animals, rabies virus localizes at the neuromuscular spindles and in proximity to the neuromuscular junction (NMJ), and is then subsequently transported via nerves to reach the CNS (Murphy, 1977). It has been suggested that the acetylcholine receptor (AChR) serves as a rabies virus receptor at the NMJ. Pretreatment of cultured chick myotubes with α-bungarotoxin, which binds to the AChR, or with antibodies to the α-subunit of the AChR, blocks binding of rabies virus to these cells and inhibits infection (Lentz *et al.*, 1982; Burrage, Tignor and Smith, 1985).

The suggestion that the AChR serves as a rabies virus receptor has generated considerable controversy. Cells lacking AChR can still be infected with rabies virus *in vitro* clearly indicating that the AChR is not the exclusive receptor for rabies virus. A considerable body of experimental evidence has been accumulated that indicates that gangliosides, including phosphatidyl serine, may be important components of a rabies virus receptor distinct from the AChR (Wunner, Reagan and Koprowski, 1984).

The rabies virus genomic RNA is enclosed in a protein shell (nucleocapsid) that is surrounded by a lipoprotein envelope that is derived as the virus buds from the plasma membrane of infected cells. A virally encoded glycoprotein ('G') is inserted into this envelope. The G glycoprotein appears to function as the viral cell attachment protein and specific regions of this protein have been identified that appear to be major determinants of viral virulence *in vivo*. N-MAbs directed against the G glycoprotein have been used to generate rabies virus variants with alterations in the amino acid sequence of the G glycoprotein. Comparison of the amino acid sequences of the glycoprotein of avirulent or attenuated variants with that of fully pathogenic variants and the wild-type virus has indicated that an arginine residue at amino acid position 333 is an important marker of viral virulence (Dietzschold *et al.*, 1983; Seif *et al.*, 1985). Variants with any one of a number of amino acid substitutions at this position (for example, glutamine, glycine, isoleucine) are avirulent or have attenuated virulence, whereas variants which preserve ARG 333 (but have substitutions at nearby positions such as 330, 336, 338) remain virulent. The mechanism(s) by which changes in the rabies glycoprotein at position 333 result in altered neurovirulence are being actively investigated. Studies of several of the avirulent variants indicate that they compete for the same cellular receptor as the wild-type virus on cultured cells, and that they do not have altered CNS tropism *in vivo*. The variants appear to spread less quickly within the CNS than the wild-type virus and appear to be restricted in their capacity to spread via certain specific neural pathways, when compared to the wild-type virus (Dietzschold *et al.*, 1985; Kucera *et al.*, 1985). Thus, alterations in the rabies glycoprotein may result in changes in the capacity of certain variant viruses to spread both to and within the CNS.

Reoviruses have also provided an important experimental system for studying the nature of viral and host factors that determine neurotropism. Reoviruses are non-enveloped viruses whose genetic information is contained in the form of ten discrete segments of double-stranded (ds) RNA. As discussed earlier, it is possible to generate reassortment viruses containing combinations of genes derived from different serotypes of virus, or even from different strains of the same serotype. After intracerebral inoculation into neonatal mice reovirus serotype 1 infects the ependymal cells lining the ventricles, but does not infect neurons. Conversely, reovirus serotype 3 infects neurons and produces acute lethal encephalitis, but does not infect ependymal cells. Using reassortants derived from types 1 and 3 it was shown that the viral S1 dsRNA segment, which encodes the outer capsid protein σ1, was the major determinant of neuronal or ependymal tropism of reoviruses (Weiner, Powers and Fields, 1980). Subsequent studies have extended this observation by documenting the importance of the same viral gene for reovirus tropism within the pituitary (Onodera *et al.*, 1981), and the retina (Tyler *et al.*, 1985).

In order to define the portions of the σ 1 protein involved in biological functions mapped to the S1 gene, variants of reovirus T3 Dearing with altered σ 1 proteins have been generated using anti-σ 1 N-MAbs (Spriggs and Fields, 1982). Comparison of the predicted amino acid sequence of the σ 1 protein of five of the variant viruses with that of their wild-type parent revealed that in each case a single amino acid substitution was present at one of two locations (amino acid 340 or 419) (Bassel-Duby *et al.*, 1986). The T3 σ 1 variants have attenuated neurovirulence and altered CNS tropism compared to the wild-type virus. For example, although the variants still infect and injure the hippocampus, septal nuclei, and mamillary bodies, they no longer injure the cortex (Spriggs, Bronson and Fields, 1983; Bassel-Duby *et al.*, 1986). A reassortant virus containing only the S1 dsRNA segment derived from a variant virus shows the same pattern of restricted neurotropism and attenuated neurovirulence as the original variants, indicating that the single amino acid substitution found in the σ 1 protein is indeed responsible for the altered phenotype of these viruses (Kaye *et al.*, 1986).

A receptor for reovirus type 3 has been identified on cultured neuronal and lymphoid cells (Co *et al.*, 1985a, 1985b). The receptor is a glycoprotein of about 67 kD. A variety of structural and pharmacological criteria have been used to demonstrate that the reovirus receptor is structurally similar to the mammalian β-adrenergic receptor. It remains to be seen whether other cellular structures can also be utilized by reovirus type 3 as cellular receptors.

An increasing body of evidence has also accumulated concerning the cell attachment proteins of both mouse hepatitis virus (MHV), a coronavirus, and lymphocytic choriomeningitis (LCM) virus. The coronaviruses are enveloped RNA viruses which have two proteins (E1, E2) inserted into their envelope. Variants of MHV have been selected with N-MAbs directed against the E2 glycoprotein (Fleming *et al.*, 1986; Dalziel *et al.*, 1986). Many of the E2 variants have attenuated neurovirulence and an altered pattern of neuropathogenicity compared to their wild-type parents. The wild-type viruses typically produce a severe acute encephalitis, although some of the survivors go on to develop a chronic demyelinating disease. Many of the E2 variant viruses produce a high incidence of demyelinating disease with little or no evidence of preceding encephalitis. Passive immunization of mice with certain anti-E2 MAbs can also convert the pattern of illness from encephalitis to demyelinating disease (Buchmeier *et al.*, 1984). These findings strongly suggest that the E2 glycoprotein of coronaviruses is a major determinant of viral pathogenicity, and that it may be involved in the tropism of these virus for neurons as variation in this protein can result in diminished neuronotropism and enhanced tropism for oligodendroglia.

The receptor for neurotropic coronaviruses such as MHV on neuronal cells has not yet been identified. A 100–110 kD protein has been found on hepatocytes and intestinal epithelial cells that appears to serve as the viral receptor in these tissues (Boyle, Weismiller and Holmes, 1987). This receptor appears to be encoded on mouse chromosome 7 (J. Boyle, personal communication), at the same locus that has previously been shown to encode a gene that mediates susceptibility to MHV infection within certain strains of inbred mice (Knobler *et al.*, 1984).

Lymphocytic choriomeningitis virus (LCMV) is an enveloped virus whose genetic information is contained in two segments of RNA. Some strains of mice infected with specific strains of LCMV develop a runting syndrome. Mice with this syndrome have high titres of LCMV in the growth hormone producing cells of the anterior pituitary. Infection of these cells results in markedly reduced levels of circulating growth hormone and this in turn accounts for the runting syndrome (Oldstone *et al.*, 1985). Reassortant viruses have been generated between strains of LCMV that produce runting syndrome in susceptible mice (such as Armstrong) and strains of LCMV that do not (such as WE). It appears that the LCMV small (S) RNA segment, which encodes the two viral envelope glycoproteins, determines the capacity of these viruses to infect growth-hormone producing pituitary cells and produce runting syndrome (Riviere *et al.*, 1985). LCMV thus provides another example of the importance of viral surface proteins in determining specific patterns of neurotropism. Interestingly, the infected growth-hormone producing cells typically do not show obvious evidence of injury at either the light or electron-microscopic level. This exemplifies a point frequently stressed by Dr Oldstone and his colleagues that the so-called 'luxury functions' of cells such as neurons can be dramatically affected by viral infection, in the absence of either alterations in more basic 'housekeeping' functions or detectable morphological changes.

Mumps virus provides another illustration of the importance of envelope glycoproteins in determining viral neurotropism. Variants of a neuroadapted strain of mumps virus (Kilham) have been selected by their resistance to N-MAbs directed against the haemagglutinin–neuraminidase (HN) glycoprotein of the viral envelope. Suckling hamsters inoculated intracerebrally with mumps virus develop lethargy, muscular wasting and incoordination followed later by hydrocephalus. The mumps virus infects the ependymal cells lining the ventricles and the aqueduct of Sylvius. Hydrocephalus develops as a result of the obstruction to CSF flow that follows aqueduct stenosis. The virus also infects neurons and glial cells throughout the cerebrum and brainstem producing a severe meningoencephalitis. By contrast, one of the mumps HN variants (M13) does not produce the early illness described above, but still causes the later developing hydrocephalus. Examination of the brains of M13 infected hamsters shows evidence of ependymal cell infection identical to that seen with wild-type virus, but only minimal signs of neuronal injury (Love *et al.*, 1985). When the structure of individual polypeptides from strains of mumps virus that differ in their neuropathogenicity are compared, the HN protein appears to show the greatest degree of variation, although no one particular HN glycoprotein peptide map correlates exactly with viral virulence (Wolinsky and Server, 1985). These results support the hypothesis that the mumps virus HN protein plays an important role in determining viral neurotropism and virally induced neuronal injury.

Some of the most dramatic advances in understanding the nature of viral–host interactions have come from investigations of the human immunodeficiency virus (HIV), the agent responsible for the acquired immunodeficiency syndrome (AIDS). Studies using a variety of techniques including culture, electron microscopy, immunocytochemistry, and *in situ* hybridization, have provided clear evidence that HIV can be found in the CNS, CSF, peripheral nervous system and muscle. The evidence accumulated to date indicates that the brunt of detectable virus, viral antigen, or viral nucleic acid is found is macrophages, endothelial cells, and to a much lesser extent, glial cells. There is, as yet, no unequivocal evidence for neuronal infection (Ho *et al.*, 1985; Shaw *et al.*, 1985; Gabuzda *et al.*, 1986; Koenig *et al.*, 1986; Wiley *et al.*, 1986).

The nature of the cell attachment protein and the cellular receptor for HIV have been extensively investigated. The HIV envelope glycoprotein (gp120) serves as the viral cell attachment protein. Comparison of the amino acid sequence of this protein from a variety of HIV isolates indicates that there are both 'constant' and 'variable' regions within the protein (Hahn *et al.*, 1986; Starcich *et al.*, 1986). To date there has been no convincing correlation between variations in the amino acid sequence of the HIV glycoprotein and specific biological properties of HIV such as virulence, neurotropism, etc. Another intriguing area of investigation involves the search for amino acid sequence homologies between the HIV glycoprotein and other biologically important molecules. A number of these have been reported, but their significance remains speculative. Some of the tantalizing homologies involve regions of immunoglobulin heavy chains (Maddon *et al.*, 1986), neuroleukin (Gurney *et al.*, 1986), and vasoactive intestinal peptide (Pert *et al.*, 1986).

The CD4 antigen has been identified as the cellular receptor for HIV on human lymphoid cells (Dalgleish *et al.*, 1984; Klatzman *et al.*, 1984). The introduction of cDNA encoding this antigen into a variety of previously HIV-resistant ($CD4^-$) human lymphoid and epithelial cell lines renders them both $CD4^+$ and susceptible to HIV infection, although the same does not appear to be true for non-human (for example, mouse) cells (Maddon *et al.*, 1986). Competition binding experiments using a large number of anti-CD4 MAbs suggest that the viral binding site is relatively large as MAbs binding to at least four distinct (non-competing) epitopes can interfere with viral infection (Sattentau *et al.*, 1986).

The observations that the HIV is neurotropic and that the CD4 antigen serves as a viral receptor have prompted an extensive search for the presence of this antigen on cells within the human CNS. Homogenates of brain tissue contain detectable CD4 mRNA sequences, indicating that this antigen is present in the CNS, assuming the mRNA is translated (Maddon *et al.*, 1986). However, knowledge concerning the precise anatomical regions within which this antigen is located and the cell types on which it is expressed remains rudimentary.

Viral enhancers, promoters, and transcriptional activators

The studies described in the preceding section provide ample evidence for the importance of viral cell attachment proteins and host cell receptors in determining the tissue tropism of a number of viruses. However, recent investigations with papovaviruses and retroviruses have provided new insights into the importance of specific elements within viral genomes, which serve to enhance transcription of viral genes or translation of viral proteins in determining both the tissue and species specificity of viruses. Binding of a virus via a specific cell attachment protein to a specific host–cell receptor may be a necessary step in the infectious cycle but should not be seen, *de facto*, as either the exclusive or even principal determinant of tropism for all viruses.

The human papovavirus, JCV, is one of the agents responsible for the disease progressive multifocal leukoencephalopathy (PML). In this disease virus infects and kills oligodendroglia, resulting in extensive demyelination throughout the CNS. JCV infection also results in dramatic and often bizarre alterations in astrocyte morphology, although infection of these cells appears to be non-permissive (that is, does not result in the production of infectious virus). Neurons are not infected by JCV. Recent studies have suggested that an 'enhancer' region present in the JCV genome may be the major determinant of this pattern of viral tropism within the CNS. Enhancers are short nucleotide sequences, often multiply repeated in tandem, that act to stimulate the transcription of associated genes (*see* Brady, Feigenbaum and Khoury, 1986, for review). In order to study the tissue specificity of the JCV enhancer region, this portion of the JCV genome was attached to an indicator gene, chloramphenicol acetyltransferase (CAT), and this complex was transfected into a variety of cell lines (Kenney *et al.,* 1984). The experimental design was based on the assumption that if the JCV enhancer was active in a particular cell line then transcription of the CAT gene would occur, and following translation of the resulting mRNA into protein, an increase in CAT enzymatic activity would be detectable. Using this assay system it was shown that the JCV enhancer is specifically active in human fetal glial cells and relatively inactive in CV-1 and HeLa cells, an *in vitro* analogue to the previously discussed *in vivo* tropism of this virus. By contrast, the enhancer region of another papovavirus, SV40, is 'promiscuous', that is, active in a wide variety of cell lines. These experiments indicate that enhancers can potentially, but do not invariably, serve as determinants of cell tropism.

Fertilized mouse eggs can be microinjected with foreign DNA. Mice which develop from these eggs contain both native and foreign DNA and have been referred to as 'transgenic' mice (*see* Messing, Chapter 9 of this volume). Transgenic mice containing the 'early region' of the SV40 genome frequently develop choroid plexus epithelial cell tumours (Brinster *et al.,* 1984). The early region of the SV40 genome contains an enhancer region as well as regions encoding a protein implicated in transformation and oncogenesis (T-antigen). If the enhancer region is omitted from the microinjected foreign DNA constructs, the frequency of choroid plexus tumours drops significantly. These results have led to the suggestion that the SV40 enhancer may determine the targeting of the T-antigen to choroid plexus epithelial cells, and that the expression of T-antigen in these cells leads to tumour formation (Palmiter *et al.,* 1985).

Transgenic mice have also been produced following microinjection of the early region of JCV (Small *et al.,* 1986). Offspring of these mice inherit JCV genomic sequences, and develop a syndrome characterized pathologically by dysmyelination within the CNS, and clinically by tremor, incoordination and seizures. The development of CNS dysmyelination correlates with the inheritance of JCV DNA sequences and the expression of high levels of JCV mRNA within the brain. It has been postulated that the JCV enhancer sequences are responsible for targeting the associated T-antigen sequences to oligodendroglia, where their expression results in cellular dysfunction leading to disordered myelination.

At least four genes that are involved in the regulation of gene transcription and in some cases mRNA translation have been identified within the HIV genome. Two of these genes (*sor* and 3′ *orf*) do not appear to be essential for HIV replication or cytopathogenicity (CPE). The remaining two genes (*tat* and *art/trs*) are required for both viral replication and CPE. Despite accumulating evidence of the mechanism(s) by which these genes and their products act (*see* Chen, 1986, for review), there is no evidence to date that they function in either a cell, tissue, or species-specific fashion. It would thus appear that the interaction of the HIV cell glycoprotein and host–cell receptors (for example, CD4 antigen) serves as the major determinant of HIV tropism, but the issue must be considered a potentially open one.

CONCLUSIONS

In this brief review the author has attempted to highlight some of the recent insights that have been gained into the molecular and genetic basis for viral tropism, with specific emphasis on those factors which appear particularly relevant to understanding the basis for the tropism of viruses for the nervous system. We have seen how host genes, acting through a variety of mechanisms, can influence the susceptibility or resistance of animals to neurotropic viruses. We have also reviewed investigations concerning the role played by individual viral genes and the proteins they encode in determining specific pathways of viral spread to the CNS in the infected host. Several examples illustrating the current state of knowledge concerning the nature of viral cell attachment proteins and host cell receptors for neurotropic viruses have been reviewed. Finally, we have seen examples of how specific viral genetic elements, such as enhancers, can act to determine the cell-specificity of certain neurotropic viruses.

Acknowledgements

Bernard Fields, John Boyle, Steven Lynn, Skip Virgin, Francisco Gonzalez-Scarano, and Dale McPhee all provided helpful discussion of the issues addressed in this review. Dr Tyler is supported by a NIAID Physician-Scientist Award (K11 AI00610) and an Alfred P. Sloan Research Fellowship. Research support includes NIH Program Project Grant 5 P50 NS16998 and additional support from the Shipley Institute of Medicine at Harvard Medical School.

References

ANDERSON, J. R. and FIELD, H. J. (1983) The distribution of herpes simplex type 1 antigen in mouse central nervous system after different routes of inoculation. *Journal of the Neurological Sciences,* **60,** 181–195

BASSEL-DUBY, R., SPRIGGS, D. R., TYLER, K. L. and FIELDS, B. N. (1986) Identification of attenuating mutations on the reovirus type 3 S1 double-stranded RNA segment using a rapid sequencing technique. *Journal of Virology,* **60,** 64–67

BECKER, Y., HADAR, J., TABOR, E., BEN-HUR, T., RABSTEIN, I., ROSEN, A. and DARAI, G. (1986) A sequence in Hpa 1-P fragment of herpes simplex virus-1 DNA determines intraperitoneal virulence in mice. *Virology,* **149,** 255–259

BOYLE, J. F., WEISMILLER, D. G. and HOLMES, K. V. and HOLMES, K. V. (1987) Genetic resistance to mouse hepatitis virus correlates with absence of virus-binding activity on target tissues. *Journal of Virology,* **61,** 185–189

BRADY, J., FEIGENBAUM, L. and KHOURY, G. (1986) Viral enhancer elements. In: *Concepts in Viral Pathogenesis,* edited by A. L. Notkins and M. B. A. Oldstone, pp. 61–67. New York: Springer-Verlag

BRINSTER, R. L., CHEN, H. Y., MESSING, A., VAN DYKE, T., LEVINE, A. J. and PALMITER, R. D. (1984) Transgenic mice harboring SV40 T-antigen genes develop characteristic brain tumors. *Cell,* **37,** 367–379

BRINTON, M. A. and NATHANSON, N. (1981) Genetic determinants of virus susceptibility: epidemiologic implications of murine models. *Epidemiologic Reviews,* **3,** 115–139

BRINTON, M. A., BLANK, K. J. and NATHANSON, N. (1984) Host genes that influence susceptibility to viral diseases. In: *Concepts in Viral Pathogenesis,* edited by A. L. Notkins and M. B. A. Oldstone, pp. 71–78. New York: Springer-Verlag

BUCHMEIER, M. J., WELSH, R., DUTKO, F. and OLDSTONE, M. B. A. (1980) The virology and immunobiology of lymphocytic choriomeningitis virus infection. *Advances in Immunology,* **30,** 275–331

BUCHMEIER, M. J., LEWICKI, H. A., TALBOT, P. J. and KNOBLER, R. L. (1984) Murine hepatitis virus-4 (strain JHM)-induced neurologic disease is modulated by monoclonal antibody. *Virology,* **132,** 261–270

BURRAGE, T. G., TIGNOR, G. H. and SMITH, A. L. (1985) Rabies virus binding at neuromuscular junctions. *Virus Research,* **2,** 273–289

CARLSON, G. A., KINGSBURY, D. T., GOODMAN, P. A. *et al.* (1986) Linkage of prion protein and scrapie incubation time genes. *Cell,* **46,** 503–511

CHEN, I. S. Y. (1986) Regulation of AIDS virus expression. *Cell,* **47,** 1–2

CLATCH, R. J., MELVOLD, R. W., MILLER, S. D. and LIPTON, H. L. (1985) TMEV-induced demyelinating disease in mice is influenced by the H-2D region: correlation with TMEV-specific delayed type hypersensitivity. *Journal of Immunology,* **135,** 1408–1413

CO, M. S., GAULTON, G. N., FIELDS, B. N. and GREENE, M. I. (1985a) Isolation and biochemical characterization of the mammalian reovirus type 3 cell-surface receptor. *Proceedings of the National Academy of Sciences USA,* **82,** 1494–1498

CO, M. S., GAULTON, G. N., TOMINAGA, A., HOMCY, C. J., FIELDS, B. N. and GREENE, M. I. (1985b) Structural similarities between the mammalian beta-adrenergic and reovirus type 3 receptors. *Proceedings of the National Academy of Sciences USA,* **82,** 5315–5318

DALGLEISH, A. G., BEVERLY, P. C. L., CLAPHAM, P. R., CRAWFORD, D. H., GREAVES, M. F. and WEISS, R. A. (1984) The CD4 (T4) antigen is an essential component of the receptor for the AIDS retrovirus. *Nature (London),* **312,** 763–767

DALZIEL, R. G., LAMPERT, P. W., TALBOT, P. J. and BUCHMEIER, M. J. (1986) Site-specific alteration of murine hepatitis virus type 4 peplomer glycoprotein E2 results in reduced neurovirulence. *Journal of Virology,* **59,** 463–471

DIAMOND, D. C., JAMESON, B. A., BONIN, J. *et al.* (1985) Antigenic variation and resistance to neutralization in poliovirus type 1. *Science,* **229,** 1090–1093

DIETZSCHOLD, B., WUNNER, W. H., WIKTOR, T. J., LOPES, A., LAFON, M., SMITH, C. and KOPROWSKI, H. (1983) Characterization of an antigenic determinant of the glycoprotein that correlates with pathogenicity of rabies virus. *Proceedings of the National Academy of Sciences USA,* **80,** 70–74

DIETZSCHOLD, B., WIKTOR, T. J., TROJANOWSKI, J. Q., MACFARLAN, R. I., WUNNER, W. H., TORRES-ANJEL, M. J. and KOPROWSKI, H. (1985) Differences in cell-to-cell spread of pathogenic and apathogenic rabies virus *in vivo* and *in vitro. Journal of Virology,* **56,** 12–18

FLEMING, J. O., TROUSDALE, M. D., EL-ZAATARI, F. A. K., STOHLMAN, S. A. and WEINER, L. P. (1986) Pathogenicity of antigenic variants of murine coronavirus JHM selected with monoclonal antibodies. *Journal of Virology,* **58,** 869–875

GABUZDA, D. H., HO, D. D., DE LA MONTE, S., HIRSCH, M. S., ROTA, T. R. and SOBEL, R. A. (1986) Immunohistochemical identification of HTLV-III antigen in brains of patients with AIDS. *Annals of Neurology,* **20,** 289–295

GRESSER, I., TOVEY, M. G., MAURY, C. and BANDU, M.-T. (1976) Role of interferon in the pathogenesis of virus diseases in mice as demonstrated by the use of anti-interferon serum. II. Studies with herpes simplex, Maloney sarcoma, vesicular stomatitis, Newcastle disease and influenza viruses. *Journal of Experimental Medicine,* **144,** 1316–1323

GURNEY, M. E., HEINRICH, S. P., LEE, M. R. and YIN, H.-S. (1986) Molecular cloning and expression of neuroleukin, a neurotropic factor for spinal and sensory neurons. *Science,* **234,** 566–574

HAHN, B. H., SHAW, G. M., TAYLOR, M. E. *et al.* (1986) Genetic variation in HTLVIII/LAV over time in patients with AIDS or at risk for AIDS. *Science,* **232,** 1548–1553

HO, D. D., ROTA, T. R., SCHOOLEY, R. T. *et al.* (1985) Isolation of HTLV-III from cerebrospinal fluid and neural tissues of patients with neurologic syndromes related to the acquired immunodeficiency syndrome. *New England Journal of Medicine,* **313,** 1493–1497

HOGLE, J. M., CHOW, M. and FILMAN, D. J. (1985) Three-dimensional structure of poliovirus at 2.9 Angstrom resolution. *Science,* **229,** 1358–1365

HOPE, J. and KIMBERLIN, R. H. (1987) The molecular biology of scrapie: the last two years. *Trends in Neuroscience,* **10,** 149–151

HOWE, H. A. and BODIAN, D. B. (1942) *Neural Mechanisms in Poliomyelitis.* New York: Commonwealth Fund

JOHNSON, R. T. (1982) *Viral Infections of the Nervous System.* New York: Raven Press

KASTRUKOFF, L. F., LAU, A. and PUTERMAN, M. (1986) Genetics of natural resistance to herpes simplex virus type 1 latent infection of the peripheral nervous system. *Journal of General Virology,* **67,** 613–621

KAYE, K. M., SPRIGGS, D. R., BASSEL-DUBY, R., FIELDS, B. N. and TYLER, K. L. (1986) Genetic basis for altered pathogenesis of an immune-selected antigenic variant of reovirus type 3 (Dearing). *Journal of Virology,* **59,** 90–97

KENNEY, S., NATARAJAN, V., STRIKE, D., KHOURY, G. and SALZMAN, N. P. (1984) JC virus enhancer-promoter active in human brain cells. *Science,* **226,** 1337–1339

KLATZMAN, D., CHAMPAGNE, E., CHAMARET, S. *et al.* (1984) T Lymphocyte T4 molecule behaves as the receptor for human retrovirus LAV. *Nature (London),* **312,** 767–768

KNOBLER, R. L., TAYLOR, B. A., WOODDELL, M. K., BEAMER, W. G. and OLDSTONE, M. B. A. (1984) Host genetic control of mouse hepatitis virus type-4 (JHM strain) replication. *Experimental and Clinical Immunogenetics,* **1,** 217–222

KOENIG, S., GENDELMAN, H. E., ORENSTEIN, J. M. *et al.* (1986) Detection of AIDS virus in macrophages in brain tissue from AIDS patients with encephalopathy. *Science,* **233,** 1089–1093

KRAH, D. and CROWELL, R. L. (1985) Properties of the deoxycholate-solubilized HeLa cell plasma membrane receptor for binding group B coxsackieviruses. *Journal of Virology,* **53,** 867–870

KUCERA, P., DOLIVO, M., COULON, P. and FLAMAND, A. (1985) Pathways of the early propagation of virulent and avirulent rabies strains from the eye to the brain. *Journal of Virology,* **55,** 158–162

LENTZ, T. L., BURRAGE, T. G., SMITH, A. L., CRICK, J. and TIGNOR, G. H. (1982) Is the acetylcholine receptor a rabies virus receptor? *Science,* **215,** 182–184

LODMELL, D. L. and EWALT, L. C. (1985) Pathogenesis of street rabies virus infections in resistant and susceptible strains of mice. *Journal of Virology,* **55,** 788–795

LOPEZ, C. (1975) Genetics of natural resistance to herpesvirus infections in mice. *Nature (London),* **258,** 152–153

LOPEZ, C. (1980) Resistance to HSV 1 in the mouse is governed by two major, independently segregating, non-H-2 loci. *Immunogenetics,* **11,** 87–92

LOVE, A., RYDBECK, R., KRISTENSSON, K., ORVELL, C. and NORRBY, E. (1985) Hemagglutinin-neuraminidase glycoprotein as a determinant of pathogenicity in mumps virus hamster encephalitis: Analysis of mutants selected with monoclonal antibodies. *Journal of Virology,* **53,** 67–74

LUO, M., VRIEND G., KAMER, G. *et al.* (1987) The atomic structure of mengovirus at 3.0 Angstrom resolution. *Science,* **235,** 182–191

MADDON, P. J., DALGLEISH, A. G., McDOUGAL, J., CLAPHAM, P., WEISS, R. and AXEL, R. (1986) The T4 gene encodes the AIDS virus receptor and is expressed in the immune system and the brain. *Cell,* **47,** 333–348

MARTIN, X. and DOLIVO, M. (1983) Neuronal and transneuronal tracing in the trigeminal system of the rat using the herpes virus suis. *Brain Research,* **273,** 253–276

MENDELSOHN, C., JOHNSON, B., LIONETTI, K. A., NOBIS, P. WIMMER, E. and RACANIELLO, V. R. (1986) Transfection of a human poliovirus receptor gene into mouse cells. *Proceedings of the National Academy of Sciences USA,* **83,** 7845–7849

MILLER, D. A., MILLER, O. J., DEV, V. G., HASHAMI, S. and TANTRAVAHI, R. (1974) Human chromosome 19 carries a poliovirus receptor gene. *Cell,* **1,** 167–173

MURPHY, F. A. (1977) Rabies pathogenesis: brief review. *Archives of Virology,* **54,** 279–297

NOTKINS, A. L. and OLDSTONE, M. B. A. (1985) (eds) *Concepts in Viral Pathogenesis.* New York: Springer-Verlag

NOTKINS, A. L. and OLDSTONE, M. B. A. (1986) (eds) *Concepts in Viral Pathogenesis II.* New York: Springer-Verlag

OAKES, J. E., GRAY, W. L. and LAUSCH, R. N. (1986) Herpes simplex virus type 1 DNA sequences which direct spread of virus from cornea to central nervous system. *Virology,* **150,** 513–517

OLDSTONE, M. B. A., AHMED, R., BUCHMEIER, M. J., BLOUNT, P. and TISHON, A. (1985) Perturbation of differentiated functions during viral infection *in vivo.* I. Relationship of lymphocytic choriomeningitis virus and host strains to growth hormone deficiency. *Virology,* **142,** 158–174

ONODERA, T., TONIOLO, A., RAY, U. R., HENSEN, A. B., KNAZEK, R. A. and NOTKINS, A. L. (1981) Virus-induced diabetes mellitus, XX: Polyendocrinopathy and autoimmunity. *Journal of Experimental Medicine,* **153,** 1457–1473

PALMITER, R. D., CHEN, H. Y., MESSING, A. and BRINSTER, R. L. (1985) SV40 enhancer and large-T antigen are instrumental in development of choroid plexus tumors in transgenic mice. *Nature,* **316,** 457–460

PERT, C. B., HILL, J. M., RUFF, M. R. *et al.* (1986) Octapeptides deduced from the neuropeptide receptor-like pattern of antigen T4 in brain potently inhibit human immunodeficiency virus receptor binding and T-cell infectivity. *Proceedings of the National Academy of Sciences USA,* **83,** 9254–9258

RIVIERE, Y., AHMED, R., SOUTHERN, P. and OLDSTONE, M. B. A. (1985) Perturbation of differentiated functions during viral infection *in vivo.* II: Viral reassortants map growth hormone defect to the S RNA of the lymphocytic choriomeningitis virus genome. *Virology,* **142,** 175–182

ROSENSTREICH, D. L., WEINBLATT, A. C. and O'BRIEN, A. D. (1982) Genetic control of resistance to infection in mice. *CRC Critical Reviews in Immunology,* **3,** 263–330

ROSSMAN, M. G., ARNOLD, E., ERICKSON, J. W. *et al.* (1985) Structure of a common cold virus and functional relationship to other picornaviruses. *Nature (London),* **317,** 145–153

SABIN, A. B. and OLITSKY, P. K. (1937, 1938) Influence of host factors on neuroinvasiveness of vesicular stomatitis virus. *Journal of Experimental Medicine,* **66,** 15–34, 35–57, and **67,** 201–228, 229–249

SATTENTAU, Q. J., DALGLEISH, A. G., WEISS, R. A. and BEVERLY, P. C. (1986) Epitopes of the CD4 antigen and HIV infection. *Science,* **234,** 1120–1123

SEIF, I., COULON, P., ROLLIN, P. E. and FLAMAND, A. (1985) Rabies virulence: effect on pathogenicity and sequence characterization of rabies virus mutations affecting antigenic site III of the glycoprotein. *Journal of Virology,* **53,** 926–934

SHARPE, A. and FIELDS, B. N. (1983) Pathogenesis of reovirus infection. In: *The Reoviridae,* edited by W. K. Joklik, pp. 229–285. New York: Plenum Press

SHARPE, A. and FIELDS, B. N. (1985) Pathogenesis of viral infections: Basic concepts derived from the reovirus model. *New England Journal of Medicine,* **316,** 486–497

SHAW, G. M., HARPER, M. E., HAHN, B. *et al.* (1985) HTLV III infection in brains of children and adults with AIDS encephalopathy. *Science,* **227,** 177–182

SMALL, J. A., SCANGOS, G. A., CORK, L., JAY, G. and KHOURY, G. (1986) The early region of human papovavirus JC induces dysmyelination in transgenic mice. *Cell,* **46,** 13–18

SPRIGGS, D. R. and FIELDS, B. N. (1982) Attenuated reovirus type 3 strains generated by selection of hemagglutinin antigenic variants. *Nature (London),* **297,** 68–70

SPRIGGS, D. R., BRONSON, R. T. and FIELDS, B. N. (1983) Hemagglutinin variants of reovirus type 3 have altered central nervous system tropism. *Science,* **220,** 505–507

STAEHLI, P., HALLER, O., BOLL, W., LINDEMANN, J. and WEISSMANN, C. (1986) Mx protein: constitutive expression in 3T3 cells with cloned Mx cDNA confers selective resistance to influenza virus. *Cell,* **44,** 147–158

STANWAY, G., HUGHES, P., MOUNTFORD, R., REEVE, P., MINOR, P., SCHILD, P. and ALMOND, J. (1984) Comparison of the complete nucleotide sequences of the genomes of the neurovirulent poliovirus P3/Leon/37 and its attenuated Sabin vaccine derivative P3/Leon 12a,b. *Proceedings of the National Academy of Sciences USA,* **81,** 1539–1543

STARCICH, B., HAHN, B., SHAW, G. *et al.* (1986) Identification and characterization of conserved and variable regions in the envelope gene of HTLV III/LAV, the retrovirus of AIDS. *Cell,* **45,** 637–648

TEMPLETON, J. W., HOLMBERG, C., GARBER, T. and SHARP, R. M. (1986) Genetic control of serum neutralizing-antibody response to rabies vaccination and survival after a rabies challenge infection in mice. *Journal of Virology,* **59,** 98–102

THOMPSON, R. L., DEVI-RAO, G. V., STEVENS, J. G. and WAGNER, E. K. (1985) Rescue of a herpes simplex virus type 1 neurovirulence function with a cloned DNA fragment. *Journal of Virology,* **55,** 504–508

TYLER, K. L., BRONSON, R. T., BYERS, K. B. and FIELDS, B. N. (1985) Molecular basis of viral neurotropism: experimental reovirus infection. *Neurology,* **35,** 88–92

TYLER, K. L. and FIELDS, B. N. (1985) Reovirus and its replication. In: *Virology,* edited by B. N. Fields, pp. 823–862. New York: Raven Press

TYLER, K. L., McPHEE, D. and FIELDS, B. N. (1986) Distinct pathways of viral spread in the host determined by reovirus S1 gene segment. *Science,* **233,** 770–774

VOGEL, S. N. and FERTSCH, D. (1987) Macrophages from endotoxin-hyporesponsive (Lps^d) C3H/HeJ mice are permissive for vesicular stomatitis virus because of reduced levels of endogenous interferon: possible mechanism for natural resistance to virus infection. *Journal of Virology,* **61,** 812–818

WEINER, H. L., POWERS, L. and FIELDS, B. N. (1980) Absolute linkage of virulence and central nervous system tropism of reoviruses to viral hemagglutinin. *Journal of Infectious Disease,* **141,** 609–616

WILEY, C. A., SCHRIER, R. D., NELSON, J. A., LAMPERT, P. W. and OLDSTONE, M. B. A. (1986) Cellular localization of human immunodeficiency virus infection within the brains of acquired immunodeficiency syndrome patients. *Proceedings of the National Academy of Science USA,* **83,** 7089–7093

WOLINSKY, J. S. and SERVER, F. (1985) Mumps virus. In: *Virology,* edited by B. N. Fields, pp. 1255–1284. New York: Raven Press

WUNNER, W. H., REAGAN, K. J. and KOPROWSKI, H. (1984) Characterization of the saturable binding sites for rabies virus. *Journal of Virology,* **50,** 691–697

YUEN, P. H., TZENG, E., KNUPP, C. and WONG, P. K. Y. (1986) The neurovirulent determinants of ts1, a paralytogenic mutant of Moloney murine leukemia virus TB, are localized in at least two functionally distinct regions of the genome. *Journal of Virology,* **59,** 59–65

8
Neuro-oncogenesis: recessive genes, activated oncogenes, and chromosome abnormalities in the development of neuroectodermal cancers

Fred Gilbert

INTRODUCTION

The role of genetic factors in the development of cancer was evident as long ago as 1866, with the publication by the French neurologist Broca of the pedigree of his wife's family showing a high frequency of breast cancer in female relatives in multiple, successive generations (Broca, 1866). Since then, and most particularly in the last 30 years, a considerable body of epidemiological and experimental evidence from human and animal studies has pointed to the accumulation of single gene changes as responsible for tumorigenesis (Lynch, 1976). Exactly which genes are involved in the transformation of particular cell types, and how these genes act, has eluded analysis until recently, with the introduction of approaches utilizing recombinant DNA technology. The results of these studies have led to a new understanding of possible roles for single gene changes, and for chromosome abnormalities as well, in the development of different cancers, including those of neuroectodermal origin.

ONCOGENES

Introducing oncogenes

Before we can discuss the role of single genes in cancer, the term 'oncogene' must be clarified. This term was initially used by virologists to refer to a class of genes in both RNA and DNA tumour viruses whose active expression (that is, the synthesis of specific mRNAs) was required for the initiation and maintenance of transformation following the infection of susceptible cells (Bishop, 1983). Examples of DNA virus oncogenes include E1A in adenovirus and middle T and large T in polyoma; examples of RNA virus oncogenes include *src* in Rous sarcoma virus and *myc* in avian myelocytomatosis virus (Bishop, 1983; Varmus, 1984).

The oncogenes carried by one class of RNA tumour viruses, the acutely transforming retroviruses, have been separately designated '*v-onc*' (viral oncogene)

(Coffin *et al.*, 1981). The latter distinction became necessary when it was shown that the sequences of individual *v-onc* were very similar to those contained within the genomes of most, if not all, eukaryotic species, from yeast to man (Varmus and Bishop, 1986). These cellular genes have been termed 'proto-oncogenes' or '*c-onc*' (cellular oncogene); it is most likely that *v-onc* were derived from *c-onc* by a series of steps, including the recombination of cellular and viral sequences following virus integration into the cellular genome, and subsequent mutations and/or rearrangements within the viral genome. Each *c-onc* is a fully functional gene in the normal host (that is, one capable of serving as a template for mRNA synthesis); *c-onc* corresponding to virtually all of the approximately 20 *v-onc* isolated from various rodent, avian, feline, and simian retroviruses have now been identified in the human genome (Varmus and Bishop, 1986) (*Table 8.1*).

Recently, the cellular oncogene category has been broadened to include sequences with no direct relationship to any described tumour virus. The category now includes, in addition to *c-onc*:

(1) Functional genes with limited sequence homology to the previously defined *c-onc*, examples of which include *N-myc* and *L-myc*, genes similar in sequence and presumed function to *c-myc*, the cellular homologue of *v-myc* in avian myelocytomatosis virus (Alt *et al.*, 1986).
(2) Functional genes with no *v-onc* homology but capable of inducing, or contributing to the induction of, transformation of certain rodent cell lines in culture, when introduced into the cell in an 'activated' form; the mechanisms of 'activation' will be described below: examples include *met, mel,* and *trk*, which induced transformation of the mouse cell line NIH 3T3 (the NIH 3T3 Transfection Assay) (Tabin *et al.*, 1982).
(3) Genes whose mRNA synthetic rates are increased following the adjacent integration into the host cell genome of retroviruses lacking *v-onc* of their own; this process is called 'promoter insertion' (Hayward, Neel and Astrin, 1981); one example is *int-1*, identified following the integration of mouse mammary tumour virus (MMTV) into mouse cell DNA (Nusse *et al.*, 1984).
(4) Transcriptionally active sequences which occur adjacent to certain consistent breakpoints in chromosomal rearrangements in particular tumour types, such as *bcl-2* on chromosome 18, identified in follicular lymphomas carrying reciprocal translocations between chromosome 14 and 18 (Tsujimoto and Croce, 1986); the enlarged cellular oncogene family now includes some 40 members (Verma, 1986) (*see Table 8.1*).

Oncogene function

A number of oncogenes, both *v-onc* and cellular oncogenes, have been cloned and sequenced. When their nucleotide sequences, and/or the consensus amino acid sequences of their putative products were compared with those available in various sequence databanks, it became clear that several bore homology to previously defined protein-coding sequences (Varmus and Bishop, 1986). These include at least one growth factor, platelet derived growth factor (PDGF), for which an altered component (the beta chain) appears to be specified by *v-sis* (Leal *et al.*, 1985) and several growth factor receptors, both identifed and unidentified as to ligand. Examples of the former include the epidermal growth factor (EGF), thyroid hormone, and macrophage colony stimulating factor (CSF-1) receptors, altered

Table 8.1 Oncogenes and chromosome localization (compiled from Verma (1986) and McKusick (1986). Oncogene abbreviations as per McKusick (1986). Column (1): *c-onc*; column (2): oncogene family members: column (3): oncogene identified by NIH 3T3 transfection assay; column (4): oncogene identified by consistent breakpoint in cancer-specific translocation or retroviral integration site

		Oncogene			
Chromosome	*Localization*	(1)	(2)	(3)	(4)
1	1p36	*fgr*			
	p36		*src*-2		
	p32			*Blym*	
	p32		*L-myc*		
	p22			*N-ras*	
	q12-qter	*ski*			
2	2p24		*N-myc*		
	p11-14	*rel*			
3	3p25	*raf*-1			
4			*raf*-2		
		kit			
5	5q33	*fms*			
6	6p12	*K-ras*-1			
	q22-q24	*myb*			
	q16-q22			*mcf*-3	
			yes-2		
7	7p14-p12	*erb-B*			
	q21-q31			*met*	
8	8q22	*mos*			
	q24	*myc*			
9	9q34	*abl*			
11	11p14	*H-ras*-1			
	q13				*bcl*-1
	q23-q24	*ets*-1			
12	12pter-q14				*int*-1
	p12		*K-ras*-2		
14	14q21-q31	*fos*			
	q32	*akt*-1			
15	15q26	*fes*			
17	17q11-q21	*erb-A*-1			
	q22			*neu*	
18	18q21	*yes*-1			
	q21				*bcl*-2
20	20q13	*src*-1			
21	21q22		*ets*-2		
22	22q12	*sis*			
X	X pter-q28		*H-ras*-2		

forms of which are coded for by *v-erbB*, *v-erbA*, and *c-fms* respectively (Ullrich *et al.*, 1984; Sap *et al.*, 1986; Sherr *et al.*, 1985); examples of the latter include the *neu* and *trk* oncogenes (Martin-Zanca, Hughes and Barbacid, 1986; Shih *et al.*, 1981).

Although the majority of proteins specified by individual oncogenes have not been identified, clues as to the functional significance of at least some has come from partial sequence analysis. Some (for example, *src, abl, raf*) contain sequences previously associated with protein kinase activity, either tyrosine or serine/threonine kinases (Varmus and Bishop, 1986). At least one, *ras*, appears to direct the synthesis of a protein similar in sequence to a subunit of a G protein, a category of proteins involved in the regulation of adenylcyclase and known to influence cellular differentiation and cell division (Stryer and Bourne, 1986). There are also several oncogenes (for example, *myc, myb,* and *fos*) whose proteins are found in the cell nucleus and are presumed to bind directly to DNA and to induce the synthesis of particular mRNA species (Rohrschneider and Gentry, 1984) (*Table 8.2*).

Table 8.2 Oncogenes, cell locations, and biochemical properties (compiled from McKusick (1986); Varmus and Bishop (1986); Verma (1986)). Oncogene abbreviations as per McKusick (1986). PDGF: platelet derived growth factor. EGF-R: epidermal growth factor receptor. CSF-1-R: macrophage colony stimulating factor receptor. G protein: GTPase subunit of G protein. Thyroid hormone-R: thyroid hormone receptor. Protein kinase (tyr) or (ser/thr): tyrosine or serine/threonine protein kinases

Cell location	*Oncogene*	*Biochemical property*
Secreted	*sis*	PDGF
Transmembrane	*erb-B*	EGF-R
	fms	CSF-1-R
Plasma membrane	*src*	Protein kinase (tyr)
	abl	Protein kinase (tyr)
	Ha-ras	G protein subunit
Cytoplasm	*raf*	Protein kinase (ser/thr)
	mos	Protein kinase (ser/thr)
Nucleus	*myc*	DNA binding
	myb	DNA binding
	erb-A	Thyroid hormone-R

Review of the above findings by several investigators led to the proposal that all cellular oncogenes code for proteins involved in the regulation of cell growth. One can then imagine an 'oncogene cascade' in which the several steps at which individual oncogenes might act include the following:

(1) extracellular growth factors; that bind to
(2) growth factor receptors (both on the cell surface, as with the EGF receptor, and within the cytoplasm, as with the thyroid hormone receptor), as well as protein kinases (tyrosine or serine/threonine kinases), and G proteins, that will lead to the activation of step (3), and/or step (4);

(3) involves the induction (by protein phosphorylation, the synthesis of G proteins, etc.) of so-called 'second messengers' (for example protein kinase C), whose role it is to carry metabolic signals from membrane receptors to nucleus, resulting in
(4) the induction of DNA binding proteins (for example the products specified by *myc, myb,* etc.) that are presumed to induce the transcription of specific mRNAs from multigene subsets, in turn affecting cell division and differentiation.

This hypothesis suggests that the coordinated expression of individual cellular oncogenes contributes to (or is responsible for) the normal control of cell growth and division.

Oncogenes and differentiation

An extensive literature now exists documenting possible relationships between the expression of individual oncogenes and differentiation or dedifferentiation in animal, as well as tissue culture, models. In whole mouse embryos, for example, levels of *N-myc*-specific mRNA are highest in early and mid-gestation, then fall as the fetus approaches term (Jakobovits *et al.*, 1985). Enhanced expression of *N-myc* in the mouse appears to be limited largely to embryonic and newborn brain and kidney tissue, detectable up to 2 weeks postnatally in the former, and 1 week postnatally in the latter (Zimmerman *et al.*, 1986).

When cultured mouse teratocarcinoma cells are exposed to retinoic acid and dibutyryl cyclic AMP, they differentiate to endoderm; the levels of *N-myc*-specific mRNA in the differentiated cell population are two-fold to five-fold less than those determined in the undifferentiated progenitor cells (Jakobovits *et al.*, 1985). The retinoic acid-induced differentiation of human neuroblastoma cells in culture, measured by the extension of neuritic processes by the cells, is also associated with a decrease in *N-myc*-specific mRNA levels (Thiele, Reynolds and Israel, 1985). *N-myc*-specific mRNA, while detectable in mouse pre-B cells, could not be detected in fully differentiated plasma cells (Zimmerman *et al.*, 1986).

Expression of the differentiated phenotype in the Friend mouse erythroleukemia cell is accompanied by a decrease in the level of *c-myc* mRNA, while the serum-induced stimulation of cell division results in a substantial increase in the level of *c-myc* mRNA (Lachman *et al.*, 1985). As a general rule, there is an inverse relationship between the transcription of individual *myc* gene family members (*N-myc, c-myc,* or *L-myc*) and the differentiation of particular cell types, and a direct relationship between the expression of the same oncogenes and cell division. If, for example, a particular *myc* family member is actively transcribed when cells of a defined cell type are dividing, then levels of the mRNA specified by that oncogene will be reduced when the cells in question stop dividing and/or differentiate (Kelly *et al.*, 1983; Lachman *et al.*, 1985).

The modulation of expression of at least three other oncogenes – *fos, fms,* and *myb* – during the differentiation of particular cell types, has been reported as well (Muller and Verma, 1984; Mitchell *et al.*, 1986). However, it remains to be determined whether changes in the transcriptional activity of one or more of these oncogenes are directly responsible for altering the growth parameters of the cell (and inducing differentiation), or are the consequence of other genetic or epigenetic events.

Oncogene activation

When the nucleotide sequences of *v-onc* are compared with those of comparable *c-onc*, differences – some slight, some quite extensive – are often evident (Varmus and Bishop, 1986). When the transfection-induced transformation of NIH 3T3 has been used to identify oncogenes, the sequences responsible for transformation are almost always non-identical with those of their counterparts in the normal human genome. The alterations in cellular oncogenes that contribute to tumorigenicity reflect different mechanisms of oncogene 'activation'. These mechanisms include the following.

Point mutation

Single base substitutions altering the amino acid composition of the protein are specified by a particular oncogene; the first described examples involved amino acid substitutions at positions 12 and 61 in *ras* sequences in human bladder carcinoma DNAs capable of inducing transformation in the NIH 3T3 transfection assay (Tabin *et al.*, 1982; Lacal *et al.*, 1986).

Molecular rearrangement

These are alterations in either the sequence transcribed as mRNA or in the level of transcription, as the result of chromosomal breaks and rearrangements that lead to the enhanced transcription of normal genes, as with the translocations in Burkitt lymphoma transposing *c-myc* from chromosome 8 to chromosome 14 (Leder *et al.*, 1983). Rearrangement of DNA sequences may lead to production of novel oncogene proteins, as with the chimeric *abl*-bcr protein resulting from the transposition of *c-abl* from chromosome 9 to a segment on chromosome 22 designated the 'break cluster region' (bcr) in chronic myelogenous leukemia [with t(9;22)] (Konopka, Watanabe and Witte, 1984; Heisterkamp *et al.*, 1985), or the activated *met* formed from the fusion of sequences derived from chromosome 1 and the *met* proto-oncogene on chromosome 7 (Park *et al.*, 1986).

Promoter insertion

These are increased levels of *c-onc* transcription, as the result of the integration of retroviruses lacking *v-onc* of their own but carrying transcription enhancer sequences, next to *c-onc* in the host genome; an example is the integration of avian leukosis virus in the chicken genome next to *c-myc*, leading to the induction of synthesis of *c-myc* mRNA (Hayward, Neel and Astrin, 1981).

Gene amplification

This is the formation of two cytogenetic markers unique to cancer cells, the intrachromosomal homogeneously staining region (HSR) and extrachromosomal double minutes (DMS), that contain multiple copies of either drug resistance genes or cellular oncogenes (Biedler and Spengler, 1986); HSR and DMS have been identified in examples of virtually every tumour type, and amplification of more than one oncogene in different examples of individual tumour types has also been documented. The only association between amplification of a single oncogene in multiple examples of a particular tumour type has been the amplification of *N-myc* in neuroblastoma (Kohl *et al.*, 1983; Schwab *et al.*, 1983).

It is likely that point mutations resulting in amino acid substitutions in the oncogene proteins alter substrate affinity or otherwise affect protein activity. An example is *ras*, in which at least some amino acid substitutions are postulated to reduce a specific function of the *ras* protein, its ability to hydrolyse GTP, ultimately prolonging or enhancing the activity of the G protein, of which *ras* is a component (Lacal *et al.*, 1986; Varmus and Bishop, 1986).

Molecular rearrangements combining sequences from different parts of the genome may affect mRNA transcription rates by transposing the functional portion of the gene away from adjacent *cis*-acting regulatory elements on its chromosome of origin or inserting a quiescent, non-transcribing gene near enhancers or promoters of active transcription. Alternatively, rearrangements to form new compound mRNAs, by altering transcript length, may affect mRNA stability and/or the functional activity of the resultant protein.

An example of active transcription of a previously silent gene as the result of translocation is *myc* in Burkitt's lymphoma, wherein the non-transcribing gene is moved from chromosome 8 to a region containing promoters of transcription within the immunoglobulin heavy chain gene complex on chromosome 14 (Leder *et al.*, 1983). An example of a new oncogene protein activity resulting from molecular rearrangement is *trk*, a fusion of sequences from tropomyosin and a tyrosine kinase membrane receptor whose liquid specificity is unknown (Martin-Zanca, Hughes and Barbacid, 1986).

Active transcription of amplified oncogenes is also very likely to be the result of movement away from *cis*-acting regulatory sequences. This conclusion is consistent with the fact that DMS are extrachromosomal and the observation that HSR are generally on chromosomes other than those to which the original single copy gene, that has been amplified, has been mapped. For example, *N-myc* has been assigned to the short arm (p) of chromosome 2 (band 2p23), but in none of the 20 or more karyotyped neuroblastomas in which *N-myc* is amplified within HSR, does the HSR develop at this site (Gilbert, unpublished data).

Oncogenes in normal and cancer cells

Cellular oncogenes are, therefore, normal constituents of the eukaryotic genome, responsible for the synthesis of different growth control elements. Aberrant expression of cellular oncogenes, as the result of active mRNA transcription at inappropriate times in the cell cycle or at inappropriate stages of a differentiated cell lineage, or as the result of the production of a variant protein, is presumed to alter the cell's proliferative capacity and to contribute to its transformation and/or to tumour progression (that is, the acquisition of a more aggressive phenotype, including rapid growth and spread in the host).

CHROMOSOMES AND CANCER

Chromosomal abnormalities in human cancers fall into three major classes: (1) rearrangements resulting in the equivalent exchange of structural material between chromosomes (reciprocal translocations); (2) rearrangements resulting in the loss of structural material from the karyotype (primarily as deletions); and (3) rearrangements or non-disjunctional events (abnormal chromosome segregation into daughter cells at mitosis) resulting in the duplication of whole chromosomes of chromosome segments (Gilbert, 1983). An example of class (1) would be the translocations between chromosome 22 and 9 that are commonly seen in chronic myelogenous leukaemia. An example of class (2) would be the deletions involving the short arm of chromosome 11 that result in the loss of structural material from band 11p13, in association with a childhood kidney tumour, Wilms' tumour, and aniridia. Examples of class (3) are trisomy for the long arm of chromosome 1 (1q) and the HSR and DMS.

Abnormalities of classes (1) and (2) tend to be unique markers of individual tumour types. For example, over 85% of chronic myelogenous leukaemias contain reciprocal translocations involving chromosome 22, and over 90% of meningiomas have either lost one chromosome 22 (monosomy 22) or carry deletions/rearrangements resulting in the loss of structural material from chromosome 22 (Yunis, 1983). Abnormalities of the third class are less specific and have been reported in individual examples of many different tumour types. As discussed previously, chromosome abnormalities of class (1) and at least some of class (3) (the HSR and DMS) are often associated with the activation of particular cellular oncogenes. A possible role for the loss of genetic information, a consequence of the abnormalities of class (2) in the development of cancer will be discussed below.

RECESSIVE GENES AND CANCER

Since Broca's original report of familial breast cancer, heritable forms of many different human cancers have been described (Lynch, 1976). These genetic cancers fall into several distinct categories:

(1) presumed repair defect syndromes (such as Fanconi anaemia, Bloom syndrome, ataxia telangiectasia, xeroderma pigmentosum, that have been associated with an increased incidence of malignancies, primarily leukaemias and lymphomas;
(2) familial cancers in which recurrences of one or more tumour types are identified in successive generations; examples include recurrences of specific tumour types, such as those of breast, eye (retinoblastoma), and kidney (renal cell carcinoma) within individual pedigrees, as well as the so-called 'cancer family syndromes' each of which is characterized by a high incidence of cancers, generally limited to a defined group of tumour types (Lynch and Krush, 1973); and
(3) genetic syndromes in which an increased predisposition to the development of particular cancers is seen in association with distinct patterns of congenital malformations and/or mental retardation, examples include neurofibromatosis (with an increased incidence of neuroblastoma, neurofibrosarcoma, and glioblastoma), and von Hippel–Lindau syndrome, in which renal carcinomas have been reported in a significant fraction of patients with the characteristic angiomata of retina and cerebellum (Lynch, 1976).

The disorders of DNA repair are inherited in an autosomal recessive manner; the disorders in categories (2) and (3) are most commonly inherited as autosomal dominants. However, what appears to be transmitted in these two categories is a predisposition to tumorigenesis; the dominant gene mutation is not sufficient in itself for the development of cancer. This is because not everyone who inherits the mutant gene in these pedigrees develops a tumour, and not every susceptible cell becomes transformed. In familial retinoblastoma, for example, an individual who inherits the gene predisposing to retinoblastoma will develop on average three discrete foci or retinoblastoma, in one or both eyes; however, the vast majority of susceptible retinoblasts (cells also carrying the mutation predisposing to retinoblastoma) will differentiate normally and never become transformed.

The latter observation, along with other epidemiological evidence, led to the proposal, some years ago, of the 'two hit' hypothesis (Knudson, 1971). This postulates that at least two gene changes are required for the development of retinoblastoma. One of these changes presumably involves a gene on the long arm (q) of chromosome 13, because of the frequent finding of deletions/rearrangements resulting in the loss of structural material from 13q in retinoblastoma karyotypes and the association between retinoblastoma and constitutional deletions (that is, deletions present in somatic and tumour cells) of band 13q14 (Yunis and Ramsay, 1978; Potluri *et al.*, 1986). The 'second hit' could involve a gene locus distinct from 13q14 in the karyotype or the same gene on the second chromosome 13 in the cell. Evidence supporting the latter possibility is discussed in Chapter 15.

Epidemiological data supporting the applicability of the 'two hit' hypothesis to explain tumorigenesis in neuroblastoma, Wilms' tumour, and phaeochromocytoma has also been presented (Knudson, 1985). Experimental studies demonstrating the loss of heterozygosity for chrosome specific markers, consistent with the homozygous involvement of single gene loci in tumorigenesis, has recently been reported in Wilms' tumour (with the loss of sequences from chromosome 11p), uveal melanoma (chromosome 2), and acoustic neuroma (chromosome 22q) (Koufos *et al.*, 1984; Mukai and Dryja, 1986; Seizinger, Martuza and Gusella, 1986).

The genes whose loss or inactivation have been postulated to be necessary events in tumorigenesis have recently been termed 'anti-oncogenes' by several authors (for example, Green and Wyke, 1985; Knudson, 1985). This is an unfortunate term, as it implies a direct relationship between the genes in question and the control of transcription or translation of one or more cellular oncogenes. Until such time as the functional significance of each of these genes has been determined, it would be most appropriate to refer to each by an abbreviation reflecting the tumour type involved – for example, Rb for the putative retinoblastoma locus at 13q14, WAGR for the locus at 11p13 implicated in the Wilms' tumour/aniridia/gonadoblastoma/retardation syndrome.

SOMATIC CELL HYBRIDS, TRANSFECTION, AND TRANSGENIC MICE

Additional arguments against dominant single gene changes sufficient for tumorigenesis have come from studies of somatic cell hybrids, DNA transfection, and transgenic mice. In the first, reports from a number of laboratories have documented that fusion of malignant cells with non-malignant cells of the same or different species initially results in non-malignant hybrids (Stanbridge, 1987).

Presumably this is a consequence of the 'correction' of the malignant phenotype by the introduction of the normal gene(s) from the non-malignant partner in the fusion that have been deleted or inactivated in the malignant parent. With the subsequent loss of chromosomes from the non-malignant partner, as the hybrid cells divide in culture the malignant potential can be re-expressed (Stanbridge, 1987).

An example specifically relating a gene change postulated to be responsible for tumorigenesis in a particular tumour type has recently been presented. When human Wilms' tumour cells were fused with a rodent hybrid cell containing a portion of a normal human 11p chromosome, the resulting hybrid cells were no longer capable of forming tumours in athymic mice (Weissman *et al.*, 1987). The loss of the normal human 11p from the hybrids over time in culture was associated with re-expression of the tumorigenic phenotype, the ability of the hybrids to form tumours in appropriate test animals.

Studies of DNA transfection, in which tumour cell DNAs and/or cloned oncogenes have induced transformation in cultured cell lines, have generally utilized one of two cell systems as substrates, either NIH 3T3 or rat embryo fibroblasts (REF). The former is an immortal (permanently dividing), aneuploid, contact inhibited mouse cell line that does not readily form tumours in athymic mice. The latter are recent explants from rat embryos, with a finite lifespan in culture, diploid chromosome number, contact inhibited phenotype, and an inability to form tumours in athymic mice. Transfection of NIH 3T3 by a number of human tumour DNAs has identified several activated oncogenes, most frequently of the *ras* family, that are by themselves capable of inducing transformation (Tabin *et al.*, 1982; Land, Parada and Weinberg, 1983). In the REF system, no single transfected oncogene is sufficient to induce transformation; the minimum number is two (Land, Parada and Weinberg, 1983). These results are consistent with the hypothesis that NIH 3T3 already carries one or more 'hits' that confer immortality and only needs a single, additional gene change (transfected activated *ras*, for example) for full transformation. The diploid, normal rat embryo fibroblasts require the interaction of at least two gene changes (such as activated *c-* or *N-myc* and *c-Ha-ras*) before transformation is achieved.

Transgenic mice represent genetic chimera, in which one or more foreign genes are directly introduced into the nucleus of a fertilized egg, integrated into nuclear DNA, and expressed in the mouse after birth (Palmiter and Brinster, 1986). Transgenic mice carrying *c-myc* linked to the mouse mammary tumour virus (MMTV) promoter develop mammary tumours at high frequency (Sinn *et al.*, 1987). However, while *c-myc* was actively expressed throughout mammary tissue, the tumours were discrete and presumably derived from single cells; this implies that gene changes in addition to the activated *c-myc* were required for malignant transformation.

Transgenic mice carrying an MMTV/*v-Ha-ras* construct will also express the transgene in mammary tissue and develop mammary carcinomas at high frequency (Sinn *et al.*, 1987). Matings between carriers of MMTV/*c-myc* and MMTV/*v-Ha-ras* produce offspring carrying both transgenes. The transgenic hybrids develop tumours earlier and at a frequency higher than that of carriers of either transgene alone (Sinn *et al.*, 1987). While expression of both transgenes was enhanced throughout mammary tissue in the hybrid animals, the tumours were again apparently monoclonal in origin. Therefore, the presence of activated *ras* and *myc* in the same cell was insufficient for malignant transformation; additional somatic events were required for the development of tumours.

TUMOUR REGRESSION, MATURATION, AND PROGRESSION

The existence of tumour regression and maturation also argues against single, dominant gene changes as responsible for the development of cancer. While most cancers will continue to grow and, without surgical or therapeutic intervention, eventually result in the death of the host, a fraction of cancers spontaneously disappear (regress) or mature to benign forms. The phenomena of regression and maturation are well known in retinoblastoma and neuroblastoma, though the frequency of either or both in these tumours is unknown.

In retinoblastoma pedigrees for example, individuals have been described who had no history of retinoblastoma themselves, but were presumed to be carriers of the predisposing 'first hit' because a parent and/or multiple children had the tumour. When carefully examined, such individuals were often found to have scars on their retinas, consistent with regressed retinoblastoma. In the case of familial neuroblastoma, parents who have had more than one child with a tumour, have been reported with ganglioneuroma, a benign, more differentiated neuroectermal tumour (Kushner, Gilbert and Helson, 1986).

Two other anomalous forms of neuroblastoma exist: neuroblastoma-in-situ and stage IV-S neuroblastoma. In a published series of autopsies of infants whose deaths could not be attributed to cancer, some 10% were found to have small foci of neuroblastoma-like change in their adrenal glands (Beckwith and Perrin, 1963). These foci were termed 'neuroblastoma-in-situ'. As the frequency of neuroblastoma in most population surveys is approximately one in 20 000 children, the implication of this finding is that these foci do not progress to clinically significant disease in the vast majority of instances.

Neuroblastoma itself has been subdivided into five clinical stages, based on age at diagnosis and extent of disease (Evans, D'Angio and Randolph, 1971). Stage IV-S, the rarest form, occurs in children under 1 year of age and is characterized by disease in adrenal gland, liver, bone marrow and skin. In some 85% of cases, the disease will spontaneously disappear without any treatment; the remaining 15% will progress to metastatic stage IV.

Alternative explanations can be presented for each of these observations: it is possible, for example, that neuroblastoma-in-situ is not a true preneoplastic lesion, merely a normal stage in the lineage of differentiated cells in the adrenal medulla, stage IV-S may be aetiologically distinct (and genetically different) from classical neuroblastoma, and tumour regression and/or maturation can result from back mutation, or from epigenetic or non-genetic events. However, these observations are also consistent with the hypothesis that while one or more gene change(s) may produce a premalignant alteration in cellular proliferative capacity, resulting perhaps in stage IV-S or neuroblastoma-in-situ, full malignant transformation of a particular cell type requires at least two specific gene changes.

In addition to the gene changes responsible for transformation, other gene changes may contribute to the ability of a tumour to expand and spread within the host (metastasis). As noted previously, chromosome abnormalities of classes (1) and (2) tend to be unique to particular tumour types; they are presumed to point to the sites of genes that play direct or primary roles in tumorigenesis. Chromosome abnormalities of class (3) tend to be less specific and have often been reported in independent examples of multiple tumour types. It is possible that the gene changes resulting from chromosome abnormalities of class (3) further enhance proliferative capacity or contribute in other ways to an already malignant cell's metastatic ability. Karyotypes of leukaemias and lymphomas prepared over time from individual patients have documented an association between the accumulation of chromosome abnormalities of class (3) and poor prognosis (Rowley, 1977). A similar association has also been made between poor prognosis and the ability to metastasize, and the presence of several class (3) changes, specifically HSR/DMS and extra 1q, in neuroblastoma (Feder and Gilbert, 1983; Brodeur *et al.*, 1984).

NEURO-ONCOGENESIS

With the above as an introduction, is it possible to combine what is known about cancer associated chromosome abnormalities, recessive cancer genes, and activated cellular oncogenes, into a unifying hypothesis that explains the role of single gene changes in the development of individual cancers. Selected tumours of neuroectodermal origin are used as model systems.

First what is known about chromosome abnormalities and oncogene expression in this family of malignancies, is reviewed. Embryonic neuroectoderm gives rise to cells populating a number of adult tissues, including the central and peripheral nervous systems and retina. Among the tumours of neuroectodermal origin are medulloblastomas, meningiomas, and various glial tumours of the central nervous system, neuroblastomas and phaeochromocytomas of the peripheral nervous system, and retinoblastomas of retina. Karyotypes have been prepared from multiple examples of each of these tumours. In virtually every instance, abnormalities of chromosome number (aneuploidy) and organization (structural rearrangements) have been demonstrated (Sandberg, 1980). However, recurrent aberrations unique to individual tumour types have been reported only in meningiomas (deletions/translocations involving the loss of structural material from chromosome 22q and monosomy 22; Sandberg, 1980); retinoblastoma (deletions/translocations involving the loss of structural material from 13q and monosomy 13; Potluri *et al.*, 1986); neuroblastoma (deletions/translocations involving the loss of structural material from 1p; Gilbert *et al.*, 1984); and a subcategory of neuroblastoma, histologically similar but clinically distinguishable by site of involvement, termed peripheral neuroepithelioma (balanced translocation involving the long arms of chromosome 11 and 22; Whang-Peng *et al.*, 1984).

Besides class (1) and (2) abnormalities that are unique to individual tumour types, a number of class (3) changes, particularly extra 1q, 17q, and 6p, as well as HSR and DMS, have been reported in multiple examples of the neuroectodermal malignancies (Gilbert *et al.*, 1984; Potluri *et al.*, 1986). For example, the development of trisomy for a 1q marker chromosome in neuroblastoma over time was documented in a patient whose tumour was sampled twice, at the time of original diagnosis and 11 months later at autopsy (Feder and Gilbert, 1983).

While the particular genes on 1q, 17q, or 6p that contribute to tumour development when present in extra copy (the result of trisomy for these chromosome segments) are unknown, the HSR and DMS in both neuroblastomas and retinoblastomas have been shown to contain amplified *N-myc* (Kohl *et al.*, 1983; Schwab *et al.*, 1983). Amplified *N-myc* has also been reported in at least one astrocytoma, and amplified *c-myc* has been demonstrated in several peripheral neuroepitheliomas (Garson, McIntyre and Kemshead, 1985; M. Israel, personal communication).

Two other activated cellular oncogenes have also been described in tumours of neuroectodermal origin. Among the tumour cell DNAs found to be capable of inducing transformation in the NIH 3T3 transfection assay were those of the human neuroblastoma cell lines, SK-N-SH and SK-N-AS (Taparowsky *et al.*, 1983; H. Rovigatti, personal communication). In each instance the responsible cellular oncogene was activated *N-ras*. Activated *neu*, present in a series of carcinogen-induced neural tumours of rats, was also found to be positive in the NIH 3T3 transfection assay (Shih *et al.*, 1981). Although *neu* has also been identified in the human genome, and mapped to 17q21, its abnormal expression in human neuroectodermal tumours has not been demonstrated.

Tying together the above observations with the requirement for a minimum of two gene changes for tumorigenesis leads to the following speculation. By analogy with the findings in retinoblastoma and Wilms' tumour, it can be argued that the recurrent deletions in neuroblastoma and meningioma also point to the sites of genes whose homozygous loss is a prerequisite for tumorigenesis in each instance. By analogy with Burkitt's lymphoma and its translocation between chromosomes 8 and 14, resulting in activation of the transposed *c-myc*, one would predict that the t(11;22) in peripheral neuroepithelioma is likely to result in the activation of a similarly transposed oncogene. The particular oncogene involved has not been identified. One possible candidate is *c-sis*, mapped to 22q12, but evidence of increased or altered transcription has yet to be presented.

How might the loss of genetic information, the consequence of a class (2) chromosome change, or the activation of an oncogene, as in a class (1) transcription, contribute to tumorigenesis? If it is accepted that cellular differentiation involves a progression through a distinct series of increasingly more differentiated steps or stages, extending from minimally differentiated embryonic stem cells to non-dividing, fully differentiated adult cells, and that the transition between each stage is controlled by single genes, then the homozygous loss of activity of one such gene would block this transition, fixing the cell in question at a particular stage in its lineage. Such a blocked cell might divide permanently; alternatively the block might be 'leaky', allowing the cell to go through a limited number of divisions, resulting perhaps in the local accumulation of cells that characterize neuroblastoma-in-situ or stage IV-S neuroblastoma, before additional gene changes occurred that allowed the cells to continue their progression through the differentiated lineage, or resulted in regression, maturation or immortality and full malignant transformation.

In accord with the 'oncogene cascade' described earlier, one could postulate that every dividing cell, and possibly each stage of a differentiated lineage, will be marked by the active synthesis of a particular DNA binding oncogene, such as a *myc* gene family member. By this hypothesis, the enhanced expression of *N-myc* reported in tumour types as diverse as neuroblastoma, Wilms' tumour, and retinoblastoma (even in tumours without *N-myc* amplification; Lee, Murphree and Benedict, 1984; Nisen *et al.*, 1986) could represent a consequence of a block to progression through a differentiated lineage, rather than its cause.

Conversely, one might be able to achieve a similar block within a differentiated lineage by constitutively activating a specific DNA binding oncogene, as, for example, in the transposition of an oncogene away from *cis*-acting suppression of transcription by chromosome translocation. The requirement for a minimum of 'two hits' for tumorigenesis is not precluded by a class (1) change; one can hypothesize that the translocation represents the 'second hit', following a specific submicroscopic structural or molecular change that predisposes to chromosome breakage and rearrangement (the 'first hit').

As the amplification of individual oncogenes cannot be demonstrated in every example of any one tumour type, amplification *per se* cannot be a necessary prerequisite for tumorigenesis. Amplification presumably develops after the primary transformation events have occurred, as is the case with the other class (3) chromosomal abnormalities that accumulate over time *in vivo* (Rowley, 1977; Gilbert, 1983). However, amplification of *N-myc* in neuroblastoma does correlate with severity of clinical disease, with virtually no examples of amplification in localized stage I (good prognosis), but a high frequency of amplification in metastatic stage IV (poor prognosis) (Brodeur *et al.*, 1984). As with the reported association between amplification of *neu* and poor prognosis in human mammary carcinoma (Slamon *et al.*, 1987) this finding is consistent with the hypothesis that oncogene amplification, and the accumulation of other class (3) chromosomal abnormalities, contributes to the ability of a tumour to expand and spread or metastasize in the host (Gilbert, 1983).

CONCLUSIONS

The cytogenic and molecular genetic analysis of different neuroectodermal malignancies has therefore enabled us to conclude that the aberrant expression of cellular oncogenes, either as an enhanced transcription of specific mRNAs or the production of altered proteins, plays a significant role in both the genesis and progression of cancer. The characterization of individual tumours in terms of the particular activated oncogenes each contains will almost certainly be of prognostic significance and could in future have therapeutic implications as well, as agents targeted against the products of each oncogene are developed.

References

ALT, F. W., DePINHO, R., ZIMMERMAN, K. *et al.* (1986) The human *myc* gene family. *Cold Spring Harbor Symposium on Quantitative Biology*, **LI:** 931–941

BECKWITH, J. B. and PERRIN, E. V. (1963) *In situ* neuroblastomas: a contribution to the natural history of neural crest tumors. *American Journal of Pathology*, **43**, 1089–1100

BIEDLER, J. and SPENGLER, B. (1986) Metaphase chromosome anomaly: association with drug resistance and cell-specific products. *Science*, **191**, 185–187

BISHOP, J. M. (1983) Cellular oncogenes and retroviruses. *Annual Review of Biochemistry*, **52**, 301–354

BROCA, P. P. (1866) *Traité des Tumeurs*. Paris: Passelin

BRODEUR, G. M., SEEGER, R. C., SCHWAB, M., VARMUS, H. E. and BISHOP, J. M. (1984) Amplification of *N-myc* in untreated human neuroblastoma correlates with advanced disease stage. *Science*, **224**, 1121–1124

COFFIN, J. M., VARMUS, H. E., BISHOP, J. M. *et al.* (1981) Proposal for naming host cell-derived inserts in retrovirus genomes. *Journal of Virology*, **40**, 953–957

EVANS, A. E., D'ANGIO, G. J. and RANDOLPH, J. (1971) A proposed staging for children with neuroblastoma. *Cancer*, **27**, 374–378

FEDER, M. and GILBERT, F. (1983) Clonal evolution in a human neuroblastoma. *Journal of National Cancer Institute*, **70**, 1051–1056

GARSON, J. A., McINTYRE, P. G. and KEMSHEAD, J. T. (1985) *N-myc* amplification in malignant astrocytoma. *Lancet*, **ii**, 718–719

GILBERT, F. (1983) Chromosomes, genes and cancer: a classification of chromosome abnormalities in cancer. *Journal of National Cancer Institute*, **71**, 1107–1114

GILBERT, F., FEDER, M., BALABAN, G. *et al.* (1984) Human neuroblastomas and abnormalities of chromosomes 1 and 17. *Cancer Research*, **44**, 5444–5449

GREEN, A. R. and WYKE, J. A. (1985) Anti-oncogenes. A subset of regulatory genes involved in carcinogenesis. *Lancet*, **ii**, 475–477

HAYWARD, W. S., NEEL, B. G. and ASTRIN, S. M. (1981) Activation of a cellular *onc* gene by promoter insertion in ALV-induced lymphoid leukosis. *Nature (London),* **290,** 475–480

HEISTERKAMP, N., STAM, K., GROFFEN, J., deKLEIN, A. and GROSVELD, G. (1985) Structural organization of the *bcr* gene and its role in the Ph[1] translocation. *Nature (London),* **315,** 758–761

JAKOBOVITS, A., SCHWAB, M., BISHOP, J. M. and MARTIN, G. R. (1985) Expression of *N-myc* in teratocarcinoma stem cells and mouse embryos. *Nature (London),* **318,** 188–191

KELLY, K., COCHRAN, B. H., STILES, C. D. and LEDER, P. (1983) Cell-specific regulation of the *c-myc* gene by lymphocyte mitogens and platelet-derived growth factor. *Cell,* **35,** 603–610

KNUDSON, A. G. (1971) Mutation and cancer: statistical study of retinoblastoma. *Proceedings of the National Academy of Sciences USA,* **68,** 820–823

KNUDSON, A. G. (1985) Hereditary cancer, oncogenes and anti-oncogenes. *Cancer Research,* **45,** 1437–1443

KOHL, N. E., KANDA, N., SCHRECK, R. R., BRUNS, G., LATT, S. A., GILBERT, F. and ALT, F. W. (1983) Transposition and amplification of oncogene-related sequences in human neuroblastoma. *Cell,* **35,** 359–367

KONOPKA, J. B., WATANABE, S. M. and WITTE, O. N. (1984) An alteration of the human *c-abl* protein in K562 leukemia cells unmasks associated tyrosine kinase activity. *Cell,* **37,** 1035–1042

KOUFOS, A., HANSEN, M. F., LAMPKIN, B. C., WORKMAN, M. L., COPELAND, N. G., JENKINS, N. A. and CAVENEE, W. K. (1984) Loss of alleles at loci on human chromosome 11 during genesis of Wilms' tumor. *Nature (London),* **309,** 170–172

KUSHNER, B., GILBERT, F. and HELSON, L. (1986) Familial neuroblastoma: case reports, literature review, and etiologic considerations. *Cancer,* **57,** 1887–1893

LACAL, J. C., SRIVASTAVA, S. K., ANDERSON, P. S. and AARONSON, S. A. (1986) *ras* p21 proteins with high or low CTPase activity can efficiently transform NIH 3T3 cells. *Cell,* **44,** 609–617

LACHMAN, H. M., HATTON, K. S., SKOULTCHI, A. I. and SCHILDKRAUT, C. L. (1985) *c-myc* mRNA levels in the cell cycle change in mouse erythroleukemia cells following inducer treatment. *Proceedings of the National Academy of Sciences USA,* **82,** 5323–5327

LAND, H., PARADA, L. F. and WEINBERG, R. A. (1983) Tumorigenic conversion of primary embryo fibroblasts requires at least two co-operating oncogenes. *Nature (London),* **304,** 596–602

LEAL, F., WILLIAMS, L. T., ROBBINS, C. and AARONSON, S. A. (1985) Evidence that the *v-813* gene product transforms by interaction with the receptor for platelet-derived growth factor. *Science,* **230,** 327–330

LEDER, P., BATTEY, J., LENOIR, G. *et al.* (1983) Translocations among antibody genes in human cancer. *Science,* **222,** 765–771

LEE, W.-H., MURPHREE, A. L. and BENEDICT, W. F. (1984) Expression and amplification of the *N-myc* gene in primary retinoblastoma. *Nature (London),* **309,** 458–460

LYNCH, H. T. (1976) *Cancer Genetics.* Springfield, Illinois: C. C. Thomas

LYNCH, H. T., KRUSH, A. J. (1973) Differential diagnosis of the cancer family syndrome. *Surgery, Gynecology and Obstetrics,* **136,** 222–224

McKUSICK, V. A. (1986) The human gene map. *Clinical Genetics,* **29,** 545–588

MARTIN-ZANCA, D., HUGHES, S. H. and BARBACID, M. (1986) A human oncogene formed by the fusion of truncated tropomyosin and protein tyrosine kinase sequences. *Nature (London),* **319,** 743–748

MITCHELL, R. L., HENNING-CHUBB, C., HUBERMAN, E. and VERMA, I. M. (1986) *c-fos* expression is neither sufficient nor obligatory for differentiation of monomyelocytes to macrophages. *Cell,* **45,** 497–504

MUKAI, S. and DRYJA, T. P. (1986) Loss of alleles at polymorphic loci on chromosome 2 in uveal melanoma. *Cancer Genetics and Cytogenetics,* **22,** 45–53

MULLER, R. and VERMA, I. M. (1984) Expression of cellular oncogenes. *Current Topics in Microbiology and Immunology,* **112,** 73–116

NISEN, P., ZIMMERMAN, K., COTTER, S., GILBERT, F. and ALT, F. (1986) Enhanced expression of the *N-myc* gene in Wilms' tumor. *Cancer Research,* **46,** 6217–6222

NUSSE, R., VAN OOYEN, A., COX, D., KAI, T., FUNG, Y. and VARMUS, H. E. (1984) Mode of proviral activation of a putative mammary oncogene (*int*-1) on mouse chromosome 15. *Nature (London),* **307,** 131–136

PALMITER, R. and BRINSTER, R. (1986) Germ-line transformation of mice. *Annual Review of Genetics,* **20,** 465–500

PARK, M., DEAN, M., COOPER, C. S., SCHMIDT, M., O'BRIEN, S. J., BLAIR, D. G. and VANDE WOUDE, G. F. (1986) Mechanism of *met* oncogene activation. *Cell,* **45,** 895–904

POTLURI, V. R., HELSON, L., ELLSWORTH, R. M., REID, T. and GILBERT, F. (1986) Chromosomal abnormalities in human retinoblastoma: a review. *Cancer,* **58,** 663–671

ROHRSCHNEIDER, L. R. and GENTRY, L. E. (1984) Subcellular location of retroviral transforming proteins define multiple mechanisms of transformation. *Advances in Viral Oncology,* **4,** 269–306

ROWLEY, J. D. (1977) Mapping of human chromosomal regions related to neoplasia. Evidence from chromosome 1 and 17. *Proceedings of the National Academy of Sciences USA,* **74,** 5729–5733

SANDBERG, A. A. (1980) *The Chromosomes in Human Cancer and Leukemia*. New York: Elsevier North-Holland

SAP, J., MUNOZ, A., DAMM, K. *et al.* (1986) The *c-erb-A* protein is a high-affinity receptor for thyroid hormone. *Nature (London)*, **324**, 635–640

SCHWAB, M., ALITALO, K., KLEMPNAUER, K.-H. *et al.* (1983) Amplified DNA with limited homology to *myc* cellular oncogene is shared by human neuroblastoma cell lines and a neuroblastoma tumor. *Nature (London)*, **305**, 245–248

SEIZINGER, B. R., MARTUZA, R. L. and GUSELLA, J. F. (1986) Loss of genes on chromosome 22 in tumorigenesis of human acoustic neuroma. *Nature (London)*, **322**, 644–649

SHERR, C. J., RETTENMIER, C. W., SACCA, R., ROUSSEL, M. F., LOOK, A. T. and STANLEY, E. R. (1985) The *c-fms* proto-oncogene product is related to the receptor for the mononuclear phagocyte growth factor, CSF-1. *Cell*, **41**, 665–676

SHIH, C., PADHY, L., MURRAY, M. and WEINBERG, R. A. (1981) Transforming genes of carcinomas and neuroblastomas introduced into mouse fibroblasts. *Nature (London)*, **290**, 261–264

SINN, E., MULLER, W., PATTENGALE, P., TEPLER, I., WALLACE, R. and LEDER, P. (1987) Co-expression of MMTV/*v-Ha-ras* and MMTV/*c-myc* genes in transgenic mice: synergistic action of oncogenes *in vivo*. *Cell*, **49**, 465–475

SLAMON, D. J., CLARK, G. M., WONG, S. G., LEVIN, W. J., ULLRICH, A. and McGUIRE, W. L. (1987) Human breast cancer: correlation of relapse and survival with amplification of the HER-2/*neu* oncogene. *Science*, **235**, 177–182

STANBRIDGE, E. J. (1987) Genetic regulation of tumorigenic expression in somatic cell hybrids. *Advances in Viral Oncology*, **6**, 83–101

STRYER, L. and BOURNE, H. R. (1986) G proteins: a family of signed transducers. *Annual Review of Cell Biology*, **2**, 391–419

TABIN, C. J., BRADLEY, S. M., BARGMANN, C. I. *et al.* (1982) Mechanism of activation of a human oncogene. *Nature (London)*, **300**, 143–149

TAPAROWSKY, E., SHIMIZU, K., GOLDFARB, M. and WIGLER, M. (1983) Structure and activation of the human *N-ras* gene. *Cell*, **34**, 581–586

THIELE, C. J., REYNOLDS, C. P. and ISRAEL, M. A. (1985) Decreased expression of *N-myc* precedes retinoic acid-induced morphological differentiation of human neuroblastoma. *Nature (London)*, **313**, 404–406

TSUJMOTO, Y. and CROCE, C. M. (1986) Analysis of the structure, transcripts, and protein products of *bc*-1-2, the gene involved in human follicular lymphoma. *Proceedings of the National Academy of Sciences USA*, **83**, 5214–5218

ULLRICH, A., COUSSENS, L., HAYFLICK, J. S. *et al.* (1984) Human epidermal growth factor receptor cDNA sequence and aberrant expression of the amplified gene in A431 epidermoid carcinoma cells. *Nature (London)*, **309**, 418–425

VARMUS, H. E. (1984) The molecular genetics of cellular oncogenes. *Annual Review of Genetics*, **18**, 553–612

VARMUS, H. E. and BISHOP, J. M. (1986) Biochemical mechanisms of oncogenic activity: proteins encoded by oncogenes. *Cancer Surveys*, **5**, 153–434

VERMA, R. S. (1986) Oncogenetics. *Molecular and General Genetics*, **205**, 385–389

WEISSMAN, B. E., SAXON, P. J., PASQUALE, S. R., JONES, G. R., GEISER, A. G. and STANBRIDGE, E. J. (1987) Introduction of a normal chromosome 11 into a Wilms' tumor cell line controls its tumorigenic expression. *Science*, **236**, 175–180

WHANG-PENG, J., TRICHE, T. J., KNUTSEN, T., MISER, J., DOUGLASS, E. and ISRAEL, M. A. (1984) Chromosome translocation in peripheral neuroepithelioma. *New England Journal of Medicine*, **311**, 584–585

YUNIS, J. J. (1983) The chromosomal basis of human neoplasia. *Science*, **221**, 227–236

YUNIS, J. J. and RAMSAY, N. (1978) Retinoblastoma and sub-band deletion of chromosome 13. *American Journal of Diseases of Childhood*, **132**, 161–163

ZIMMERMAN, K. A., YANCOPOULOS, G. D., COLLUM, R. G. *et al.* (1986) Differentiated expression of *myc* family genes during murine development. *Nature (London)*, **319**, 780–783

9
Transgenic mice and neurological disease

Albee Messing

INTRODUCTION

The ability to introduce defined sequences of DNA into the germline of animals opened a new era in studies on the regulation of gene expression and the pathological consequences of atypical gene expression. Genes subjected to experimental manipulation can now be exposed to a wide variety of cell types undergoing normal developmental processes in an intact animal. Extensive reviews have recently appeared (Gordon and Ruddle, 1985; Palmiter and Brinster, 1986). This chapter focuses on how 'transgenic' animals have been used in neuroscience, and in particular on their potential use in studying the pathogenesis of neurological diseases.

METHODS OF GENE TRANSFER INTO ANIMALS AND GENERAL CONSIDERATIONS ABOUT EXPRESSION

Several methods exist for introducing foreign genes into the mammalian germline: nuclear transplantation; direct microinjection of DNA into pronuclei of fertilized eggs; retroviral infections of embryos; or colonization of blastocyst-stage embryos by cultured embryonic stem cells or teratocarcinomas. Each has advantages and disadvantages that will not be discussed here (for details of techniques, *see* Gordon and Ruddle, 1983; Brinster *et al.*, 1985; Hogan, Costantini and Lacy, 1986). The most widely used method to date is direct microinjection into pronuclei. Most work has been done in the mouse, although progress is being made with other species. Briefly, single-cell fertilized eggs are injected with a solution of DNA, and the microinjected eggs are then transferred into the oviduct of a pseudopregnant foster mother who carries them to term. The DNA spontaneously integrates into an apparently random site in the host genome. With the microinjection technique, multiple copies of the genes often combine in head-to-tail tandem arrays, but usually integrate at a single site. In general, there is very poor correlation between gene copy number and levels of expression. When integration occurs prior to the first cellular division, the gene should be present in all cells of the developing fetus, including the germ line, and be passed to the progeny. Mosaicism occurs in

Table 9.1 Transgenic mice with expression or pathology in CNS, PNS or skeletal muscle

*Gene**	*Site of expression or pathology*	*Reference*
(*a*) *Viral genes causing pathology*		
SV40	Choroid plexus papilloma	Brinster *et al.* (1984)
SV40	Choroid plexus papilloma	Palmiter *et al.* (1985)
SV40	Peripheral hypomyelination	Messing *et al.* (1985)
SV40 and *v-myc*	Choroid plexus papilloma	Small *et al.* (1985)
Adeno E1A/SV40	Glioblastoma	Kelly *et al.* (1986)
JC virus	Adrenal neuroblastoma	Small *et al.* (1986a)
	CNS hypomyelination	Small *et al.* (1986b)
(*b*) *Genes normally expressed in nervous system*		
mMT–hHPRT	Brain (cell type unknown)	Stout *et al.* (1985)
mMBP	Brain – oligodendrocytes	Readhead *et al.* (1987)
		Popko *et al.* (1987)
mGnRH	CNS neurons	Mason *et al.* (1986b)
(*c*) *Genes with ectopic expression in nervous system*		
mMT–GH	CNS neurons	Swanson *et al.* (1985)
mMT–rSS	Anterior pituitary – gonadotrophs	Low *et al.* (1986)
(*d*) *Genes normally expressed in muscle*		
Rat α-actin	Skeletal and cardiac muscle	Shani (1986)
Rat myosin light-chain 2	Skeletal muscle	Shani (1985)
(*e*) *Genes with low-level or erratic expression in brain or muscle*		
mMT–hGH	Brain	Palmiter *et al.* (1983)
Mouse α_1-globin	Brain	Rusconi (1984)
Retroviral Eco gpt	Brain	Stuhlmann *et al.* (1984)
Hepatitis surface antigen	Brain	Chisari *et al.* (1985)
mMT-hGH-β-globin	Brain	Townes *et al.* (1985)
Retroviral	Brain	van der Putten *et al.* (1985)
Mouse E_α^d MHC class II	Brain	Yamamura (1985)
Mouse α_2 (I) collagen-CAT	Brain	Khillan *et al.* (1986)
Mouse H-2K/hGH MHC class I	Brain	Morello *et al.* (1986)
Rabbit β-globin	Muscle	Lacy *et al.* (1983)
Mouse E_α^d MHC class II	Muscle	Pinkert *et al.* (1985)
Chicken transferrin	Brain and muscle	McKnight *et al.* (1983)
Retroviral neo	Brain and muscle	Stewart *et al.* (1985)
Mouse H-2D^d MHC class I	Brain and muscle	Bieberich *et al.* (1986)
Retroviral-CAT	Brain and muscle	Overbeek *et al.* (1986)
Retroviral-β-globin	Brain and muscle	Soriano *et al.* (1986)

*Gene abbreviations: SV40: simian virus 40; mMT: mouse metallothionein promotor; hHPRT: human hypoxanthine phosphoribosyltransferase; mMBP: mouse myelin basic protein; mGnRH: mouse gonadotropin-releasing hormone; GH: rat or human growth hormone; rSS: rat somatostatin; CAT: chloramphenicol acetyltransferase

approximately 30% of transgenic mice produced by microinjection, presumably due to integrations after the first division. With present techniques of gene transfer by retroviruses or embryonic stem cell/teratocarcinoma colonization, the founder animals are always mosaic.

Although earlier attempts at obtaining expression of the foreign genes were less successful, the general experience now is that many are expressed in an appropriate tissue-specific manner. This occurs despite the varying influence of different chromosomal locations. In some cases proper developmental patterns and quantitative levels of expression are also achieved. The tissue-specificity appears to depend upon regulatory sequences (enhancers/promotors) separate from the structural sequences, and hybrid genes can be constructed in which expression of a structural gene is redirected to a different cell type due to the substitution of an alternate enhancer/promotor sequence. In still other cases entirely unexpected patterns of expression occur, the reasons for which are not clear. Transgenic animals offer unique opportunities for studying the consequences of atypical gene expression. A comprehensive listing of transgenic mice that have expressed foreign genes or shown pathological changes in the nervous system or skeletal muscle is shown in *Table 9.1*.

EXPRESSION OF VIRAL GENES WITH PATHOLOGICAL EFFECTS IN THE NERVOUS SYSTEM

Neoplasia

Transgenic mice are ideally suited for testing hypotheses that expression of exogenous genes, or overexpression of endogenous genes such as oncogenes, are involved in particular disease processes. The first dramatic example of an effect on the nervous system occurred when the partial genome from an oncogenic DNA virus, simian virus 40 (SV40), was introduced into mice (Brinster *et al.*, 1984; *see also* Small *et al.*, 1985). Mice carrying the SV40 early region, which codes for two proteins (large T and small t antigens), consistently succumbed by 5–12 months of age to a characteristic syndrome which included papillomas of the choroid plexus and non-neoplastic lesions of the thymus and kidney (*Figure 9.1*). This phenotype was passed on to the progeny, and several breeding lines have now been established

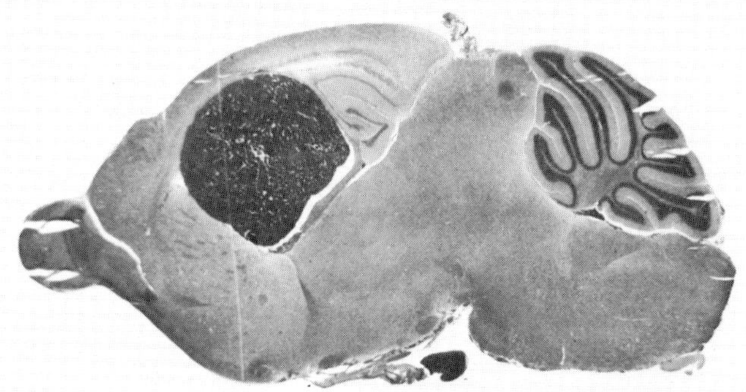

Figure 9.1 Haematoxylin and eosin stained section through the brain of a 5 month old mouse carrying the SV40–MGH gene construction (Brinster *et al.*, 1984), showing a large choroid plexus papilloma arising in one of the lateral ventricles

in which the phenotype has remained consistent through 15 generations. Subsequent study showed that when the SV40 enhancer sequence was deleted the incidence of choroid plexus tumours was markedly reduced (Palmiter *et al.*, 1985). The basis for the SV40 enhancer's apparent specificity for the choroid plexus has not yet been established. It is also not certain whether other oncogenes can similarly transform the choroid plexus epithelium, since an SV-*myc* construction failed to produce mice with any noticeable phenotype (Adams *et al.*, 1985). However, this gene construction may not have been expressed at all. Recently, Yagaloff *et al.* (1986) demonstrated high-affinity binding of ^{125}I-LSD to the transgenic choroid plexus tumours. These tumours turn out to have the highest known density of serotonin 5-HT$_{1C}$ receptors, a subtype specific to the choroid plexus. Induction of such neoplasms may thus have unexpected benefits by providing large quantities of tissues (or sources of cell lines) not otherwise available for biochemical or molecular analysis.

Other cell types within the central nervous system (CNS) are also susceptible to transformation by SV40, as illustrated by transgenic mice carrying the SV40 early region genes in association with an enhancer/promotor region from adenovirus E1A (type 5) (Kelly *et al.*, 1986). Two such mice showed signs of altered equilibrium and hind-leg paralysis at 3–4 months of age, and necropsy revealed that both had developed cerebral glioblastomas. The basis for this gene construction's specificity for a glial cell type within the central nervous system is not clear.

Defective myelination

In contrast to the neoplastic disorders described above, two types of myelination defects have been described in transgenic mice carrying papovavirus early region genes, either from SV40 or JC virus. The first type arose unexpectedly when the SV40 enhancer (72 base pair repeat) was deleted from a construction which also contained a metallothionein-growth hormone fusion gene (Messing *et al.*, 1985). Instead of developing choroid plexus papillomas, these mice died at 2–6 months from hypomyelinating peripheral neuropathy, hepatocellular carcinomas, and insulinomas. This pattern of pathology may reflect control by the metallothionein sequences (Palmiter *et al.*, 1985). The peripheral neuropathy shows morphological similarity to the Trembler mouse mutant and congenital hypomyelination syndromes in humans. Preliminary evidence suggests that large T antigen is being expressed in the peripheral nerves of these mice, presumably in the Schwann cells (Messing *et al.*, unpublished observations).

The second group of transgenic mice with myelination defects carried portions of the JC virus, a human papovavirus associated with progressive multifocal leukoencephalopathy, a demyelinating disease of the CNS. The JC virus has substantial homology to SV40, and the JC early region also codes for large T and small t antigens. When the JC early region was introduced into mice along with the JC enhancer/promotor, there was a high incidence of perinatal mortality, and four of the five remaining mice developed metastatic adrenal neuroblastomas by the ages of 4–10 months (Small *et al.*, 1986). In addition, progeny from two of the founder animals displayed a remarkably different phenotype, consisting of tremors, seizures and death (Small *et al.*, 1986b). These clinical signs resemble those seen in the jimpy and quaking mouse mutants. Examination of the CNS showed oligodendrocytes with large bizarre nuclei, marked hypomyelination, and myelin debris in macrophages. Analysis of mRNA levels indicated high expression in the brain relative to other tissues, except for one group where there was also high expression in the lung with no apparent effect. In one line, inbreeding to produce mice homozygous for the JC genes yielded animals with increased levels of message and a more severe phenotype.

Among the interesting questions posed by these studies are why the cell specificity differs so much between SV40 and JC virus, and why, among cells that do express these genes, some are transformed, some show dysfunction but not transformation, and some apparently are not affected at all. In the case of oligodendrocytes in the CNS or Schwann cells in the PNS, expression of a large T antigen of either virus may somehow be uniquely disruptive to the process of myelination or the maintenance of myelin.

EXPRESSION OF GENES NORMALLY ACTIVE IN THE NERVOUS SYSTEM OR SKELETAL MUSCLE

Five genes of vertebrate origin which are normally expressed in the nervous system or skeletal muscle have been tested to date (*see Table 9.1*(*a*) and (*d*). One of the primary aims in these studies is to delineate sequences capable of directing appropriate tissue-specific expression. For example, the hypoxanthine phosphoribosyl-transferase (HPRT) gene is normally expressed in all tissues, but at four-fold higher levels in the CNS with particularly high levels in the basal ganglia. Stout *et al.* (1985) introduced a cDNA from the HPRT gene attached to the murine metallothionein promotor (mMT), a promotor which in many other gene constructs confers the metallothionein pattern of expression (liver, pancreas, kidney, intestine). Despite the mMT promotor, several of these mice showed preferential expression in the CNS. There were no obvious phenotypic effects of elevating CNS HPRT levels four-fold. Presumably there are sequences within the HPRT cDNA responsible for high CNS expression, although the expressing cell type(s) is still not known.

The expression of two other genes, for myelin basic protein (MBP, expressed in oligodendrocytes) and gonadotropin-releasing hormone (GnRH, expressed in various CNS neurons), would have been more difficult to study in transgenic mice (Mason *et al.,* 1986b; Popko *et al.,* 1987; Readhead *et al.,* 1987), were murine mutants not available which were defective in the endogenous genes. In both cases the phenotypic defect in the spontaneous mutant was corrected by introduction of the foreign genes, indicating substantial preservation of normal tissue, cell, and developmental specificity in the regulation of these transgenes (*see below*).

Two muscle-specific genes have also shown relatively normal tissue-specific expression when introduced into mice: rat α-actin (Shani, 1986) and rat myosin light-chain 2 (Shani, 1985). No overt phenotype resulted from the possible overexpression of these two gene products.

One of the major benefits of these studies will be the ability to direct gene expression to particular cell types in attempts to either correct or induce models of disease processes.

ECTOPIC EXPRESSION IN THE NERVOUS SYSTEM

The consequences of expression of a structurally normal protein in an inappropriate cell type are difficult to predict. Such ectopic expression has occasionally taken place unexpectedly in the central nervous system of transgenic animals after the introduction of hybrid genes. Two examples involved the mMT promotor in association with structural genes for growth hormone (Swanson *et al.*, 1985) or somatostatin (Low *et al.*, 1986). In the case of the mMT–GH fusion genes, the highest levels of expression were in the liver, as expected, due to control by the mMT promotor, and lower levels were detected in the brain (Palmiter *et al.*, 1983) where astrocytes are a normal site of mMT-I synthesis. However, in the mMT-GH transgenic mice, expression of the growth hormone was taking place not in astrocytes, but in neurons in multiple areas of the brainstem and cerebrum, with highest levels in the paraventricular and supraoptic nuclei of the hypothalamus. Since expression was now taking place in a cell type in which neither the promotor nor structural sequences were known to be active before, it was proposed that some 'combinatorial specificity' had been generated with this particular fusion gene. The subsequent demonstration that mMT–somatostatin fusion genes were expressed in gonadotrophs of the anterior pituitary may provide another example of combinatorial specificity (Low *et al.*, 1986). An obvious phenotype has not been recognized due to ectopic expression of these genes. Numerous other genes have shown erratic or low-level ectopic expression in the nervous system or muscle. Even though the cells responsible for expression in each of these cases have not been identified, these mice may become useful resources for studies on the consequences of ectopic expression on neural development and function, and they have been listed in *Table 9.1*(*e*).

INSERTIONAL MUTAGENESIS

In any type of gene transfer where the foreign DNA integrates into the host genome, occasional host genes will be disrupted by the integration itself. Although the effects of such insertional mutations can be difficult to detect, a minimum estimate is that they occur in approximately 7% of transgenic mice. Since the integration site can often be characterized through probes for the foreign DNA, such mice offer a powerful new approach towards the identification of mutated genes associated with particular phenotypes. Several lethal and non-lethal mutations have been studied so far (reviewed in Palmiter and Brinster, 1986), though none to date has had a primary effect on the nervous system. Nevertheless, insertional mutations with interesting neurological phenotypes are likely to be found.

CORRECTION OF GENETIC DEFECTS

Germ line introduction of functional genes offers an attractive approach towards the correction of certain phenotypes associated with loss of function in mutated genes. Reversal of mutant phenotypes has already been accomplished with addition of the growth hormone genes to the dwarf little mouse (Hammer, Palmiter and Brinster, 1984), major histocompatibility complex (MHC) genes to immune-deficient mice (Le Meur *et al.*, 1985; Pinkert *et al.*, 1985; Yamamura *et al.*, 1985), globin genes to β-thalassaemic mice (Costantini, Chada and Magram, 1986), gonadotropin-releasing hormone genes to hypogonadal mice (Mason *et al.*, 1986b), and myelin basic protein (MBP) genes to myelin deficient mice (Popko *et al.*, 1987; Readhead *et al.*, 1987). While germ line manipulations are unlikely to be an appropriate therapeutic strategy for human disorders (Anderson, 1984), these studies illustrate the value of utilizing mutants with a deficiency in function of a particular gene as a background against which to evaluate the effects of experimentally introduced genes.

The shiverer mouse mutant displays defective myelination of the CNS associated with defects in the myelin basic protein (MBP) gene. Readhead *et al.* (1987) have used this mutant to study the effects of introducing new copies of the murine MBP gene. By injecting the gene first into wild-type mice and then back-crossing to shiverers so as to transfer the foreign gene to a shiverer background, a dramatic restoration of neurological function took place. Shiverers hemizygous for the foreign MBP gene showed an extended life span compared to normal shiverers, and mice made homozygous for the foreign MBP gene had only subtle deficits, did not have seizures, and were still alive at 7 months of age. MBP mRNA was detected only in brain, and showed a normal developmental pattern of expression with a peak at 18 days postnatally. Ultrastructurally there was increased myelination of CNS axons, although not to normal levels. This agreed well with the biochemical evidence that mRNA and MBP protein levels were only restored to 20–25% of wild types. These investigators have developed several lines of transgenic mice with remarkable correlations between levels of expression of MBP, degree of myelination, and correction of the mutant phenotype (Popko *et al.*, 1987).

The hypogonadal mouse was shown to have a large deletion in the gene for the common precursor of gonadotropin-releasing hormone (GnRH) and GnRH-associated peptide, proteins normally produced by neurons in the septopreoptic area of the hypothalamus (Mason *et al.*, 1986a). After introduction of an intact GnRH gene, gonadal size, hormone levels, reproductive behaviour, and litter sizes were all restored to normal, indicating correct cell-specific and regulated expression of the transgene (Mason *et al.*, 1986b).

By utilizing a background of absence or deficiency of a particular gene product against which to study the effects of an experimentally introduced gene, these studies have now set the stage for making deliberate alterations in the foreign genes and being able to study the effects of these changes. In the case of myelin basic protein, we may begin to get a glimpse of how this protein contributes to the process of myelination in the central nervous system.

TRANSGENIC MICE AS BIOASSAYS FOR EFFECTS OF DEFECTIVE GENES

Recent efforts in gene mapping have allowed remarkable progress in the identification of loci for genetic diseases in which the gene product or its function are not known. As mentioned above, transgenic mice are uniquely suited for testing the effects of overexpression of a particular gene product. Such mice may be useful in fully testing the consequences of expression of candidate mutant genes, at least for dominant mutations in which a gain of function is likely to have taken place.

Expression need not take place in the nervous system itself, and one of the primary advantages of transgenic animals is the ability to study systemic disorders with secondary effects on the nervous system. As an example, familial amyloidotic polyneuropathy is an autosomal dominant condition in which a variant plasma transthyretin may lead to deposition of amyloid in peripheral nerves. Sasaki *et al.* (1986) have produced mice carrying the abnormal transthyretin gene attached to the mMT promotor in an attempt to see whether expression in the liver could produce a murine model of the polyneuropathy.

Similarly, transgenic mice will prove useful as test systems once candidate genes are identified for other autosomal dominant conditions such as Huntington's disease and familial Alzheimer's disease (Gusella *et al.*, 1983; St George-Hyslop *et al.*, 1987). The potential role of the β-amyloid gene in the pathogenesis of Alzheimer's disease (Delabar *et al.*, 1987; Goldgaber *et al.*, 1987; Tanzi *et al.*, 1987) can be explored by forcing overexpression in neurons or other cell types.

A more difficult problem arises when one wishes to study the effects of a recessive gene, since without a previously identified deficiency mutant, as in the shiverer or hypogonadal mouse studies described above, one would have to overcome the functioning of the normal endogenous genes. An ideal situation would be site-specific integration of the foreign DNA into the homologous endogenous gene, in effect substituting one for the other. Such site-specific integration has not yet been achieved with transgenic animals, but recent success with mammalian cell culture systems indicates progress towards this goal (Smithies *et al.*, 1985). Alternatively, specific induction or selection of mutations at particular loci prior to introduction of the foreign DNA would also produce a suitable background, as with the recent derivation of $HPRT^-$ murine mutants through use of embryonal stem cell cultures and subsequent colonization of embryos (Hooper *et al.*, 1987; Kuehn *et al.*, 1987). It will be interesting to use the $HPRT^-$ mice to study the effects of the HPRT gene construction described above (Stout *et al.*, 1985).

In general, one limitation in using transgenic mice to model the effects of defective genes will be that some phenotypes may be species-specific, perhaps due to anatomical or biochemical variations that cannot be fully duplicated in the mouse system. In addition, there is currently only limited control over the range, levels, and developmental onset of expression in transgenic mice. Nevertheless, our increasing ability to target gene expression to particular cell types in the nervous system will offer numerous opportunities for probing the molecular pathogenesis of neurological disorders.

References

ADAMS, J. M., HARRIS, A. W., PINKERT, C. A. *et al.* (1985) The *c-myc* oncogene driven by immunoglobulin enhancers induces lymphoid malignancy in transgenic mice. *Nature (London)*, **318**, 533–538

ANDERSON, W. F. (1984) Prospects for human gene therapy. *Science*, **226**, 401–409

BIEBERICH, C., SCANGOS, G., TANAKA, K. and JAY, G. (1986) Regulated expression of a murine class I gene in transgenic mice. *Molecular and Cellular Biology*, **6**, 1339–1342

BRINSTER, R. L., CHEN, H. Y., MESSING, A., VAN DYKE, T., LEVINE, A. J. and PALMITER, R. D. (1984) Transgenic mice harboring SV40 T-antigen genes develop characteristic brain tumors. *Cell*, **37**, 367–379

BRINSTER, R. L., CHEN, H. Y., TRUMBAUER, M. E., YAGLE, M. K. and PALMITER, R. D. (1985) Factors affecting the efficiency of introducing foreign DNA into mice by microinjecting eggs. *Proceedings of the National Academy of Sciences USA*, **82**, 4438–4442

CHISARI, F. V., PINKERT, C. A., MILICH, D. R. *et al.* (1985) A transgenic mouse model of the chronic hepatitis B surface antigen carrier state. *Science*, **230**, 1157–1160

COSTANTINI, F., CHADA, K. and MAGRAM, J. (1986) Correction of murine β-thalassemia by gene transfer into the germ line. *Science*, **233**, 1192–1194

DELABAR, J. M., GOLDGABER, D., LAMOUR, Y. *et al.* (1987) β-Amyloid gene duplication in Alzheimer's disease and karyotypically normal Down syndrome. *Science*, **235**, 1390–1392

GOLDGABER, D., LERMAN, M. I., McBRIDE, O. W., SAFFIOTTI, U. and GAJDUSEK, D. C. (1987) Characterization and chromosomal localization of a cDNA encoding brain amyloid of Alzheimer's disease. *Science*, **235**, 877–880

GORDON, J. W. and RUDDLE, F. H. (1983) Gene transfer into mouse embryos: production of transgenic mice by pronuclear injection. *Methods in Enzymology,* **101,** 411–433

GORDON, J. W. and RUDDLE, F. H. (1985) DNA-mediated genetic transformation of mouse embryos and bone marrow – a review. *Gene,* **33,** 121–136

GUSELLA, J. F., WEXLER, N. S., CONNEALLY, P. M. *et al.* (1983) A polymorphic DNA marker genetically linked to Huntington's disease. *Nature (London),* **306,** 234–238

HAMMER, R. E., PALMITER, R. D. and BRINSTER, R. L. (1984) Partial correction of murine hereditary growth disorder by germ-line incorporation of a new gene. *Nature (London),* **311,** 65–67

HOGAN, B., COSTANTINI, F. and LACY, E. (1986) *Manipulating the Mouse Embryo: a Laboratory Manual.* Cold Spring Harbor: Cold Spring Harbor Laboratory

HOOPER, M., HARDY, K., HANDYSIDE, A., HUNTER, S. and MONK, M. (1987) HPRT-deficient (Lesch–Nyhan) mouse embryos derived from germline colonization by cultured cells. *Nature (London),* **326,** 292–295

KELLY, F., KELLERMANN, O., MECHALI, F., GAILLARD, J. and BABINET, C. (1986) Expression of simian virus 40 oncogenes in F9 embryonal carcinoma cells, in transgenic mice, and in transgenic embryos. In: *DNA Tumor Viruses: Control of Gene Expression and Replication,* edited by M. Botchan, T. Grodzicker and P. A. Sharp, pp. 363–372. Cold Spring Harbor: Cold Spring Harbor Laboratory

KHILLAN, J. S., SCHMIDT, A., OVERBEEK, P. A., DE CROMBRUGGHE, B. and WESTPHAL, H. (1986) Developmental and tissue-specific expression directed by the α_2 type I collagen promotor in transgenic mice. *Proceedings of the National Academy of Sciences USA,* **83,** 725–729

KUEHN, M. R., BRADLEY, A., ROBERTSON, E. J. and EVANS, M. J. (1987) A potential animal model for Lesch–Nyhan syndrome through introduction of HPRT mutations into mice. *Nature (London),* **326,** 295–298

LACY, E., ROBERTS, S., EVANS, E. P., BURTENSHAW, M. D. and COSTANTINI, F. D. (1983) A foreign β-globin gene in transgenic mice: integration at abnormal chromosomal positions and expression in inappropriate tissues. *Cell,* **34,** 343–358

LE MEUR, M., GERLINGER, P., BENOIST, C. and MATHIS, D. (1985) Correcting an immune-response deficiency by creating E_α gene transgenic mice. *Nature (London),* **316,** 38–42

LOW, M. J., LECHAN, R. M., HAMMER, R. E. *et al.* (1986) Gonadotroph-specific expression of metallothionein fusion genes in pituitaries of transgenic mice. *Science,* **231,** 1002–1004

MASON, A. J., HAYFLICK, J. S., ZOELLER, R. T. *et al.* (1986a) A deletion truncating the gonadotropin-releasing hormone gene is responsible for hypogonadism in the *hpg* mouse. *Science,* **234,** 1366–1371

MASON, A. J., PITTS, S. L., NIKOLICS, K. *et al.* (1986b) The hypogonadal mouse: reproductive functions restored by gene therapy. *Science,* **234,** 1372–1378

McKNIGHT, G. S., HAMMER, R. E., KUENZAL, E. A. and BRINSTER, R. L. (1983) Expression of the chicken transferrin gene in transgenic mice. *Cell,* **34,** 335–341

MESSING, A., CHEN, H. Y., PALMITER, R. D. and BRINSTER, R. L. (1985) Peripheral neuropathies, hepatocellular carcinomas and islet cell adenomas in transgenic mice. *Nature (London),* **316,** 461–463

MORELLO, D., MOORE, G., SALMON, A. M., YANIV, M. and BABINET, C. (1986) Studies on the expression of an H-2K/human growth hormone fusion gene in giant transgenic mice. *EMBO Journal,* **5,** 1877–1883

OVERBEEK, P. A., LAI, S. P., VAN QUILL, K. and WESTPHAL, H. (1986) Tissue-specific expression in transgenic mice of a fused gene containing RSV terminal sequences. *Science,* **231,** 1574–1577

PALMITER, R. D., NORSTEDT, G., GELINAS, R. E., HAMMER, R. E. and BRINSTER, R. L. (1983) Metallothionein-human GH fusion genes stimulate growth of mice. *Science,* **222,** 809–814

PALMITER, R. D., CHEN, H. Y., MESSING, A. and BRINSTER, R. L. (1985) The SV40 enhancer and large T-antigen are instrumental in development of choroid plexus tumors in transgenic mice. *Nature (London),* **316,** 457–460

PALMITER, R. D. and BRINSTER, R. L. (1986) Germ-line transformation of mice. *Annual Review of Genetics,* **20,** 465–499

PINKERT, C. A., WIDERA, G., COWING, C. *et al.* (1985) Tissue-specific, inducible and functional expression of the E^d_α MHC class II gene in transgenic mice. *EMBO Journal,* **4,** 2225–2230

POPKO, B., PUCKETT, C., LAI, E. *et al.* (1987) Myelin deficient mice: expression of myelin basic protein and generation of mice with varying levels of myelin. *Cell,* **48,** 713–721

READHEAD, C., POPKO, B., TAKAHASHI, N. *et al.* (1987) Expression of a myelin basic protein gene in transgenic shiverer mice: correction of the dysmyelinating phenotype. *Cell,* **48,** 703–712

RUSCONI, S. (1984) Gene transfer in living organisms. In: *The Impact of Gene Transfer Techniques in Eukaryotic Cell Biology,* edited by J. S. Schell and R. Starlinger, pp. 134–152. Berlin: Springer-Verlag

SASAKI, H., TONE, S., NAKAZATO, M. *et al.,* (1986) Generation of transgenic mice producing a human transthyretin variant: a possible model for familial amyloidotic polyneuropathy. *Biochemical and Biophysical Research Communications,* **139,** 794–799

SHANI, M. (1985) Tissue-specific expression of rat myosin light chain 2 gene in transgenic mice. *Nature (London),* **314,** 283–286

SHANI, M. (1986) Tissue-specific and developmentally regulated expression of a chimeric actin-globin gene in transgenic mice. *Molecular and Cellular Biology,* **6,** 2624–2631

SMALL, J. A., BLAIR, D. G., SHOWALTER, S. D. and SCANGOS, G. (1985) Analysis of a transgenic mouse containing simian virus 40 and *v-myc* sequences. *Molecular and Cellular Biology,* **5,** 642–648

SMALL, J. A., KHOURY, G., JAY, G., HOWLEY, P. M. and SCANGOS, G. A. (1986a) Early regions of JC virus and BK virus induce distinct and tissue-specific tumors in transgenic mice. *Proceedings of the National Academy of Sciences USA,* **83,** 8288–8292

SMALL, J. A., SCANGOS, G. A., CORK, L., JAY, G. and KHOURY, G. (1986b) The early region of human papovavirus JC induces dysmyelination in transgenic mice. *Cell,* **46,** 13–18

SMITHIES, O., GREGG, R. G., BOGGS, S. S., KORALEWSKI, M. and KUCHERLAPATI, R. S. (1985) Insertion of DNA sequences into the human chromosomal β-globin locus by homologous recombination. *Nature (London),* **317,** 230–234

SORIANO, P., CONE, R. D., MULLIGAN, R. C. and JAENISCH, R. (1986) Tissue-specific and ectopic expression of genes introduced into transgenic mice by retroviruses. *Science,* **234,** 1409–1413

ST GEORGE-HYSLOP, P., TANZI, R. E., POLINSKY, R. J. *et al.* (1987) The genetic defect causing familial Alzheimer's disease maps on chromosome 21. *Science,* **235,** 885–890

STEWART, C. L., VANEK, M. and WAGNER, E. F. (1985) Expression of foreign genes from retroviral vectors in mouse teratocarcinoma chimaeras. *EMBO Journal,* **4,** 3701–3709

STOUT, J. T., CHEN, H. Y., BRENNAND, J., CASKEY, C. T. and BRINSTER, R. L. (1985) Expression of human HPRT in the central nervous system of transgenic mice. *Nature (London),* **317,** 250–252

STUHLMANN, H., CONE, R., MULLIGAN, R. C. and JAENISCH, R. (1984) Introduction of a selectable gene into different animal tissue by a retrovirus recombinant vector. *Proceedings of the National Academy of Sciences USA,* **81,** 7151–7155

SWANSON, L. W., SIMMONS, D. M., ARRIZA, J. *et al.* (1985) Novel developmental specificity in the nervous system of transgenic animals expressing growth hormone fusion genes. *Nature (London),* **317,** 363–366

TANZI, R. E., GUSELLA, J. F., WATKINS, P. C. *et al.* (1987) Amyloid β protein gene: cDNA, mRNA distribution, and genetic linkage near the Alzheimer locus. *Science,* **235,** 880–884

TOWNES, R. M., CHEN, H. Y., LINGREL, J. B., PALMITER, R. D. and BRINSTER, R. L. (1985) Expression of human β globin genes in transgenic mice: effects of a flanking metallothionein-human growth hormone fusion gene. *Molecular and Cellular Biology,* **5,** 1977–1983

VAN DER PUTTEN, H., BOTTERI, F. M., NUKKER, A. D. *et al.* (1985) Efficient insertion of genes into the mouse germ line via retroviral vectors. *Proceedings of the National Academy of Sciences USA,* **82,** 6148–6152

YAGALOFF, K. A., LOZANO, G., VAN DYKE, T., LEVINE, A. J. and HARTIG, P. R. (1986) Serotonin 5-HT_{1C} receptors are expressed at high density on choroid plexus tumors from transgenic mice. *Brain Research,* **385,** 389–394

YAMAMURA, K., KIKUTANI, H., FOLSOM, V. *et al.* (1985) Functional expression of a microinjected E^{d}_{α} gene in C57BL/6 transgenic mice. *Nature (London),* **316,** 67–69

10
Messenger RNA levels in neurological disease

Marcelle R. Morrison

INTRODUCTION

Many questions about the structure and function of the human nervous system can be addressed by studying appropriate animal models. For example, much of our rudimentary understanding of the molecular events underlying simple forms of memory (Goelet *et al.*, 1986) or neuronal development (Harris, 1985), comes from the study of *Aplysia* and *Drosophila*. Animal research is limited in its usefulness for studying topics such as the molecular basis for human neurological disease, since few animal models for these diseases are available. For this reason, researchers have investigated the postmortem stability of different human brain constituents. Such studies have shown that useful comparisons of many brain metabolites in diseased and normal postmortem human brain can be made (Bowen *et al.*, 1977a, 1977b; Mann, Barton and Davies, 1978; Perry, Perry and Tomlinson, 1982; Perry and Perry, 1983). The development of techniques to quantitate messenger RNA (mRNA) concentrations in postmortem human brain now allows us to determine if differences seen at the protein level in diseased brain reflect altered mRNA levels and hence altered rates of protein synthesis. This chapter reviews the evidence for the postmortem stability of brain mRNAs and describes recent progress in quantitating mRNA levels in normal human brain and in the brains of patients with Alzheimer's disease.

MESSENGER RNA ANALYSIS IN POSTMORTEM BRAIN

Messenger RNA stability

Early studies of postmortem human brain compared total RNA levels in different cell types using a number of histological procedures (Mann, 1978; Mann, Yates and Stamp, 1978; Mann and Yates, 1979). These studies measured levels of the most abundant intracellular RNAs, ribosomal RNAs (rRNAs). The integrity of mRNA was first analysed when a variety of established techniques (Kaplan, Bernstein and Gioio, 1979; Marotta, Strocchi and Gilbert, 1979; Morrison *et al.*, 1980) were used to isolate total RNA from human postmortem brain. Several groups showed, by *in vitro* translation of these RNA isolates, that despite some overall degradation, translationally active mRNAs were still present in RNA isolated from postmortem human brain. Analysis of *in vitro* translation products by gel electrophoresis showed that many moderately abundant mRNAs remained intact up to 16 h post mortem (Gilbert *et al.*, 1981; Marotta *et al.*, 1981; Morrison and Griffin, 1981; Morrison, Griffin and White, 1985, 1986; Sajdel-Sulkowska, Coughlin and Marotta, 1983; Sajdel-Sulkowska *et al.*, 1983; Whatley *et al.*, 1984) and that even mRNAs encoding proteins of high molecular weight were still translationally active (Sajdel-Sulkowska, Coughlin and Marotta, 1983; Morrison, Griffin and White, 1985, 1986). The postmortem degradation pattern thus differs markedly from that obtained when mRNA is adventitiously degraded during isolation. In the latter case, even mRNAs of average molecular weight are selectively degraded compared to those of low molecular weight (Morrison, unpublished observations). The relative levels of different mRNAs in RNA isolates from human brain is discussed below.

Although intact mRNAs were present, the yields of total RNA in these early studies were quite variable. The lowest recoveries (40 μg total RNA/g wet weight tissue) were obtained when RNA was isolated from polysomes (Gilbert *et al.*, 1981), probably because the majority of mRNAs were no longer polysomal (Gilbert *et al.*, 1981; Marotta *et al.*, 1981). Recoveries were 8–10-fold higher when total cellular RNA was isolated by homogenizing the tissues directly in denaturing agents (Morrison and Griffin, 1981). In different laboratories, the recovery of total cellular RNA ranges from 162 μg to > 500 μg/g wet weight of tissue (Marotta *et al.*, 1981; Morrison and Griffin, 1981; Sajdel-Sulkowska *et al.*, 1983; Whatley *et al.*, 1984; Guillemette *et al.*, 1986; Johnson, Morgan and Finch, 1986; Morrison, Griffin and White, 1986; Taylor *et al.*, 1986; May *et al.*, 1987). Part of the variation in mRNA yield, seen even in different RNA isolates from the same brain region (Morrison, unpublished data), is probably due to a variable ratio of white to grey matter in the dissected tissue, since grey matter contains a higher proportion of cell bodies and hence RNA. Another variable may be the extraction procedure itself. We obtain consistently higher yields and better quality RNA if we isolate RNA by a series of precipitations in a guanidinium/ethanol mix (Ilaria *et al.*, 1985), as described by Cox (1968), rather than by pelleting the RNAs through a caesium chloride gradient (Chirgwin *et al.*, 1979; Morrison and Griffin, 1981).

The *in vitro* translation analyses described above can be used to determine the overall quality of an RNA isolate, but two-dimensional gel analysis of the *in vitro* translation products is only sensitive enough to determine whether relatively abundant mRNAs are still translationally active. The integrity of less abundant mRNAs in RNA isolates from human brain has been assayed in other laboratories by immunoprecipitation of translation products or by functional analysis of proteins synthesized in frog oocytes after injection of mRNA. The results show that frog oocytes injected with RNA isolated from fetal human brain synthesize serum proteins (Dziegielewska *et al.*, 1986), cholinesterases (Soreq, Zevin-Sonkin and Razon, 1984), and a number of functional ion channels (Gunderson, Miledi and Parker, 1984a, 1984b), thus documenting that intact mRNAs for these proteins are still present in the RNA isolates.

Denaturing gel electrophoresis and Northern hybridization of mRNAs to radiolabelled recombinant DNA probes can be used to determine the integrity of rRNAs as well as the integrity of individual mRNA species. We have shown, by denaturing gel analysis, that the 28S and 18S rRNAs are still present in total RNAs isolated from postmortem brain (Ilaria *et al.*, 1985; Morrison, Griffin and White, 1986). The ratio of 28S to 18S RNA is often less than 2:1 showing that there is some preferential degradation of the 4.7 kb 28S RNA during the postmortem interval. Northern hybridization of total RNAs isolated from postmortem human brain to α and β tubulin cDNA probes showed that most of the α and β tubulin mRNAs migrate at 1.8 kb, as expected (Morrison, Griffin and White, 1986; Pardue *et al.*, unpublished data). However, a variable proportion of the mRNAs were no longer full length and the intensity of hybridization was less than to RNA isolated from adult rat brain. These results demonstrate that some of the tubulin mRNAs in the RNA isolated from human brain are degraded and that the extent of degradation varies from one RNA preparation to another. The results are in accord with the variable decrease in levels of translationally active mRNAs in postmortem RNA isolates (Morrison and Griffin, 1981).

Northern hybridization analysis with a number of other cDNA probes has shown that relatively intact mRNAs for thy 1 (Johnson, Morgan and Finch, 1986), the human prion protein analogue (Johnson, Morgan and Finch, 1986; Robakis *et al.*, 1986a, 1986b), several myelin basic proteins (Kamholz *et al.*, 1986), the 70 kDa heat shock protein (Guillemette *et al.*, 1986), transthyretin (Herbert *et al.*, 1986), acetylcholinesterase (Soreq *et al.*, 1985) and sequences hybridizing to several anonymous DNA probes (Wood *et al.*, 1986), are also still present in RNAs isolated from adult human brain up to 36 h post mortem. Although most of these mRNAs are of small to average molecular weight (2 kb or less), one study showed that glutamic acid decarboxylase mRNA (3.7 kb) can be detected in total RNA isolated from postmortem human brain (Wood *et al.*, 1986). Most of the mRNA is intact, but the relative level in human brain is approximately 2-fold less than in cat brain (Wood *et al.*, 1986). A recent study also documents that at least a proportion of the 6 kb mRNA encoding tau is still intact in RNA isolates from human brain (Tanzi *et al.*, 1987).

In most of these studies, hybridization to lower molecular weight degradation products shows that the RNAs are partially degraded. Johnson, Morgan and Finch (1986) have documented that this degradation is greater than in rat brain, even when rat brain is cooled postmortem at a rate analogous to that of human brain. Any amount of RNA degradation in tissue samples, especially if variable, means that individual mRNA levels cannot be directly compared in different RNA isolates without reference to an internal standard (*see below*, p. 139).

There are several measurements that should be used as general indicators of RNA integrity in RNA isolates from human postmortem brain. The simplest is to evaluate the integrity of the rRNAs after denaturing gel electrophoresis. However, it has been our experience that an occasional sample will have relatively intact rRNAs, but that its mRNAs will be significantly degraded. Several other methods are therefore more reliable:

(1) Total *in vitro* translational activity can be used to compare the level of translationally active mRNA in different samples and the size of the translation products can be analysed on one dimensional gels (Morrison and Griffin, 1981; Morrison, Griffin and White, 1986, 1987);
(2) The amount of polyadenylated mRNA in the sample can be independently measured by hybridization with poly(U) (Morrison and Lingrel, 1976; Morrison, Griffin and White, 1986; Morrison *et al.*, 1986; 1987a, 1987b); and
(3) The levels of individual mRNAs can be compared to those in RNA isolated from animal brain.

This comparison gives an indication of how much mRNA degradation has taken place post mortem, if the assumption is made that mRNA levels in human and animal brain are equivalent (Morrison and Griffin, 1981; Morrison, Griffin and White, 1986; Morrison *et al.*, 1986; Wood *et al.*, 1986).

We find that approximately 40% (of brains) obtained through routine autopsy (14 brains, 4–16 h postmortem at room temperature) do not contain significant levels of intact mRNA (Morrison *et al.*, 1987a; Pardue *et al.*, unpublished data). This RNA degradation is not simply a function of postmortem interval (Johnson, Morgan and Finch, 1986; Pardue *et al.*, unpublished data). Dr W. S. T. Griffin (personal communication) also finds that approximately 50% of brains obtained through routine autopsy are unsuitable either for *in situ* hybridization or for immunohistochemistry. We believe that among the most important factors influencing RNA stability are the agonal state of the patient and the degree and duration of hypoxia (Morrison, Griffin and White, 1986; Morrison *et al.*, 1986, 1987a, 1987b; Roses *et al.*, 1987). Studies on the postmortem stability of other brain constituents also show that agonal state and hypoxia are factors of prime importance (Bowen *et al.*, 1977a, 1977b; Perry, Perry and Tomlinson, 1982; Wester *et al.*, 1985).

Dr Allen Roses at Duke University has initiated a rapid autopsy programme (RAP) to obtain brains for studying the molecular biology of Alzheimer's disease. This programme minimizes possible premortem hypoxia as well as the length of the postmortem interval (Roses *et al.*, 1987). In contrast to the brains we have obtained through routine autopsy, we have shown in collaborative studies with Drs Roses and Gilbert that intact RNA can be isolated from 100% (7/7) of the brains we obtain through the RAP (Morrison *et al.*, 1987a, 1987b). This reproducibility is important in subsequent mRNA quantitation.

Messenger RNA quantitation

It would be of little practical value to quantitate specific mRNA levels in RNA isolated from postmortem human brain if the relative amounts of different mRNAs in the RNA isolates were not the same as *in vivo* amounts. We have addressed the issue of relative yield and quantitation in several ways. First, we compared the levels of *in vitro* translation of RNAs isolated from control rat brains subjected to different postmortem conditions. As noted above (p. 136) the lowered RNA recovery and *in vitro* translation showed that there was a significant loss both of rRNA and translationally active mRNAs during the postmortem interval, especially if brains were maintained at room temperature postmortem (Morrison and Griffin, 1981). The relative levels of individual mRNA species in the RNA isolates were then compared by two-dimensional gel electrophoresis of the *in vitro* translation products. Surprisingly, the results showed that many of the abundant medium-sized mRNAs, isolated from rat cerebellum maintained for 16 h post mortem at room temperature, were present at the same relative concentrations in these RNA isolates as in RNAs isolated from fresh rat cerebellum. Similar results were obtained for RNAs isolated from postmortem human cerebellum (Morrison and Griffin, 1981). These results show that postmortem RNA degradation is random for abundant mRNAs of low to medium molecular weight.

To determine whether the relative mRNA amounts in RNA isolated from postmortem human brain can be used to predict relative mRNA levels *in vivo*, we next analysed the mRNA content of RNA isolates from different human brain regions. We found that RNA isolated from optic nerve is, as expected, highly enriched in mRNA encoding the glial marker, glial fibrillary acidic protein (Morrison and Griffin, 1985). The level of this mRNA is also increased in RNAs isolated from gliotic brain regions of patients with Joseph disease, whereas there is no increase in glial fibrillary protein mRNA in unaffected cortex (Morrison and Rosenberg, 1983). In another series of experiments, we compared the relative amounts of different *in vitro* translated tubulin isotypes synthesized by RNAs isolated from human cortex and cerebellum at different stages of development. We found that levels of mRNAs encoding α and β tubulins decrease developmentally in both human cortex (Morrison, Pardue and Griffin, 1983) and in human cerebellum (Morrison and Griffin, 1985). The relative levels of the mRNA encoding a basic β tubulin isotype, β_1, are much higher in fetal cortex than in early postnatal cerebellum, whereas very little of the mRNA encoding this isotype is present in either adult cortex or adult cerebellum (Morrison, Pardue and Griffin, 1983; Morrison and Griffin, 1985). Importantly, these results mirror the developmental profiles of the tubulin isotype mRNAs in rat cerebellum and cortex (Morrison, Pardue and Griffin, 1981, 1983; Morrison and Griffin, 1984, 1985, 1986). These experiments show that the relative amounts of several abundant mRNAs in RNA isolated from postmortem human brain, quantitated by *in vitro* translation, are similar to the relative amounts *in vivo*.

Quantitation by *in vitro* translation gives an immediate assessment of the relative levels of different mRNAs in an RNA isolate from postmortem brain. As documented above, changes in the levels of a particular mRNA in an RNA isolate from postmortem brain can be quantitated relative to the total amount of proteins synthesized *in vitro*, even though the absolute amount of all mRNAs is reduced. Similar comparisons have not yet been made using hybridization to cDNA probes to quantitate mRNA levels in RNA isolated from postmortem human brain. Even when a significant proportion of the mRNA remains intact, as judged by Northern hybridization, the level of hybridization in different RNA isolates cannot be taken as an indication of the *in vivo* levels of that mRNA. This is because no distinction can be made between an alteration in mRNA levels due to differences in overall mRNA degradation in different samples, and between an alteration in mRNA levels that is biologically significant. It is therefore impossible to use data obtained from one Northern hybridization to make conclusions about the relative levels of an mRNA species in different RNA isolates.

The way to determine the relative levels of an mRNA in different RNA isolates from postmortem tissue is, therefore, to quantitate the levels of several expressed mRNAs in the same RNA isolates. The levels of the experimental mRNA in these isolates can then be compared to the levels of the other mRNAs. Optimally, several other mRNAs whose size and abundance bracket the mRNA of interest would be measured, so as to eliminate any bias caused by selective degradation of higher molecular weight or less abundant mRNAs in the RNA isolates. In a recent study, hybridization to a glyceraldehyde 3 phosphodehydrogenase (G3PD) cDNA probe was used as a control for the level of hybridization of the β amyloid precursor cDNA probe to RNA isolated from different human brain regions (Tanzi *et al.*, 1987). The results showed that the levels of G3PD mRNA are relatively constant in each brain region whereas those of the β amyloid mRNA precursor vary widely from one region to another (Tanzi *et al.*, 1987).

The possibility of RNA degradation makes the correct quantitation of mRNAs in RNA isolated from human brain by dot blot hybridization particularly difficult

(Elshourbagy *et al.*, 1985), especially if the integrity of the mRNAs and the specificity of hybridization is not first checked by Northern analysis with several cDNA probes.

Human brain cDNA libraries

The preceding sections have documented that relatively intact mRNAs are still present in total RNA isolates from postmortem human brain but have not addressed the question of whether or not these mRNAs are still polyadenylated. This is an important consideration as most cDNA libraries are constructed using the poly(A) tail on the mRNA to prime first-strand cDNA synthesis. If deadenylation of mRNAs occurred during the postmortem interval, the deadenylated mRNAs would not be represented in cDNA libraries. In order to determine if mRNAs present in postmortem human brain are still adenylated, we isolated polyadenylated (poly$(A)^+$) mRNAs from adult human cortex RNA (Morrison, Pardue and Griffin, 1983). The poly$(A)^+$ RNA was purified by three successive hybridizations to oligo(dT) cellulose; rRNA still bound to the poly$(A)^+$ mRNA after the second hybridization was dissociated by heating the partially purified poly$(A)^+$ mRNA prior to the third hybridization (Morrison, Baskin and Rosenberg, 1977). We found that the content of poly$(A)^+$ mRNAs in RNA isolated from both fetal and adult human cortex (1.5%) is equivalent to that isolated from rat cortex and cerebellum as well as to that isolated from mouse neuroblastoma cells (Morrison, Baskin and Rosenberg, 1977; Morrison, Pardue and Griffin, 1983; Morrison and Griffin, unpublished data). As outlined above (p. 139), the relative amounts of individual tubulin mRNAs in the poly$(A)^+$ RNA preparations isolated from human postmortem brain were similar to their relative amounts at comparable stages of rat brain development. We could find no evidence from the two-dimensional gel profiles that abundant mRNAs of low to intermediate molecular weight had been selectively deadenylated postmortem. Sajdel-Sulkowska, Coughlin and Marotta (1983) independently showed that more than 1% of total RNA is polyadenylated in RNA isolated from human brain, and that proteins of up to 130 000 molecular weight are synthesized *in vitro* by the poly$(A)^+$ mRNA fraction. Johnson, Morgan and Finch (1986) and Guillemette *et al.* (1986) have more recently shown, by Northern blot hybridization to several different cDNA probes, that at least a fraction of the corresponding mRNAs are still present in the poly$(A)^+$ RNA fraction and are relatively intact.

Evidence that most of the poly$(A)^+$ mRNAs isolated from human brain are undegraded was obtained when we estimated the size of cDNAs reverse transcribed from adult human cortex poly$(A)^+$ RNA. The cDNAs were the same size as those made from a rat mRNA template, with an average size of approximately 2000 nucleotides (Michael *et al.*, 1982; Morrison and Griffin, 1985). The cDNA inserts in a small human cortex cDNA library made from this material were similar in size to those in a rat brain cDNA library that had been constructed using the same techniques (Michael *et al.*, 1982; Morrison and Sparkman, 1984). Several more complete human brain cDNA libraries, mostly in lambada vectors, have now been constructed in other laboratories (Kurnit *et al.*, 1982; Puckett *et al.*, 1985; Bray *et al.*, 1986; Jaye *et al.*, 1986; Kamholz *et al.*, 1986; Kretschmar *et al.*, 1986; Nathans, Thomas and Hogness, 1986; Neve *et al.*, 1986; Wood *et al.*, 1986), and sequences that encode a number of specific mRNAs have been identified. In one study, several of the myelin basic protein cDNA clones, isolated from a 1 day postnatal human brain cDNA library, were almost full length (2.2 kb; Kamholz *et al.*, 1986), showing that at least some of the mRNAs encoding myelin basic protein were undegraded in the original poly(A)$^+$ RNA isolates.

MESSENGER RNA LEVELS IN ALZHEIMER'S DISEASE

Concentrations of total RNA

The major pathological features of Alzheimer's disease (AD) are the high incidence of extracellular senile plaques, intracellular paired helical filaments (PHF) and cerebrovascular amyloid (CVA) in specific regions of AD cortex and hippocampus (Tomlinson, Blessed and Roth, 1970; Morimatsu *et al.*, 1975; Vinters and Gilbert, 1983). *In vivo* metabolic studies show that there is less blood flow to cortex in the severely demented AD population (Farkas *et al.*, 1982; Duara *et al.*, 1986), suggesting a generalized decrease in cortical metabolism. Cortical protein synthesis is also reduced (Bustani *et al.*, 1983). However, while the levels of certain cortical neuropeptides and enzymes are reduced (for example, somatostatin and choline acetyltransferase; Davies and Maloney, 1976; Davies and Terry, 1981), others are unaltered (for example, cholecystokinin, vasointestinal peptide and glutamic acid decarboxylase; Rossor *et al.*, 1980; Ferrier *et al.*, 1983). These results suggest that, as in Huntington's disease (Ferrante *et al.*, 1985; Koh, Peters and Choi, 1986), specific cell populations are more compromised than others.

If indeed cellular metabolism is compromised in Alzheimer's disease, there might be a reduction in intracellular levels of total RNA and mRNA, as either a cause or a result of the neuropathology. All early morphological studies of RNA levels in individual cell types in postmortem Alzheimer's disease brain, deduced from comparison of nucleolar volumes, showed that neurons with cytoplasmic PHF contain less intracellular RNA than do comparable neurons in the same brain region with no cytoplasmic PHF (Dayan and Ball, 1973; Mann, Yates and Stamp, 1978; Mann *et al.*, 1981a; Mann and Yates, 1981). Reduced nucleolar volume is not a feature unique to PHF-containing neurons, as a similar reduction in nucleolar volume is seen in neurons with other intracellular inclusions or in neurons with granulovacuolar degeneration (Mann, 1978; Mann and Sinclair, 1978; Mann and Yates, 1979; Mann, Yates and Stamp, 1978).

There is considerably more disagreement as to whether there is an overall decrease in intracellular RNA in Alzheimer's disease brain cells. Some reports show a decrease in nucleolar volume in all cortical neurons (Mann and Yates, 1981; Mann *et al.*, 1981a, 1981b), and others suggest that the decrease is only found in neurons containing PHF (Mann, 1978; Dayan and Ball, 1973). Direct RNA measurement in dissected single cell bodies shows that the RNA content of cells in Alzheimer's disease prefrontal cortex is the same as that in control cortex (Uemura and Hartmann, 1978) whereas cells with neurofibrillary tangles contain half the normal amount (Uemura and Hartmann, 1979). Recent cytophotometric analysis of the RNA content of hippocampal pyramidal cells demonstrated a 33–46% decrease in Alzheimer's disease hippocampus, and a statistically lower pyramidal neuron count in several regions of Alzheimer's disease hippocampus (Doebler *et al.*, 1987a). Further analysis showed that the decrease in intracellular RNA was confined to pyramidal cells containing detectable levels of the Alz 50 protein (Wolozin *et al.*, 1986; Rhoads *et al.*, 1987).

There are also reports that RNA loss may not be confined to regions, such as the hippocampus, that exhibit neuropathological lesions. Mann, Yates and Barton (1977) reported RNA loss ranging from 7 to 51% in a variety of brain cell types. Similarly, Doebler *et al.* (1987b) found that the total RNA content of individual Golgi type 1 neurons in caudate and putamen was 17–23% lower in Alzheimer's disease, whereas the RNA content of astrocytes in hippocampus and dentate gyrus was unchanged (Rhoads, personal communication). These results would suggest a more global decrease in RNA in large neurons in Alzheimer's disease. In all these studies, the influence of postmortem delay and fixation regimens on RNA preservation in individual neurons in Alzheimer's disease brain must be considered, especially in light of our results showing increased susceptibility of Alzheimer's disease RNA to degradation (*see below*). The only studies to eliminate the possibility of postmortem artefact were those of Mann *et al.* (1981a, 1981b). These authors showed that nucleolar volume is decreased by 36% and RNA content by 26% in pyramidal cells in layers 3 and 5 of biopsied Alzheimer's disease temporal cortex as compared to corresponding cells in control cortex (Mann *et al.*, 1981b). Reduction in neuronal nucleolar volume correlated with the frequency of both senile plaques and PHF in temporal cortex (Mann *et al.*, 1981a).

A number of groups have now isolated total RNA and quantitated mRNA levels in both Alzheimer's disease and control brain. In the first direct comparison of mRNA levels in Alzheimer's disease and control brain, total RNA was isolated from Alzheimer's disease and control cortex and mRNA contents were assayed by *in vitro* translation (Sajdel-Sulkowska *et al.*, 1983). RNA recovery was very low even in control cortex (approximately 100 μg/g wet weight of tissue) and 50% less total RNA was recovered in Alzheimer's disease cortex. Quantitation of translationally active mRNA by *in vitro* translation showed that this was reduced 5-fold in Alzheimer's disease cortex. Two-dimensional gel analysis of *in vitro* translation products showed that Alzheimer's disease mRNA synthesized most proteins less efficiently. Decreased synthesis of the tubulins was frequently observed (Sajdel-Sulkowska *et al.*, 1983). Taylor *et al.* (1986) and Guillemette *et al.* (1986) reported, in contrast to the results of Sajdel-Sulkowska *et al.* (1983), that total RNA levels were not reduced in Alzheimer's disease cortex. However, Taylor *et al.* (1986, 1987) found a 62% loss of ribosomal RNA in Alzheimer's disease when quantitated by Northern hybridization rather than by optical density. They suggested that the Alzheimer's disease RNA isolates contain short pieces of degraded RNA that contribute to the optical density readings but are not quantitated by hybridization. Poly(A) levels in the Alzheimer's disease total RNA, quantitated by hybridization to radiolabelled poly(U), were 66% of controls. Guillemette *et al.* (1986) directly isolated poly(A)$^+$ RNA from control and Alzheimer's disease cortex and, after correcting for rRNA contamination by hybridization with poly(U), showed that Alzheimer's disease poly(A)$^+$ RNA was 58% of control values.

In our initial studies, we found that RNA isolated from both pathologically affected and unaffected regions of Alzheimer's disease brain was consistently more degraded than RNA isolated from control brains. Further analysis showed that selective degradation of Alzheimer's RNA during isolation only took place if the tissue was allowed to thaw, even minimally, before homogenization in chaotropic agents. In contrast, thawing of control human brain or rat brain did not result in RNA degradation (Ilaria *et al.*, 1985; Morrison, Griffin and White, 1986; Morrison *et al.*, 1987a, 1987b). The increased susceptibility of Alzheimer's disease RNA to degradation during isolation could result from a defect in the alkaline ribonuclease inhibitor in Alzheimer's disease brain, as has been suggested by Sajdel-Sulkowska and Marotta (1984). However, we have found that as much alkaline ribonuclease is bound to inhibitor in our Alzheimer's disease brains as in controls (Morrison *et al.*, 1987a, 1987b; Maschhoff *et al.*, unpublished data). The reason for the increased susceptibility of Alzheimer's disease RNA to degradation is therefore still unknown. One possibility is that Alzheimer's disease patients are more hypoxic than control patients for several days prior to death (Roses *et al.*, 1987). Hypoxia may make their lysosomes more fragile, releasing acidic ribonucleases into the cytoplasm (de Duve and Beaufay, 1959). We have not yet directly tested this possibility as lysosomes are ruptured in both Alzheimer's disease and control brains after storage at −70 °C (Morrison *et al.*, 1987a, 1987b).

As was our experience with control brains (*see above,* p. 138), approximately 40% of the Alzheimer's disease brains that were obtained through routine autopsy were unsuitable for the isolation of intact RNA, whereas we isolated undegraded RNA from 100% (7/7) of the Alzheimer's disease brains obtained through the Duke rapid autopsy programme.

We found no statistical difference in the recovery of total RNA from comparable regions of control and Alzheimer's disease brain, whether regions of high (inferior temporal cortex), medium (occipital pole) or low (cerebellum) plaque content were assayed. Quantitation of mRNA content, both by *in vitro* translation in a reticulocyte lysate and by solution hybridization to radiolabelled poly(U), also showed no statistical differences between RNAs isolated from these three regions of control and Alzheimer's disease brain (Morrison, Griffin and White, 1986; Morrison *et al.*, 1986, 1987a, 1987b; Pardue *et al.*, unpublished data). Similar results were obtained when recoveries of total RNA and poly$(A)^+$ RNA from control and Alzheimer's disease hippocampus in the Los Angeles brain bank were compared (May *et al.*, 1987). Both our group and the Finch group have compared RNA recoveries and mRNA content from Alzheimer's disease and control cortex and cerebellum obtained through the Duke rapid autopsy programme. Again, there were no statistical differences between the control and Alzheimer's disease values. Our results show no dramatic reduction in total RNA or in total mRNA in either affected or unaffected regions of Alzheimer's disease cortex or cerebellum. Because of normal variation in RNA yields, we cannot exclude the possibility that there is a slight decrease in RNA content in Alzheimer's disease brain. Our results also do not exclude the likely possibility that there is a dramatic reduction in intracellular RNA in specific subpopulations of compromised neurons, while the RNA content of non-compromised neurons and of glia remains normal.

The differences in RNA yield between different groups highlights the problem of quantitating RNA levels in postmortem human brain accurately, especially if control and diseased tissue is being compared.

Levels of specific messenger RNAs

One-dimensional gel analysis show no observable difference between *in vitro* translation products of RNAs isolated from corresponding regions of Alzheimer's disease and control brain (Morrison *et al.*, 1987a, 1987b; Pardue *et al.*, unpublished data). On two-dimensional gels, the most dramatic difference between the abundant *in vitro* translation products was a 2–3-fold increase in the levels of the mRNA that encodes glial fibrillary acidic protein (GFAP) in Alzheimer's disease cortex (Morrison *et al.,* 1986). Increased levels of GFAP mRNA have also been reported in Alzheimer's disease cortex by comparing *in vitro* translation products (Sajdel-Sulkowska *et al.,* 1986) and by Northern hybridization of poly(A)$^+$ mRNAs to a GFAP cDNA probe (Goss, Morgan and Finch, 1987; Zain *et al.,* 1987). Induction of GFAP mRNA correlates with gliosis in Alzheimer's disease cortex (Brun, 1983; Griffin, personal communication). We also noted increased levels of GFAP synthesized by mRNAs isolated from cerebellum of several Alzheimer's disease brains. Although this is a brain area in which we find no plaques and tangles (Pardue *et al.,* unpublished data), there is a noticeable increase in histochemical staining with GFAP antibody in Alzheimer's disease cerebellum (Griffin, personal communication). These results suggest that, at least in some of the Alzheimer's disease brains in the Dallas Brain Bank, the cerebellum is more compromised than would be expected on the basis of routine pathological examination.

We have also done Northern hybridization comparisons between Alzheimer's disease and control brain RNAs using tubulin and actin cDNA probes. Although hybridization levels vary from one RNA isolate to another, we can see no consistent differences between the levels of these mRNAs in control and Alzheimer's disease cortex and cerebellum (Pardue *et al.,* unpublished data). McLachlan *et al.* (1987) observed, by Northern hybridization of total cortical RNA to several different cDNA probes, that some mRNA levels are not reduced in Alzheimer's disease cortex whereas that encoding the 68 kDa neurofilament protein mRNA is 27% of control values. Dot blot hybridization of total cortical RNA with preproenkephalin and preprosomatostatin probes showed levels of mRNA in Alzheimer's disease that were 84% and 70% respectively of control values (Taylor *et al.,* 1986).

Due to the increased potential for RNA degradation during isolation from Alzheimer's disease brain (*see above,* p. 142), quantitation of experimental mRNA levels with respect to constitutively expressed mRNAs, such as pIB16 (Milner and Sutcliffe, 1983), is even more important in Alzheimer's disease brain than it is for analysis of control human brain. This is illustrated in one recent study where hybridization to a tau cDNA probe was used as a standard for quantitation of the β amyloid precursor mRNA in control and Alzheimer's disease cortex and cerebellum. Both the tau and the β amyloid precursor probes hybridized equally to the cerebellar RNA isolate. There was no hybridization of the tau cDNA probe to RNA isolated from Alzheimer's disease cortex (Tanzi *et al.,* 1987). These results demonstrate that the lack of β amyloid precursor probe hybridization to the same cortical RNA isolate was most probably due to degradation of the RNA sample rather than to low levels of the β amyloid precursor mRNA in Alzheimer's disease cortex (*see below*, p. 145 for further discussion of β amyloid mRNA levels in both Alzheimer's disease and control brain).

One other mRNA that has been analysed in RNA isolates from Alzheimer's disease brain is that encoding nerve growth factor (NGF). Two lines of evidence suggest that NGF might be involved in the aetiology of Alzheimer's disease. Firstly, NGF mRNA is present in brain (Shelton and Reichardt, 1984; Korsching *et al.,* 1985) and is highest in regions innervated by the cholinergic cells of the basal forebrain that are compromised in Alzheimer's disease (Goedert *et al.,* 1986; Large *et al.,* 1986; Shelton and Reichardt, 1986; Whittemore *et al.,* 1986). Secondly, NGF is selectively taken up and retrogradely transported by these cholinergic cells

(Seiler and Schwab, 1984). In order to determine whether NGF mRNA levels were altered in Alzheimer's disease, Goedert *et al.* (1986) compared the amount of NGF mRNA in Alzheimer's disease and control cortex. Although they showed no change, this study did not compare the hybridization level of NGF mRNA to that of any control mRNAs, thus making it difficult to evaluate the results obtained. If the measurements are accurate, then NGF mRNA levels are not abnormally low in the area of Alzheimer's disease cortex measured, making it unlikely that degeneration of cholinergic cells projecting to cortex is caused by decreased synthesis of this particular trophic factor. Conversely, the results would also suggest that NGF mRNA levels were not significantly increased in Alzheimer's disease in response to neuronal cell death in the basal forebrain.

Importantly, several groups have now succeeded in making cDNA libraries of good quality from poly(A)$^+$ RNA isolated from Alzheimer's disease cortex or hippocampus (May *et al.,* 1987; Zain *et al.,* 1987; Gilbert and Roses, personal communication). Both abstracts state that cDNA sizes are normal and the Finch group reports an average insert size of 1.5 kb (May *et al.,* 1987). The libraries are being screened for sequences altered in amount in Alzheimer's disease brain (Johnson and Finch, 1987; May *et al.,* 1987).

β Amyloid precursor messenger RNA

Attempts to define the protein constituents of PHF, plaque amyloid and cerebrovascular amyloid (CVA) were initially complicated by their insolubility and impurity (Selkoe, Ihara and Salazar, 1982; Allsop, Landon and Kidd, 1983). Amyloid and CVA can now be isolated in relatively pure form and solubilized in formic acid, yielding peptides 4 kDa in size (Allsop, Landon and Kidd, 1983; Glenner and Wong, 1984; Masters *et al.,* 1985). The amino termini of the 4 kDa peptides were sequenced and were shown to have no homology to sequences in protein databanks. The amyloid and CVA peptides had identical amino termini, and the sequence of the first 28 amino acids of amyloid and CVA isolated from Alzheimer's disease brain was identical to the corresponding amino terminal sequence of amyloid isolated from Down's syndrome brain (Glenner and Wong, 1984; Wong, Quaranta and Glenner, 1985; Masters *et al.,* 1985).

Several laboratories have used oligodeoxynucleotide probes corresponding to parts of the 4 kD amino terminal sequence to identify the corresponding cDNA clones from fetal (Kang *et al.,* 1987; Tanzi *et al.,* 1987), or normal adult (Goldgaber *et al.,* 1987) human brain cDNA libraries. Sequence analysis of the cDNA clones showed that the amino terminal 27 amino acids of the amyloid peptide is present near the carboxy terminal end of a much larger precursor. This precursor has many of the characteristics of a transmembrane receptor (Kang *et al.,* 1987).

Northern blot analysis of the 3.7 kb mRNA encoding the β amyloid precursor shows that it is present in fetal (Kang *et al.,* 1987; Tanzi *et al.,* 1987), and adult (Goldgaber *et al.,* 1987; Tanzi *et al.,* 1987) human brain. Highest levels of the mRNA are found early in development and its levels are increased 2-fold in trisomy 21 fetal brain as compared to control fetal brain (Tanzi *et al.,* 1987). The mRNA is also present at high levels in a number of other 20–22 week human fetal tissues (Tanzi *et al.,* 1987), but in very low levels in adult placenta, thymus, and liver (Goldgaber *et al.,* 1987).

Tanzi *et al.* (1987) also determined the relative levels of the β amyloid precursor mRNA in different regions of adult normal brain. In these experiments, the authors used a control cDNA probe for hybridization to demonstrate that the mRNAs isolated from the different brain regions were of comparable quality (*see above*, p. 139). They showed that the levels of the β amyloid precursor mRNA were highest in the perisylvian cortex and other associated areas of cortex, as well as in the cerebellum. There was negligible hybridization to hippocampus or to motor areas of cortex (Tanzi *et al.*, 1987). Kang *et al.* showed (unpublished observations) that the β amyloid precursor mRNA in Alzheimer's disease cortex is similar in size to that in normal cortex and Tanzi *et al.* (1987) reported that this mRNA is present at similar levels to control in Alzheimer's disease cerebellum.

Concurrently with attempts to identify and characterize the β amyloid precursor mRNA, Gusella's group (St George-Hyslop *et al.*, 1987), used restriction fragment length polymorphisms of chromosome 21 markers to determine if any region of chromosome 21 could be linked to Alzheimer's disease. Linkage analysis of four families in which Alzheimer's disease was inherited showed that linkage could be demonstrated on the centromeric portion of the long arm of chromosome 21, thus establishing an inherited defect in these families. Interestingly, the amyloid gene has also been localized to chromosome 21 (Goldgaber *et al.*, 1987; Kang *et al.*, 1987; Robakis *et al.*, 1987; Tanzi *et al.*, 1987). Tanzi *et al.* (1987) have shown that the amyloid gene is linked to one of the chromosome 21 probes which is linked to the Alzheimer's disease locus. Preliminary evidence, from three cases of sporadic Alzheimer's disease, that a region of the long arm of chromosome 21 including the amyloid locus is trisomic (Delabar *et al.*, 1987), would suggest that duplication of this gene, or of the region surrounding it, is the genetic defect in both sporadic and familial Alzheimer's disease.

The identification of the gene encoding the amyloid precursor and analysis of the distribution of its mRNA raises a number of important questions about where amyloid is synthesized and about the factors regulating its deposition. If the distribution of the precusor in Alzheimer's disease brain is similar to that shown by Tanzi *et al.* (1987) in control brain, then its levels correlate neither with the distribution of plaque amyloid nor of CVA in Alzheimer's disease brain, for plaque amyloid and CVA are rare in cerebellum while plaque amyloid, but not CVA, is most prevalent in hippocampus (Mandybur, 1975; Vinters and Gilbert, 1983). The relatively high levels of the mRNA in fetal organs again does not correlate with amyloid deposition.

There are several possible explanations for these observations. If the β amyloid precursor is a member of a multigene family, as suggested by Robakis *et al.* (1987), then it is possible that the cDNA probes are cross-hybridizing with several mRNAs of similar molecular weight, only one of which encodes the precursor to β amyloid. If, on the other hand, there is only one genetic locus for β amyloid, then the precursor gene or its mRNA may be different in Alzheimer's disease and in controls. A precedent for this exists in familial amyloid polyneuropathy where a point mutation in plasma prealbumin predisposes it to form amyloid (Dwulet and Benson, 1984). Another possibility is that there is inappropriate initiation of translation on the β amyloid precursor mRNA in Alzheimer's disease, just prior to the amyloid sequence itself (Breimer and Denny, 1987). Recent *in situ* analysis of the β amyloid precursor mRNA shows that its levels are increased in particular regions of Alzheimer's disease hippocampus compared to controls (Higgins *et al.*, 1987), suggesting that increased levels of the β amyloid precursor mRNA may lead to amyloid deposition (Goldgaber *et al.*, 1987). If there is no difference between the Alzheimer's disease and normal β amyloid precursor genes or in the nature of the primary translation product, then alteration in the intracellular environment must regulate the levels of synthesis of the precursor protein, or alter either its degradation rate or degradation pathway in specific cell types.

Isolation and sequence comparison of β amyloid precursor cDNA from Alzheimer's disease brain and of its gene will be necessary to determine if a mutation or duplication is present. Comparison of the primary translation products of the mRNAs from Alzheimer's disease and normal tissues, and their post-translational processing and degradation, will show whether there are specific alterations in these pathways in regions of Alzheimer's disease brain where amyloid is deposited.

There is now evidence that the amyloid beta protein gene is not the genetic locus of the defect in familial Alzheimer's disease (Van Broeckhoven *et al.*, 1987; Tanzi *et al.*, 1987). Several groups have also shown that they cannot find evidence for duplication either of the amyloid beta protein gene or of other chromosome 21 genes in familial and sporadic Alzheimer's disease (St George-Hyslop *et al.*, 1987; Podlisny *et al.*, 1987; Tanzi *et al.*, 1987).

Acknowledgements

This work was supported by NIH grants HD14886 (MRM) and AG05537 (WST Griffin) as well as by the Leland Fikes Association. Thanks to E. Katherine Miller for editing and to Cheryl Beisert for typing the text.

References

ALLSOP, D., LANDON, M. and KIDD, M. (1983) The isolation and amino acid composition of senile plaque core protein. *Brain Research,* **259,** 348–352

BOWEN, D. M., SMITH, C. B., WHITE, P., GOODHARDT, M. J., SPILLANE, J. A., FLACK, R. H. A. and DAVISON, A. N. (1977a) Chemical pathology of the organic dementias. I. Validity of biochemical measurements on human post-mortem brain specimens. *Brain,* **100,** 397–426

BOWEN, D. M., SMITH, C. B., WHITE, P., FLACK, R. H. A., CARRASCO, L. H., GEDYE, J. L. and DAVISON, A. N. (1977b) Chemical pathology of the organic dementias. II. Quantitative estimation of cellular changes in post-mortem brains. *Brain,* **100,** 427–453

BRAY, P., CARTER, A., SIMONS, C. *et al.* (1986) Human cDNA clones for four species of $G_{\alpha s}$ signal transduction protein. *Proceedings of the National Academy of Sciences USA,* **83,** 8893–8897

BREIMER, L. H. and DENNY, P. (1987) Alzheimer amyloid aspects. *Nature (London),* **326,** 749–750

BROECKHOVEN, VAN *et al.* (1987) *Nature,* **329,** 153–156

BRUN, A. (1983) An overview of light and electron microscopic changes. In: *Alzheimer's Disease. The Standard Reference,* edited by B. Reisberg, pp. 37–47. London: The Free Press, Collier MacMillan

BUSTANI, P., HENRY, J. F., SOUSSALINE, F. and COMAR, D. (1983) In: *Functional Radionuclide Imaging of the Brain,* edited by P. G. Magistretti, pp. 319–326. New York: Raven Press

CHIRGWIN, J. M., PRZYBYLA, A. E., MacDONALD, R. J. and RUTTER, W. J. (1979) Isolation of biologically active ribonucleic acid from sources enriched in ribonuclease. *Biochemistry,* **18,** 5294–5299

COX, R. A. (1968) The use of guanidinium chloride in the isolation of nucleic acids. *Methods in Enzymology,* **12B,** 120–129

DAVIES, P. and MALONEY, A. J. F. (1976) Selective loss of central cholinergic neurons in Alzheimer's disease. *Lancet,* **ii,** 1403

DAVIES, P. and TERRY, R. D. (1981) Cortical somatostatin-like immunoreactivity in cases of Alzheimer's disease and senile dementia of the Alzheimer type. *Neurobiology of Aging,* **2,** 9–14

DAYAN, A. D. and BALL, M. J. (1973) Histometric observations on the metabolism of tangle-bearing neurons. *Journal of the Neurological Sciences,* **19,** 433–436

DE DUVE, C. and BEAUFAY, H. (1959) Tissue fractionation studies: 10 Influence of ischemia on the state of some bound enzymes in rat liver. *Biochemical Journal,* **73,** 610–616

DELABAR, J.-M., GOLDGABER, D., LAMOUR, Y., NICOLE, A., HURET, J.-L., DE GROUCHY, J. *et al.* (1987) β amyloid gene duplication in Alzheimer's disease and karyotypically normal Down syndrome. *Science,* **235,** 1390–1392

DOEBLER, J. A., MARKESBERY, W. R., ANTHONY, A. and RHOADS, R. E. (1987a) Neuronal RNA in relation to neuronal loss and neurofibrillary pathology in the hippocampus in Alzheimer's disease. *Journal of Neuropathology and Experimental Neurology,* **46,** 28–39

DOEBLER, J. A., MARKESBERY, W. R., ANTHONY, A. and RHOADS, R. E. (1987b) *In situ* analysis of neostriatal RNA and chromatin in Alzheimer's disease. *Neurology,* **37,** 309–313

DUARA, R., GRADY, C., HAXBY, J., SUNDARAM, M. *et al.* (1986) Positron emission tomography in Alzheimer's disease. *Neurology,* **36,** 879–887

DWULET, F. E. and BENSON, M. D. (1984) Primary structure of an amyloid and its plasma precursor in a heredofamilial polyneuropathy of Swedish origin. *Proceedings of the National Academy of Sciences USA,* **81,** 694–698

DZIEGIELEWSKA, K. M., SAUNDERS, N. R., SCHEJTER, E. J., ZAKUT, H., ZEVIN-SONKIN, D., ZISLING, R. and SOREQ, H. (1986) Synthesis of plasma proteins in fetal, adult, and neoplastic human brain tissue. *Developmental Biology,* **115,** 93–104

ELSHOURBAGY, N. A., LIAO, W. S., MAHLEY, R. W. and TAYLOR, J. M. (1985) Apoprotein E mRNA is abundant in the brain and adrenals, as well as in the liver, and is present in other peripheral tissues of rats and marmosets. *Proceedings of the National Academy of Sciences USA,* **82,** 203–207

FARKAS, T., FERRIS, S. H., WOLF, A. P. *et al.* (1982) ^{18}F-2-Deoxy-2-fluoro-D-glucose as a tracer in the positron emission tomographic study of senile dementia. *American Journal of Psychiatry,* **139,** 352–353

FERRANTE, R. J., KOWALL, N. W., BEAL, M. F., RICHARDSON, E. P. JR, BIRD, E. D. and MARTIN, J. B. (1985) Selective sparing of a class of striatal neurons in Huntington's disease. *Science,* **230,** 561–563

FERRIER, I. N., CROSS, A. J., JOHNSON, J. A. *et al.* (1983) Neuropeptides in Alzheimer type dementia. *Journal of the Neurological Sciences,* **62,** 159–170

GILBERT, J. M., BROWN, B. A., STROCCHI, P., BIRD, E. D. and MAROTTA, C. A. (1981) The preparation of biologically active messenger RNA from human postmortem brain tissue. *Journal of Neurochemistry,* **36,** 976–984

GLENNER, G. G. and WONG, C. W. (1984) Alzheimer's disease and Down's syndrome: sharing of a unique cerebrovascular amyloid fibril protein. *Biochemical and Biophysical Research Communications,* **122,** 1131–1135

GOEDERT, M., FINE, A., HUNT, S. P. and ULLRICH, A. (1986) Nerve growth factor mRNA in peripheral and central rat tissues and in the human central nervous system: Lesion effects in the rat brain and levels in Alzheimer's disease. *Molecular Brain Research,* **1,** 85–92

GOELET, P., CASTELLUCCI, V. F., SCHACHER, S. and KANDEL, E. R. (1986) The long and the short of long-term memory – a molecular framework. *Nature (London),* **322,** 419–422

GOLDGABER, D., LERMAN, M. I., McBRIDE, O. W., SAFFIOTTI, U. and GAJDUSEK, D. C. (1987) Characterization and chromosomal localization of a cDNA encoding brain amyloid of Alzheimer's disease. *Science,* **235,** 877–880

GOSS, J. R., MORGAN, D. G. and FINCH, C. E. (1987) Glial fibrillary acidic protein mRNA increases during a wasting agonal state in old mice. Implications for the increased GFAP mRNA in Alzheimer's disease. *Society for Neuroscience Abstracts,* **13,** 1325

GUILLEMETTE, J. G., WONG, L., CRAPPER McLACHLAN, D. R. and LEWIS, P. N. (1986) Characterization of messenger RNA from the cerebral cortex of control and Alzheimer-afflicted brain. *Journal of Neurochemistry,* **47,** 987–997

GUNDERSEN, C. B., MILEDI, R. and PARKER, I. (1984a) Messenger RNA from human brain induces drug- and voltage-operated channels in *Xenopus* oocytes. *Nature (London),* **308,** 421–424

GUNDERSEN, C. B., MILEDI, R. and PARKER, I. (1984b) Properties of human brain glycine receptors expressed in *Xenopus* oocytes. *Proceedings of the Royal Society of London B,* **221,** 235–244

HARRIS, W. A. (1985) Genetics and development of the nervous system. *Journal of Neurogenetics,* **2,** 179–196

HERBERT, J., WILCOX, J. N., PHAM, K.-T. C. *et al.* (1986) Transthyretin: a choroid plexus-specific transport protein in human brain. *Neurology,* **36,** 900–911

HIGGINS, G. A., LEWIS, D. A., YOUNG, W. G. *et al.* (1987) Quantitative *in situ* hybridization shows differential expression of amyloid-β protein mRNA within neurons of the hippocampal formation in Alzheimer's disease. *Society for Neuroscience Abstracts,* **13,** 818

ILARIA, R., WINES, D., PARDUE, S., JAMISON, S., OJEDA, S. R., SNIDER, J. and MORRISON, M. R. (1985) A rapid microisolation procedure for isolating RNA from multiple samples of human and rat brain. *Journal of Neuroscience Methods,* **15,** 165–174

JAYE, M., HOWK, R., BURGESS, W., RICCA, G. A., CHIU, I.-M., RAVERA, M. W. *et al.* (1986) Human endothelial cell growth factor: cloning, nucleotide sequence, and chromosome localization. *Science,* **233,** 541–545

JOHNSON, S. A. and FINCH, C. E. (1987) Cloning of RNA sequences whose prevalence is increased in Alzheimer cortex. *Society for Neuroscience Abstracts,* **13,** 1325

JOHNSON, S. A., MORGAN, D. G. and FINCH, C. E. (1986) Extensive post-mortem stability of RNA from rat and human brain. *Journal of Neuroscience Research,* **16,** 267–280

KAMHOLZ, J., DE FERRA, F., PUCKETT, C. and LAZZARINI, R. (1986) Identification of three forms of human myelin basic protein by cDNA cloning. *Proceedings of the National Academy of Sciences USA,* **83,** 4962–4966

KANG, J., LEMAIRE, H.-G., UNTERBECK, A. *et al.* (1987) The precursor of Alzheimer's disease amyloid A4 protein resembles a cell-surface receptor. *Nature (London),* **325,** 733–736

KAPLAN, B. B., BERNSTEIN, S. L. and GIOIO, A. E. (1979) An improved method for the rapid isolation of brain ribonucleic acid. *Biochemistry Journal,* **183,** 181–184

KOH, J.-Y., PETERS, S. and CHOI, D. W. (1986) Neurons containing NADPH-diaphorase are selectively resistant to quinolinate toxicity. *Science,* **234,** 73–76

KORCSHING, S., AUBURGER, G., HEUMANN, R., SCOTT, J. and THOENEN, H. (1985) Levels of nerve growth factor and its mRNA in the central nervous system of the rat correlate with cholinergic innervation. *EMBO Journal,* **4,** 1389–1393

KRETZSCHMAR, H. A., STOWRING, L. E., WESTAWAY, D., STUBBLEBINE, W. H., PRUSINER, S. B. and DEARMOND, S. J. (1986) Molecular cloning of a human prion protein cDNA. *DNA,* **5,** 315–324

KURNIT, D. M., WENTWORTH, B. M., DE LONG, L. and VILLA-KOMAROFF, L. (1982) Construction of cloned libraries from RNA of human fetal tissues. *Cytogenetics and Cell Genetics,* **34,** 193–203

LARGE, T. H., BODARY, S. C., CLEGG, D. O., WESKAMP, G., OTTEN, U. and REICHARDT, L. F. (1986) Nerve growth factor gene expression in the developing rat brain. *Science,* **234,** 352–355

MANDYBUR, T. I. (1975) The incidence of cerebral amyloid angiopathy in Alzheimer's disease. *Neurology,* **25,** 120–126

MANN, D. M. A. (1978) Granulovacuolar degeneration in pyramidal cells of the hippocampus. *Acta Neuropathologica (Berlin),* **42,** 149–151

MANN, D. M. A. and SINCLAIR, K. G. A. (1978) The quantitative assessment of lipofuscin pigment, cytoplasmic RNA and nucleolar volume in senile dementia. *Neuropathology and Applied Neurobiology,* **4,** 129–135

MANN, D. M. A. and YATES, P. O. (1979) The effects of ageing on the pigmented nerve cells of the human locus caeruleus and substantia nigra. *Acta Neuropathologica (Berlin),* **47,** 93–97

MANN, D. M. A. and YATES, P. O. (1981) The relationship between formation of senile plaques and neurofibrillary tangles and changes in nerve cell metabolism in Alzheimer type dementia. *Mechanisms of Ageing and Development,* **17,** 395–401

MANN, D. M. A., YATES, P. O. and BARTON, C. M. (1977) Cytophotometric mapping of neuronal changes in senile dementia. *Journal of Neurology, Neurosurgery, and Psychiatry,* **40,** 299–302

MANN, D. M. A., BARTON, C. M. and DAVIES, J. S. (1978) Post-mortem changes in human central nervous tissue and the effects on quantitation of nucleic acids and enzymes. *Histochemical Journal,* **10,** 127–135

MANN, D. M. A., YATES, P. O. and STAMP, J. E. (1978) The relationship between lipofuscin pigment and ageing in the human nervous system. *Journal of the Neurological Sciences,* **37,** 83–93

MANN, D. M. A., NEARY, D., YATES, P. O., LINCOLN, J., SNOWDEN, J. S. and STANWORTH, P. (1981a) Neurofibrillary pathology and protein synthetic capability in nerve cells in Alzheimer's disease. *Neuropathology and Applied Neurobiology,* **7,** 37–47

MANN, D. M. A., NEARY, D., YATES, P. O., LINCOLN, J., SNOWDEN, J. S. and STANWORTH, P. (1981b) Alterations of protein synthetic capacity of nerve cells in Alzheimer's disease. *Journal of Neurology, Neurosurgery, and Psychiatry,* **44,** 97–102

MAROTTA, C. A., STROCCHI, P. and GILBERT, J. M. (1979) Biosynthesis of heterogeneous forms of mammalian brain tubulin subunits by multiple messenger RNAs. *Journal of Neurochemistry,* **33,** 231–246

MAROTTA, C. A., BROWN, B. A., STROCCHI, P., BIRD, E. D. and GILBERT, J. M. (1981) *In vitro* synthesis of human brain proteins including tubulin and actin by purified polysomes. *Journal of Neurochemistry,* **36,** 966–975

MASTERS, C. L., SIMMS, G., WEINMAN, N. A., MULTHAUP, G., McDONALD, B. L. and BEYREUTHER, K. (1985) Amyloid plaque core protein in Alzheimer disease and Down syndrome. *Proceedings of the National Academy of Sciences USA,* **82,** 4245–4249

MAY, P. C., JOHNSON, S. A., MASTERS, J. N., LAMPERT-ETCHELLS, M. and FINCH, C. E. (1987) Cloning of poly(A) RNA differentially regulated in Alzheimer's disease from a hippocampal cDNA library. *Society for Neuroscience Abstracts,* **13,** 1325

McLACHLAN, D. R., LUKIW, W. J., WONG, L. and BECH-HANSE, N. T. (1987) Selective messenger RNA reduction in Alzheimer's disease. *Journal of Cellular Biochemistry,* **Supplement 11D,** 196

MICHAEL, A. J., GAMMON, R. B., PARDUE, S., CROALL, D. E., SPARKMAN, D. R., GRIFFIN, W. S. T. and MORRISON, M. R. (1982) Construction of a recombinant DNA library containing sequences complementary to human adult cortex polyadenylated RNAs. *Neuroscience Abstracts,* **8,** 704

MILNER, R. J. and SUTCLIFFE, J. G. (1983) Gene expression in rat brain. *Nucleic Acids Research,* **11,** 5497–5520

MORIMATSU, M., HIRAI, S., MURAMATSU, A. and YOSHIKAWA, M. (1975) Senile degenerative brain lesions and dementia. *Journal of the American Geriatrics Society,* **23,** 390–406

MORRISON, M. R. and LINGREL, J. B. (1976) Molecular weight determinations on the α-globin and β-globin mRNAs. *Biochimica et Biophysica Acta,* **447,** 104–109

MORRISON, M. R. and GRIFFIN, W. S. T. (1981) The isolation and *in vitro* translation of undegraded messenger RNAs from human postmortem brain. *Analytical Biochemistry,* **113,** 318–324

MORRISON, M. R. and ROSENBERG, R. N. (1983) Specific messenger RNA changes in Joseph disease cerebella. *Annals of Neurology,* **14,** 73–79

MORRISON, M. R. and SPARKMAN, D. R. (1984) Genetic engineering approaches to disease of the nervous system. In: *Symposium on Neuronal Communications,* edited by B. J. Meyer and S. Kramer, pp. 264–271. Cape Town: Balkema Press

MORRISON, M. R. and GRIFFIN, W. S. T. (1984) Developmental expression of specific messenger RNAs in rat and human brain. In: *Symposium on Neuronal Communications,* edited by B. J. Meyer and S. Kramer, pp. 107–113. Cape Town: Balkema Press

MORRISON, M. R. and GRIFFIN, W. S. T. (1985) Molecular biology of the mammalian brain. In: *Gene Expression in Brain,* edited by C. Zomzely-Neurath and W. A. Walker, pp. 57–98. New York: Wiley Press

MORRISON, M. R. and GRIFFIN, W. S. T. (1986) Quantitation and *in situ* localization of tubulin mRNA in the mammalian nervous system In: *Dynamic Aspects of Microtubule Biology,* Vol. 466, edited by D. Soifer, pp. 51–74. New York: Annals of the New York Academy of Sciences

MORRISON, M. R., BASKIN, F. and ROSENBERG, R. N. (1977) Quantitation and characterization of poly(A)-containing messenger RNAs from mouse neuroblastoma cells. *Biochemica et Biophysica Acta,* **476,** 228–237

MORRISON, M. R., PARDUE, S. and GRIFFIN, W. S. T. (1981) Developmental alterations in the levels of translationally-active messenger RNAs in the postnatal rat cerebellum. *Journal of Biological Chemistry,* **256,** 3350–3356

MORRISON, M. R., GRIFFIN, W. S. T. and WHITE, C. L. III (1985) Brain messenger RNA in Alzheimer's disease. *Journal of Neurogenetics,* **2,** 164

MORRISON, M. R., GRIFFIN, W. S. T. and WHITE, C. L. III (1986) Brain messenger RNA in Alzheimer's disease. In: *The Role of RNA and DNA in Brain Function,* edited by A. Giuditta, B. B. Kaplan and C. Zomzely-Neurath, pp. 142–159. Boston: Martinus Nijhoff Press

MORRISON, M. R., PARDUE, S. and GRIFFIN, W. S. T. (1983) Altered expression of different tubulin electrophoretic variants during human cortex development. *Journal of Neurogenetics,* **1,** 105–111

MORRISON, M. R., PARDUE, S., PRASHAD, N., CROALL, D. E. and BRODEUR, R. (1980) Relative increase in polysomal mRNA for R1 cAMP binding protein in neuroblastoma cells treated with N-Dibutyryl-adenosine 3′,-5′-phosphate. *European Journal of Biochemistry,* **106,** 463–472

MORRISON, M. R., PARDUE, S., MASCHHOFF, K. *et al.* (1986) Brain messenger RNA levels and ribonuclease activity in Alzheimer's disease. *Biochemical Society Bulletin,* **8,** 58

MORRISON, M. R., PARDUE, S., MASCHHOFF, K. *et al.* (1987a) Brain messenger RNA levels and ribonuclease activity in Alzheimer's disease. *Biochemical Society Transactions,* **15,** 133–134

MORRISON, M. R., PARDUE, S., MASCHHOFF, K. *et al.* (1987b) Messenger RNA analysis in control and Alzheimer's disease brain: Rapid autopsy evaluation. In: *Alzheimer's Disease: Advances in Basic Research and Therapies,* edited by R. J. Wurtman, S. H. Corkin and J. H. Growdon, pp. 505–509. Cambridge, Massachusetts; Center for Brain Sciences and Metabolism Charitable Trust

NATHANS, J., THOMAS, D. and HOGNESS, D. S. (1986) Molecular genetics of human color vision: the genes encoding blue, green, and red pigments. *Science,* **232,** 193–202

NEVE, R. L., HARRIS, P., KOSIK, K. S., KURNIT, D. M. and DONLON, T. A. (1986) Identification of cDNA clones for the human microtubule-associated protein tau and chromosomal localization of the genes for tau and microtubule-associated protein 2. *Molecular Brain Research,* **1,** 271–280

PERRY, E. K. and PERRY, R. H. (1983) Human brain neurochemistry – some postmortem problems. *Life Sciences,* **33,** 1733–1743

PERRY, E. K., PERRY, R. H. and TOMLINSON, B. E. (1982) The influence of agonal status on some neurochemical activities of postmortem human brain tissue. *Neuroscience Letters,* **29,** 303–307

PODLISNY, *et al.* (1987) *Science,* **238,** 669–671

PUCKETT, C., KAMHOLZ, J., RORKE, L. and LAZZARINI, R. A. (1985) Construction and characterization of mouse and human brain cDNA 'expression' libraries. *Neurology,* **35,** 259

RHOADS, R. E., DOEBLER, J. A., MARKESBERY, W. R., ANTHONY, A., DAVIES, P. and SHEFF, S. W. (1987) RNA levels and Alz-50 immunoreactivity in Alzheimer's disease hippocampus. *Journal of Cellular Biochemistry,* **Supplement 11D,** 198

ROBAKIS, N. K., SAWH, P. R., WOLFE, G. C., RUBENSTEIN, R., CARP, R. I. and INNIS, M. A. (1986a) Isolation of a cDNA clone encoding the leader peptide of prion protein and expression of the homologous gene in various tissues. *Proceedings of the National Academy of Sciences,* **83,** 6377–6381

ROBAKIS, N. K., DEVINE-GAGE, E. A., JENKINS, E. C. *et al.* (1986b) Localization of a human gene homologous to the PrP gene on the p arm of chromosome 20 and detection of PrP-related antigens in normal human brain. *Biochemical and Biophysical Research Communications,* **140,** 758–765

ROBAKIS, N. K., WISNIEWSKI, H. M., JENKINS, E. C. *et al.* (1987) Chromosome 21q21 sublocalisation of gene encoding beta-amyloid peptide in cerebral vessels and neuritic (senile) plaques of people with Alzheimer disease and Down syndrome. *Lancet,* **i,** 384–385

ROSES, A. D., GILBERT, J. R., BARTLETT, R. J., PERICAK-VANCE, M. A., DAWSON, D. V. and MORRISON, M. R. (1987) Sib-pair linkage and subtraction hybridization strategies for Alzheimer's disease. *Banbury Reports, 27,* in press

ROSSOR, M., FAHRENKRUG, J., EMSON, P., MOUNTJOY, C., IVERSEN, L. and ROTH, M. (1980) Reduced cortical choline acetyltransferase activity in senile dementia of Alzheimer type is not accompanied by changes in vasoactive intestinal polypeptide. *Brain Research,* **201,** 249–253

SAJDEL-SULKOWSKA, E. M. and MAROTTA, C. A. (1984) Alzheimer's disease brain: alterations in RNA levels and in a ribonuclease-inhibitor complex. *Science,* **225,** 947–949

SAJDEL-SULKOWSKA, E., COUGHLIN, J. F. and MAROTTA, C. A. (1983) *In vitro* synthesis of polypeptides of moderately large size by poly(A)-containing messenger RNA from postmortem human brain and mouse brain. *Journal of Neurochemistry,* **40,** 670–680

SAJDEL-SULKOWSKA, E. M., COUGHLIN, J. F., STATON, D. M. and MAROTTA, C. A. (1983) *In vitro* protein synthesis by messenger RNA from the Alzheimer's disease brain. In: *Banbury Report 15, Biological Aspects of Alzheimer's Disease,* edited by R. Katzman, pp. 193–200. Cold Spring Harbor Laboratory, Cold Spring Harbor

SAJDEL-SULKOWSKA, E. M., MAJOCHA, R. E., SALIM, M. *et al.* (1986) Messenger RNA dependent expression of GFAP in the Alzheimer's disease brain. *Society for Neuroscience Abstracts,* **12,** 1400

SEILER, M. and SCHWAB, M. E. (1984) Specific retrograde transport of nerve growth factor (NGF) from neocortex to nucleus basalis in the rat. *Brain Research,* **300,** 33–39

SELKOE, D. J., IHARA, Y. and SALAZAR, F. J. (1982) Alzheimer's disease: Insolubility of partially purified paired helical filaments in sodium dodecyl sulfate and urea. *Science,* **215,** 1243–1245

SHELTON, D. L. and REICHARDT, L. F. (1984) Expression of the β-nerve growth factor gene correlates with the density of sympathetic innervation in effector organs. *Proceedings of the National Academy of Sciences USA,* **81,** 7951–7955

SHELTON, D. L. and REICHARDT, L. F. (1986) Studies on the expression of the β nerve growth factor (HGF) gene in the central nervous system: level and regional distribution of NGF mRNA suggest that NGF functions as a trophic factor for several distinct populations of neurons. *Proceedings of the National Academy of Sciences USA,* **83,** 2714–2718

SOREQ, H., ZEVIN-SONKIN, D. and RAZON, N. (1984) Expression of cholinesterase gene(s) in human brain tissues: translational evidence for multiple mRNA species. *EMBO Journal,* **3,** 1371–1375

SOREQ, H., ZEVIN-SONKIN, D., AVNI, A., HALL, L. M. C. and SPIERER, P. (1985) A human acetylcholinesterase gene identified by homology to the Ace region of *Drosophila. Proceedings of the National Academy of Sciences USA,* **82,** 1827–1831

ST GEORGE-HYSLOP, P. H., TANZI, R. E., POLINSKY, R. J., HAINES, J. L., NEE, L., WATKINS, P. C. (1987) The genetic defect causing familial Alzheimer's disease maps on chromosome 21. *Science,* **235,** 885–890

ST GEORGE-HYSLOP, P. H. *et al.* (1987) *Science,* **238,** 664–666

TANZI, R. E., GUSELLA, J. F., WATKINS, P. C. *et al.* (1987) Amyloid β protein gene: cDNA, mRNA distribution, and genetic linkage near the Alzheimer locus. *Science,* **235,** 880–884

TANZI, R. E. *et al.* (1987) *Nature,* **329,** 156–157

TANZI, R. E. *et al.* (1987) *Science,* **238,** 666–669

TAYLOR, G. R., CARTER, G. I., CROW, T. J. *et al.* (1986) Recovery and measurement of specific RNA species from postmortem brain tissue: a general reduction in Alzheimer's disease detected by molecular hybridization. *Experimental and Molecular Pathology,* **44,** 111–116

TAYLOR, G. R., CARTER, G. I., CROW, T. J. *et al.* (1987) Recovery and measurement of RNA in Alzheimer's disease by molecular hybridization. *Journal of Neurology, Neurosurgery, and Psychiatry,* **50,** 356

TOMLINSON, B. E., BLESSED, G. and ROTH, M. (1970) Observations on the brains of demented old people. *Journal of Neurological Sciences,* **11,** 205–242

UEMURA, E. and HARTMANN, H. A. (1978) RNA content and volume of nerve cell bodies in human brain. I. Prefrontal cortex in aging normal and demented patients. *Journal of Neuropathology and Experimental Neurology,* **37,** 487–496

UEMURA, E. and HARTMANN, H. A. (1979) Quantitative studies of neuronal RNA on the subiculum of demented old individuals. *Brain Research Bulletin,* **4,** 301–305

VINTERS, H. V. and GILBERT, J. J. (1983) Cerebral amyloid angiopathy: Incidence and complications in the aging brain. II. The distribution of amyloid vascular changes. *Stroke,* **14,** 924–928

WESTER, P., BATEMAN, D. E., DODD, P. R. *et al.* (1985) Agonal status affects the metabolic activity of nerve endings isolated from postmortem human brain. *Neurochemical Pathology,* **3,** 169–180

WHATLEY, S. A., HALL, C., DAVISON, A. N. and LIM, L. (1984) Alterations in the relative amounts of specific mRNA species in the developing human brain in Down's syndrome. *Biochemical Journal,* **220,** 179–187

WHITTEMORE, S. R., EBENDAL, T., LARKFORS, L. *et al.* (1986) Developmental and regional expression of nerve growth factor messenger RNA and protein in the rat central nervous system. *Proceedings of the National Academy of Sciences USA,* **83,** 817–821

WOLOZIN, B. L., PRUCHNICKI, A., DICKSON, D. W., DAVIES, P. L. (1986) Neuronal antigen in the brains of Alzheimer patients. *Science,* **232,** 648–650

WONG, C. W., QUARANTA, V. and GLENNER, G. G. (1985) Neuritic plaques and cerebrovascular amyloid in Alzheimer disease are antigenically related. *Proceedings of the National Academy of Sciences USA,* **82,** 8729–8732

WOOD, T. L., FRANTZ, G. D., MENKES, J. H. and TOBIN, A. J. (1986) Regional distribution of messenger RNAs in postmortem human brain. *Journal of Neuroscience Research,* **16,** 311–324

ZAIN, S. B., REHMAN, S., SAJDEL-SULKOWSKA, E. M. *et al.* (1987) Molecular cloning and *in vitro* translation studies of GFAP mRNA from normal and Alzheimer disease brain tissue. *Journal of Cellular Biochemistry,* **Supplement 11D,** 198

11
Molecular genetics of Joseph disease

Roger N. Rosenberg and Abraham Grossman

AETIOLOGY OF JOSEPH DISEASE

In 1972, two papers described dominantly inherited neurological disease in two Azorean-Portuguese families residing in Massachusetts. The first report (Nakano, Dawson and Spence, 1972) referred to 'Machado disease' to acknowledge the proband family in which there was a progressive cerebellar disease with distal atrophy and sensory loss beginning in the fifth decade. A sural nerve biopsy from one patient documented segmental demyelination. The second report (Woods and Schaumburg, 1972) described 'nigrospinodentatal degeneration with nuclear ophthalmoplegia' in the Thomas family, presenting in the third to fifth decades with gait ataxia, external ophthalmoplegia, facial-lingual fasciculations, spasticity, and limb rigidity. The neuropathological features included neuronal loss in the substantia nigra, dentate nuclei, and Clarke's column in the spinal cord. Myelin pallor was noted in the posterior columns, lateral corticospinal, and spinothalamic tracts of the spinal cord, most evident in the thoracic portion.

In 1976, Rosenberg *et al.* described a progressive motor system disease beginning in the second and third decades in another Portuguese family, the Joseph family of California, consisting of 329 persons in nine generations at that time. The clinical picture differed from that of the two previous reports in that affected members of the Joseph family had progressive spasticity, lurching unsteadiness of gait due to spasticity (without cerebellar signs), spastic dysarthria, loss of fast saccadic eye movements, ophthalmoparesis for upward gaze, and facial-lingual fasciculations, with prominent dystonia of head, face, limbs, and trunk in some individuals. The neuropathology consisted of neuronal loss in the striatum, substantia nigra, basis pontis, dentate nucleus, cerebellar cortex, and anterior horns of the spinal cord. Cerebrospinal fluid (CSF) homovanillic acid levels were reduced, correlating with neuronal loss in the zona compacta of the substantia nigra. The Joseph family disorder differed from the Massachusetts cases because: (a) there was striatonigral degeneration; (b) dystonia was severe in some patients; and (c) cerebellar signs were not evident in any family member examined in 1975. The disease was also different from olivopontocerebellar degeneration because the medullary inferior olives were unaffected. It was unlike Huntington's disease (HD) because intelligence remained normal in all Joseph patients, and the cerebral cortex was histologically normal.

Epidemiological studies were necessary to establish the spectrum of clinical and neuropathological findings. The logical place to proceed was in the Azores Islands, particularly on the island of Flores, from which both the Joseph and Thomas families originated. Other suggestions of variable clinical and neuropathological features were made as personal communications to Dr Rosenberg from Dr Coutinho and Dr Andrade of the Neurology Service, Hospital Santo Antonio, Porto, Portugal. They travelled to Flores in December 1976, where they found patients with prominent spasticity and extrapyramidal signs and some with cerebellar disorders or peripheral neuropathy. Coutinho and Andrade (1978) emphasized types I, II, and III as variants of the autosomal dominant disorder in affected individuals on Flores. Type I corresponded to the typical Joseph phenotype, with only extrapyramidal and pyramidal signs, including dystonia, athetosis, and rigidity with associated spasticity. Type II included patients with cerebellar deficits and spasticity. Type III was reserved for patients with the Machado phenotype – patients with a dominantly inherited form of motor polyneuropathy and true cerebellar deficits (Coutinho and Andrade, 1978; Rosenberg *et al.*, 1978).

Romanul *et al.* (1977) described another Massachusetts family of Portuguese ancestry with a similar clinical spectrum of gait ataxia, parkinsonian rigidity, ophthalmoparesis, muscle fasciculation, areflexia and associated nystagmus, cerebellar tremor, and extensor plantar responses. Postmortem study of one patient showed striatonigral degeneration with severe neuronal loss and gliosis in the putamen; in another patient, the basal ganglia were normal. Their observations supported the concept of clinical and neuropathological variation. It seemed likely that all the reported families of similar ancestry represented a single genetic entity with variable expression, and that a better understanding of this disease might accrue from a visit to Flores and San Miguel, from which the family by Romanul *et al.* (1977) had emigrated.

AZOREAN NEUROEPIDEMIOLOGY

In 1977, Rosenberg and Nyhan joined Coutinho in the Azores to investigate the presence of this genetic disease, especially on the island of Flores. Rosenberg *et al.* (1978) encountered typical genetic disease, as described in the Joseph and Thomas families, on Flores in the Sousa family.

Synthesizing the data from the Joseph, Thomas, and Sousa families (Rosenberg *et al.*, 1978), the data of Coutinho and Andrade (1978), and the data of Romanul *et al.* (1977), we believe we are witnessing the expression of a single gene that causes: (a) early onset disease with pyramidal and extrapyramidal signs (type I); (b) intermediate age of onset and progression, with cerebellar deficits as well as pyramidal and extrapyramidal findings (type II); or (c) later onset, with peripheral neuropathy and cerebellar disorder (type III). Dominantly inherited parkinsonism with distal atrophy in one family could constitute a type IV disease. It took a systemic evaluation of the California, Massachusetts, and Flores families, however, to document the variations of clinical expression. The additional neuropathological findings of Romanul *et al.* (1977), indicating the wider pathological involvement with striatonigral degeneration in only some patients, also implied varied expression of this mutant gene.

More recently, two reports have cited this genetic disorder in two families, one in north-eastern Portugal (Lima and Coutinho, 1980) and one in New York (Healton *et al.*, 1980) – a black family traced to North Carolina. They had clinical features similar to those of the Joseph and Thomas families, but neither family had any identifiable linkage to the Azores. Neuropathological studies are required to prove that these families have the same disease.

Three recent reports also cite the presence of Joseph disease in Japan (Goto *et al.*, 1982; Sakai, Ohta and Ishino, 1983; Yuasa *et al.*, 1986). The clinical features are the same as in the American and Portuguese families, and the neuropathological features are quite similar but have an accentuation of a dentatorubroluysian degeneration (Yuasa *et al.*, 1986). Joseph disease has also been described in a family from India by Bharucha, Bharucha and Bhabda (1986). It is suggested that Joseph disease was brought to India and Japan possibly as early as the 16th century by Portuguese navigators and clergy who established missionaries in the western Indian state of Goa and in the southern Japanese island of Kyushu where the disease is found today.

MOLECULAR GENETICS OF JOSEPH DISEASE

The primary molecular defect in autosomal dominant spinocerebellar degeneration is unknown. In general, the molecular defect in most dominantly inherited diseases has not been characterized, in contrast to the known enzyme abnormalities of recessively inherited disorders. DNA polymorphism studies have not yet been reported in human spinocerebellar degenerations, so we now describe preliminary findings of linkage studies using restriction fragment length polymorphisms (RFLP) in the Joseph disease type of spinocerebellar degeneration. The data suggest that the primary mutation for Joseph disease may reside on chromosome 1, near the amylase gene locus (1p21), from loose linkage data with the Rhesus blood group factor and pancreatic amylase polymorphism (amy-2) (Forster-Gibson *et al.*, 1984; Myers *et al.*, 1986). The data suggesting linkage between 1p loci and the mutant gene responsible for Joseph disease are weak, and efforts to look elsewhere for linkage continue (Myers *et al.*, 1987).

A modifier gene, if it cosegregates with the primary mutant gene in an individual, could explain the variation in gene expression and penetrance, as well as occasional skipped generations, that are seen so often in dominantly inherited neurological diseases. An electrophoretic variant of erythrocyte acid phosphatase (*ACP*-1) has been associated with the absence of neurological disorder in one family with Joseph disease and there may be a modifier gene close to the *ACP*-1 locus on chromosome 2 (2p23).

Linkage studies

We used two probes for the short arm of human chromosome 1 and two probes for the short arm of chromosome 2 including a rat probe pc XP-38 (from Dr R. Macdonald, University of Texas, Health Science Center, Dallas), for an amylase (*Amy*) gene sequence which maps to 1p1 (McKusick, 1983). This probe contained 1.2 kb of 5′ end of rat cDNA, and has been described in detail (MacDonald *et al.*, 1980). RFLPs of *Amy* sequences have been reported when human genomic DNAs were digested with *Msp I* or *Pst I* restriction enzymes (Skolnick, Willard and Menlove, 1984). We also found RFLP in two out of four Joseph disease (JD) families after DNA digestion with these enzymes.

Members of one of these families (W15) had either two or one *Msp I* polymorphic fragments (approximate size 4.5 and 4.6 kb) which hybridized with the *Amy* probe (*Figure 11.1*). In this family, an unaffected and a not at risk member (2) has only one 4.6 kb fragment, an unaffected and at risk member (1) has another 4.5 kb fragment, and affected members (3 and 4) have both 4.5 and 4.6 kb fragments of DNA. RFLP was not found in this family after digestion of DNA with *Pst I*.

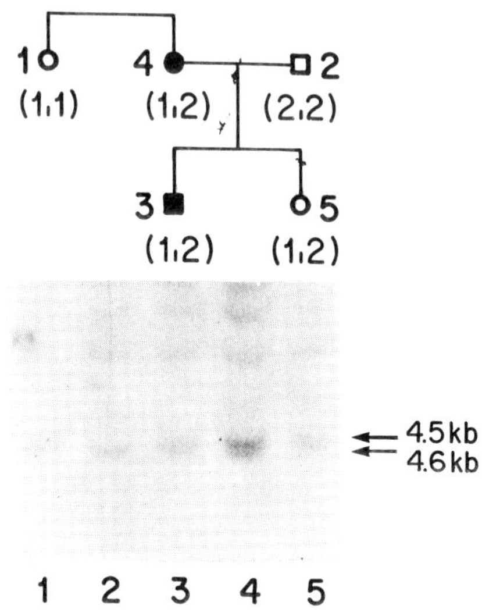

Figure 11.1 Restriction fragment length polymorphisms in a Joseph disease (RFLPs) (W15) family are shown. Solid square and circle indicate persons affected with Joseph disease. DNAs of patients have been digested with *Msp* 1 and hybridized with the *pcHPA* nick translated probe, characteristic for human *Amy* gene. Arrows and numbers indicate sizes of hybridizing polymorphic genomic DNA fragments in kilobases (kb) (*see text* for details)

However, we found RFLP after hybridization of *Amy* probes with *PstI* digested genomic DNAs from the second Joseph disease family (W2.3–7). In general, three bands with approximate sizes 6.0, 5.0, and 4.0 kb showed visible polymorphism (*Figure 11.2*). Patient 4, who is unaffected and not at risk, has only one 5.0 kb band. Her affected husband (patient 3) has two bands of 6.0 and 4.0 kb. Their affected son (patient 1) has three bands 6.0, 5.0, and 4.0 kb, similar to his affected half-brother (patient 2).

Two probes, pro-opiomelanocortin (POMC) and *N-myc*, known to be on the short arm of chromosome 2 (Skolnick, Willard and Menlove, 1984), were used in hybridization experiments with DNAs from one large family (E-2). This family demonstrates a significant reduction of affected descendants. The hybridization experiments with genomic DNAs from this family were carried out to find possible cosegregation of RFLP with the absence of disease.

The *POMC* probe showed RFLP after digestion of human genomic DNAs with *Rsa I* (Skolnick, Willard and Menlove, 1984). We also found RFLP using *POMC* probe with *Rsa I* digests from the E-2 family. Two DNA fragments of approximate size 2.5 (type 1) and 2.2 kb (type 2) are polymorphic (*Figure 11.3*). Two members of this family not at risk or unaffected, patient 3 and her daughter (patient 15), are homozygous for the 2.5 kb band; three patients, children of an affected mother (5,

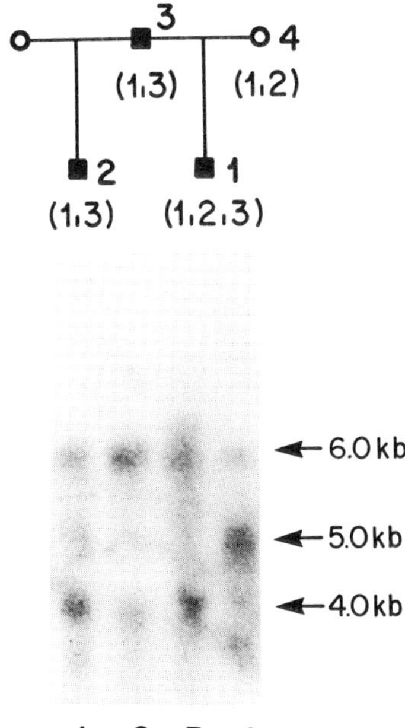

Figure 11.2 Restriction fragment length polymorphisms in Joseph disease (RFLPs) (JD) (W2.3–7) family are shown. Solid squares indicate affected Joseph disease patients. DNAs of patients have been digested with *Pst* I and hybridized with the *pcHPA* nick translated probes characteristic for the human *Amy* gene. Arrows and numbers indicate sizes of hybridizing polymorphic genomic DNA fragments in kb (*see* text for details)

6, and 8), are homozygous for the 2.2 kb band; and ten family members are heterozygous for both bands. Four members (1, 2, 7, and 9) among these ten are affected and the rest of them are potentially at risk (*see Figure 11.3*). Inheritance of 2.5 kb (type 1) and 2.2 kb (type 2) DNA fragments is mendelian and they cosegregated with A and B type of *ACP*-1. *POMC* and *ACP*-1 genes are closely linked and situated in the same region (2p23) of chromosome 2.

It is of interest to follow the cosegregation of these two genes with the appearance of Joseph disease in this family. Four affected members of the family (1, 2, 7, and 9) have identical genotypes for *ACP*-1 (BB) and for *POMC* (1,2) genes (*see Figure 11.3*).

We did not find RFLP after digestion of genomic DNAs with either *EcoRI* or *Taq I* and hybridization with the *N-myc* probe. Only one fragment of 1.9 kb hybridized with the *N-myc* probe after digestion with *Eco RI*. Two *Taq I* fragments, approximate size 4 and 3.8 kb, hybridized with the *N-myc* probe.

Modifier (*Mo*) gene

Joseph disease is clearly transmitted as an autosomal dominant trait as can be seen from the pedigrees shown in *Figures 11.1, 11.2* and *11.3*. The proportion of affected children of heterozygous (affected) parents should be close to 50%. However, in the 164 families with 694 descendants at risk, only 266 of descendants (38.3%) aged over 40 years had developed disease, which is significantly smaller ($p < 0.001$) than the expected number (347) of affected descendants. The observed reduction of the number of the affected descendants in the investigated Joseph disease families is a

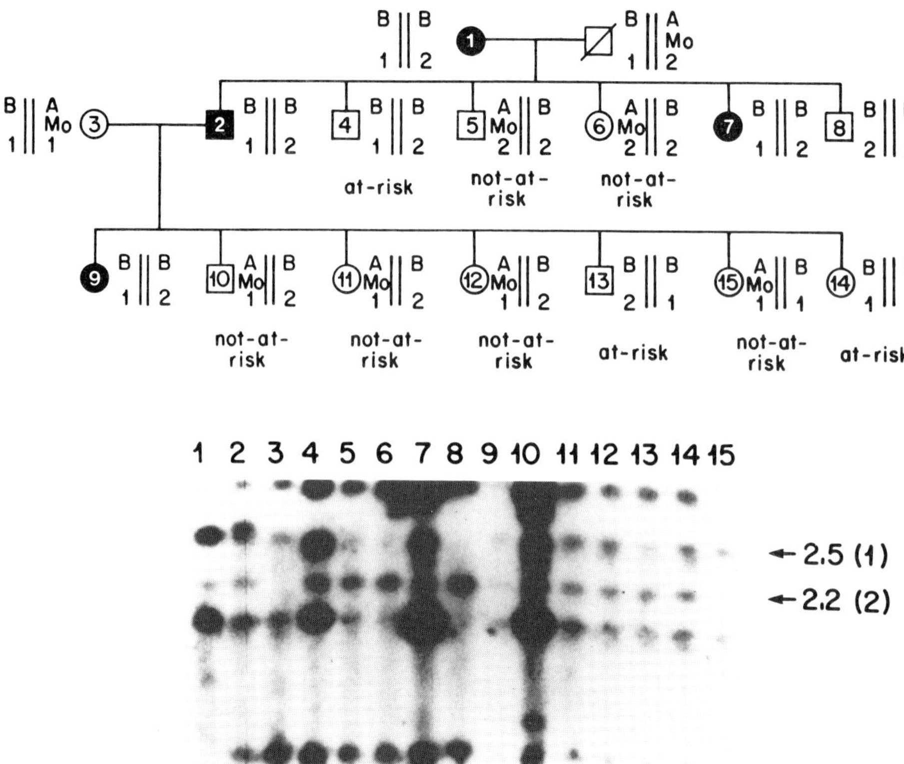

Figure 11.3 Erythrocyte acid phosphatase type and restriction fragment length polymorphisms (RFLPs) are shown for Cordeiro family members. A portion of this family's pedigree is shown here. Solid circles and squares indicate patients with Joseph disease. Patient numbers correspond to the patient's position on the Southern blot. DNAs were digested with *Rsa* 1 and hybridized with the nick translated *POMC* probe. Two DNA fragments, 2.5 kb (1) and 2.2 kb (2), are polymorphic. Combinations of alleles for *ACP*-1 (A and B) and RFLP (1 and 2) and their association with the short arm of the chromosome 2 are shown for each individual. Genotypes of *ACP*-1 and RFLPs and the combination of alleles for the deceased person have been deduced from the segregation of these alleles among his children and grandchildren (*see* text for details)

result of the reduced penetrance of the disorder in some families. Such families represented approximately 10% of the total population. The Cordeiro family is an example of a family with a statistically significant high number of non-affected children. Using the curve of cumulative distribution for age of onset, we calculate that six children of 16 at risk should have developed the disease. However, only three have developed it. Part of the Cordeiro family pedigree is shown in *Figure 11.3*.

Erythrocyte acid phosphatase-1 (*ACP*-1) was investigated for its three alleles (A, B, C) in the Cordeiro family members. Two alleles, A and B, have been found in this family as seen in *Figure 11.3*. In the Cordeiro family, four affected persons have a BB phenotype of *ACP*-1. Nine children with a BA phenotype have not yet developed Joseph disease, as well as four with a BB genotype. The actual absence of the affected descendants with BA phenotype of *EAP*-1 is apparently not a matter of chance ($0.1 > p > 0.05$), and this observation leads to the conclusion that the BA phenotype of *EAP*-1 is linked to the absence of Joseph disease expression.

CONCLUSIONS

An understanding of the molecular bases of dominantly inherited neurological diseases will require use of recombinant DNA techniques. DNA polymorphisms associated with a dominantly inherited disorder in an individual family would provide molecular markers to identify at risk persons who are affected. Major achievements in this regard have been recently made by identifying linked DNA probes, and the precise chromosomal assignments for Huntington's disease and X-linked muscular dystrophy (Gusella *et al.*, 1983; Bakker *et al.*, 1985).

The findings on Southern blots using leukocyte DNA from two families with Joseph disease hybridized to the rat or human amylase cDNA probes are interesting. They were made possible by initially finding loose linkage between the Joseph disease locus and the Rhesus factor mapped to chromosome 1p32-1 pter and the *Amy*-2 locus at 1p21 (Forster-Gibson *et al.*, 1987; Myers *et al.*, 1986). The human amylase probe assigned to the short arm of chromosome 1 (1p21), provided a pattern on Southern blots from two Joseph disease families in which there was appropriate segregation of fragments in affected individuals but not in those not at risk. In family 1 (W-15), two affected persons had amylase cDNA hybridization to 4.5 and 4.6 kb fragments and a not at risk person showed only a 4.6 kb fragment hybridization signal (*see Figure 11.1*). This implies that the Joseph disease gene resides on the chromosome which yielded the 4.5 *Msp I* fragment. In family 2 (W2.3-7), three affected persons had amylase cDNA probe hybridization as follows: (a) 6.0 and 4.0 kb; (b) 6.0, 5.0, and 4.0 kb fragments; and (c) 6.0 and 4.0 kb fragments. A not at risk family member had only a 5.0 kb hybridization signal (*see Figure 11.2*). This result suggests that in this family the mutant Joseph disease gene may reside on a chromosome that yielded 6.0 and 4.0 kb *Pst I* fragments. In both families, at risk persons had patterns like affected patients. If there is linkage, this would suggest that these at risk individuals will develop clinical disease in the future. They will be followed carefully to monitor their neurological status. The finding of this DNA polymorphism pattern utilizing the cDNA probe for amylase indicates that the Joseph disease locus may be linked to the amylase locus. These studies will be extended to additional families and additional chromosome 1 probes to establish lod scores and to determine whether the Joseph disease gene is situated in this part of genome. It is of interest to note that one family with autosomal dominant Charcot–Marie–Tooth disease has also been mapped to chromosome 1 due to linkage with the Duffy locus (Bird, Ott and Giblett, 1982).

Genetic diseases whether recessive, dominant, or X-linked refer to a single gene defect. When another gene has the ability to modify the penetrance or expressivity of the primary gene responsible for the clinical entity, then the second gene is referred to as a modifier gene. We have found evidence for a modifier gene in Joseph disease. It may explain the variation encountered in penetrance and expressivity which is typical of many common dominantly inherited neurological disorders. The occurrence of skipped generations in large pedigrees may also be due to the presence of a modifier. The concept of a modifier gene is used here to explain the full variance of disease expression seen in patients with Joseph disease.

No distinction is made here between gene modifiers affecting Joseph disease gene penetrance and expressivity and an epistatic gene which totally suppresses expression of the mutant Joseph disease gene.

An association of an electrophoretic form of erythrocyte acid phosphatase (*ACP*-1) and absence of clinical disease was seen in one of our Joseph disease families (*see Figure 11.3*), so the postulated modifier gene may be linked to the *ACP*-1 locus and cosegregate with it. This implies that those persons who inherit the modifier gene (linked and identified by a form of *ACP*-1) and the mutant Joseph disease gene will have minimal or no clinical disease. The modifier gene product, it is presumed, is able to correct or stabilize the mutant Joseph disease locus product and thus prevent the expression of the genetic disorder in genotypically affected persons. The gene map location for *ACP*-1 is 2p23 which is also the map location for the *POMC* locus. It is of considerable interest to note that at the 2p23 region in Joseph disease patients that there is polymorphism present for *ACP*-1 and *POMC*. However, the *N-myc* DNA probe gave no polymorphism. We think that this is an important internal control, as the *N-myc* gene also has a chromosome map assignment in the 2p23 region and thus emphasizes the specificity of the *ACP*-1 and *POMC* polymorphisms.

The Cordeiro family (*see Figure 11.3*) represents a Joseph disease family with a reduced number of affected descendants. Segregation of three genetically inherited traits (*JD, ACP*-1, and *POMC*) have been studied. Alleles of *ACP*-1 are designated as A and B, and 2.5 and 2.2 kb *Rsa I* fragments of DNA hybridizing with the *POMC* probe are designated as 1 and 2, respectively.

If we suppose that the *Mo* gene is located on chromosome 2, then the affected patient 9 inherited the $^{B}_{2}$ chromosome from her father (patient 2) who is affected and apparently does not have a modifier gene, and the $^{B}_{1}$ chromosome without the *Mo* gene from her healthy mother (patient 3). Similarly, the paternal chromosome $^{B}_{1}$ of the affected patient 2 does not carry the *Mo* gene. Therefore, the homologous second chromosome $^{A}_{2}$ of his deceased healthy father is the chromosome which would carry the *Mo* gene. His brother (4) and sister (7) with paternal $^{B}_{1}$ chromosomes, which is without the *Mo* gene, are either at risk or affected. His brother (5) and sister (6) with $^{A}_{2}$ paternal chromosomes are not at risk for the disease expression, since the $^{A}_{2}$ chromosome might carry a *Mo* gene, although they could be carriers of Joseph disease gene. Patient 8 carries the $^{B}_{2}$ maternal chromosome and a crossover paternal chromosome $^{B}_{2}$. Clinical status of this patient depends where the hypothetical *Mo* gene is located, that is, closer to *EAP* or to the *POMC* gene.

As we indicated above, the $^{B}_{1}$ chromosome of 3, a healthy spouse, does not carry the *Mo* gene since her daughter who inherited $^{B}_{1}$ chromosome (patient 9) is affected. However, if patient 3 does carry a *Mo* gene, it should be situated on the chromosome $^{A}_{1}$. Therefore, all her children with the $^{B}_{1}$ maternal chromosome (13 and 14) are either at risk for Joseph disease or affected (9). At the same time, her children (10, 11, 12 and 15) are not at risk to develop the disease, although they might be carriers of the Joseph disease gene. We are following descendants not affected at the present time in this family to see who will develop Joseph disease.

It is our view that the primary mutant Joseph disease gene product encoded on chromosome 1 interacts with genes at the 2p23 region. A modifier gene linked to *ACP*-1 and *POMC* has its product interact with the Joseph disease gene product to correct its altered biological phenotype and thus minimize disease expression. The presence of a modifier gene in Joseph disease to explain variation in penetrance and expressivity of disease is a concept that needs further study as applied to all dominantly inherited neurological disorders. It is a more precise and directly assayable postulate rather than explaining clinical variation as being due to normal background genetic heterogeneity involving many genes.

It is our view that the primary mutation for Joseph disease may reside on chromosome 1p and for some families with olivopontocerebellar atrophy (OPCA) on chromosome 6 near the HLA complex. We suggest the Joseph disease locus is influenced by a postulated modifier gene locus at 2p23 which is responsible in part for variation in disease expression.

It is hoped these data and ideas will provide additional approaches and insights into the pathogenesis of Joseph disease and OPCA. The use of leukocyte genomic DNA to obtain DNA polymorphisms on Southern blots for Joseph disease appears most promising. A molecular marker for this disease may become available in the near future.

References

BAKKER, E., HOFKER, M. H., GOOR, N. *et al.* (1985) Prenatal diagnosis and carrier detection of Duchenne muscular dystrophy with closely linked RFLPs. *Lancet,* **i,** 655–658

BIRD, T., OTT, J. and GIBLETT, E. (1982) Evidence for linkage of Charcot–Marie–Tooth neuropathy to the Duffy locus on chromosome 1. *American Journal of Human Genetics,* **34,** 388–394

BHARUCHA, N., BHARUCHA, E. and BHABHA, S. (1986) Machado–Joseph Azorean disease in India. *Archives of Neurology,* **43,** 142–144

COUTINHO, P. and ANDRADE, C. (1978) Autosomal dominant system degeneration in Portuguese families of the Azores Islands. *Neurology,* **28,** 703–709

FORSTER-GIBSON, C., MYERS, S., SIMPSON, N., ROSENBERG, R., SEQUEIROS, J. and MacLEOD, P. (1984) Investigation of linkage in 12 kindreds with Machado–Joseph disease. *American Journal of Human Genetics,* **36,** 9S

GOTO, I., TOBIMATSU, S., OHTA, M., HOSOKAWA, S., SHIBASAKI, H. and KUROIWA, Y. (1982) Dentatorubropallidoluysian degeneration: clinical, neuro-ophthalmologic, biochemical, and pathologic studes on autosomal dominant form. *Neurology,* **32,** 1395–1399

GUSELLA, J., WEXLER, N., CONNEALLY, P. *et al.* (1983) A polymorphic DNA marker genetically linked to Huntington's disease. *Nature (London),* **306,** 234–238

HEALTON, E., BRUST, J., KERR, D., RESOR, S. and PENN, A. (1980) Presumably Azorean disease in a non-Portuguese family. *Neurology,* **33,** 74–80

LIMA, L. and COUTINHO, P. (1980) Clinical criteria for diagnosis of Machado–Joseph disease: report of a non-Azorean Portuguese family. *Neurology,* **30,** 319–322

MacDONALD, R., CREVAR, M., SWAIN, W., PICTET, R., THOMAS, G. and RUTTER, W. (1980) Structure of a family of rat amylase genes. *Nature (London),* **287,** 1–5

McKUSICK, V. A. (1983) *Mendelian Inheritance in Man,* 6th edition, p. xl. Baltimore: Johns Hopkins University Press

MYERS, S., MacLEOD, P., FORSE, R., FORSTER-GIBSON, C. and SIMPSON, N. (1986) Machado–Joseph disease: linkage analysis between the loci for the disease and 18 protein markers. *Cytogenetics and Cell Genetics* **43,** 226–228

NAKANO, K., DAWSON, D. and SPENCE, A. (1972) Machado disease; a hereditary ataxia in Portuguese immigrants to Massachusetts. *Neurology,* **22,** 49–55

ROMANUL, F. C. A., FOWLER, H. L., RADVANY, J. *et al.* (1977) Azorean disease of the nervous system. *New England Journal of Medicine,* **296,** 1505–1508

ROSENBERG, R. N., NYHAN, W. L., BAY, C. and SHORE, P. (1976) Autosomal dominant striatonigral degeneration. *Neurology (Minneapolis),* **26,** 703

ROSENBERG, R. N., NYHAN, W. L., COUTINHO, P. and BAY, C. (1978) Joseph disease: an autosomal dominant neurological disease in the Portuguese of the United States and the Azores Islands. In: *The Inherited Ataxias,* edited by P. Kark, R. N. Rosenberg and L. Schut, pp. 33–57. New York: Raven Press

SAKAI, T., OHTA, M. and ISHINO, H. (1983) Joseph disease in a non-Portuguese family. *Neurology,* **33,** 74–80

SKOLNICK, M., WILLARD, H. and MENLOVE, L. (1984) Report of the committee on human gene mapping by recombinant DNA techniques. Human gene mapping 7. *Cytogenetics and Cell Genetics,* **37,** 210–273

WOODS, B. T. and SCHAUMBURG, H. (1972) Nigro-spino-dentatal degeneration with nuclear ophthalmoplegia. *Journal of Neurological Science,* **17,** 149–166

YUASA, T., OHAMA, E., HARAYAMA, H. *et al.* (1986) Joseph's disease: clinical and pathological studies in a Japanese family. *Annals of Neurology,* **19,** 152–157

12
Huntington's disease

T. Conrad Gilliam and James F. Gusella

INTRODUCTION

Huntington's disease is an autosomal dominant disorder affecting the central nervous system. It is characterized by abnormal movements and dementia which begin in midlife and pursue an inexorable progression leading to complete disability and death, usually 15–20 years after the initial onset of symptoms. Neuropathological changes result from premature nerve cell death which occurs throughout the brain, but is most prominent in the striatum. In recent years, anatomical and neurochemical categorization of the neuronal elements primarily affected in Huntington's disease has improved, but the underlying mechanism of cell death remains unknown. A tragic aspect of Huntington's disease is that the late age of onset often results in the defect being transmitted unknowingly to offspring prior to the onset of symptoms in the parent. The discovery of a genetic marker closely linked to Huntington's disease has now made it possible to provide reliable presymptomatic diagnosis in offspring of some affected individuals, although predictive testing raises significant psychological, ethical, and legal issues that have yet to be fully addressed. Of perhaps greater significance, the availability of a linked marker has focused molecular strategies for identification and subsequent characterization of the Huntington disease gene based on its chromosomal map location.

DESCRIPTION OF THE DISEASE

Symptoms and signs

Although the characteristic movement disorder of Huntington's disease was reflected in the former name for the disease, Huntington's chorea (Huntington, 1872; Hayden, 1981), some of the earliest signs may involve subtle emotional and psychological manifestations. Personality changes frequently precede the movement disorder by a number of years but cannot themselves be considered diagnostic of Huntington's disease (Myers and Martin, 1982; Bird, 1980). Often they are reflected in chronic depression or increased irritability first noticed by family members. The motor abnormalities in Huntington's disease usually begin as minor adventitious movements which may be hidden by incorporation into apparently purposeful acts. Over time, however, the movement disorder gradually progresses until the entire body is consumed in chorea. Huntington's disease can also manifest as a rigid form, which predominates among early onset cases but is rare in adults (Bittenbender and Quadfasel, 1962). As the disease progresses, memory and cognitive function are also impaired, accompanying the complete physical debilitation. Some 15–20 years after onset of Huntington's disease, the patient usually succumbs to a secondary complication of the disorder, such as pneumonia. The devastating loss of control of both physical and mental faculties caused by Huntington's disease is the frightening but uncertain fate that 'at-risk' individuals must confront as they ponder their future. This prospect, combined with the psychological effects of the Huntington's disease gene itself, undoubtedly contributes to the increased rate of suicide observed in this disease (Schoenfeld *et al.*, 1984).

Prevalence

The prevalence of Huntington's disease is five to ten per 100 000 in Caucasian populations, and is thought to be lower (though is less well documented) in non-Caucasians. There is a suggestion that the origin of the Huntington's disease gene in Africans and orientals may derive from a Caucasian ancestor. In some instances, Huntington's disease is reported to be very rare, while some areas have an unusually high prevalence of the disorder. Hayden and Beighton (1982) reported only three affected blacks in a population of 19 million in Africa. In Tasmania, the disease appears at a rate of 174 per million (Brothers, 1949). The most intensively studied example of a focal concentration of Huntington's disease is the Lake Maracaibo region in Venezuela. This population, first described by Negrette in 1963, formed the major resource responsible for successful linkage of a DNA marker to the Huntington's disease gene locus and is described in more detail below (p. 170).

Age on onset

The onset of symptoms of Huntington's disease is variable even among members of the same family, with a mean of 35–42 years (Brothers, 1964; Hayden, 1981; Myers and Martin, 1982) based upon the standard criterion for detecting abnormal movements. Although the majority of cases have onset in middle age, 6% present before the age of 21 and 28% after 50 years of age. Some researchers feel that age at death is more reliable than age of onset for pedigree and population comparisons in Huntington's disease due to the subjective nature of diagnosis early in the disorder. Definite differences in disease phenotype can be discerned in cases of early versus late onset of Huntington's disease. In approximately 3% of patients, the disorder begins before the age of 15 and is characterized by intellectual decline, seizures, rigidity, myoclonus, or dystonic movements, rather than chorea (Bittenbinder and Quadfasel, 1962). This form of the disease progresses rapidly, and in most instances is transmitted by the father (Bird, Caro and Pilling, 1974; Hayden *et al.*, 1982). In a minority of adult onset cases, choreiform movements begin as late as the sixth to eighth decade of life and progress slowly. Cognitive impairment is also less pronounced, and the disease may continue for 20–30 years before the patient dies from secondary complications or causes unrelated to the disease itself. In a New England study, significantly more of the 68 individuals with onset after 50 years of age inherited the disease gene from their mother (Myers *et al.*, 1983).

The variability in age of onset has engendered considerable speculation to explain the observed differences. Two of the more obvious possibilities are allelic heterogeneity (different independent disease alleles at the same locus) and non-allelic heterogeneity (multiple different gene loci causing similar symptoms). The latter is an important issue with respect to predictive testing using linked markers. If either type of heterogeneity were responsible for variability in the manifestation of the disorder, one would expect less variation within than among families. Went *et al.* (1975) described two families from the Netherlands who reside in close proximity to each other and who, over the past seven to eight generations, have produced 44 and 78 affected members, respectively. The average age at death was significantly different between the two pedigrees with a mean of 47.1 years for the first, and 63.9 years for the second. In a more extensive investigation, Went *et al.* (1983) studied 1100 patients with Huntington's disease from 102 families, comparing the variation in age at death both within and among the different families. The authors argue that variation in the mean age at death is greater between families than within families, suggesting more than one genetic form of Huntington's disease.

In recent years, the Venezuela disease pedigree, comprising over 5000 individuals from a single Huntington's disease pedigree living around Lake Maracaibo, has become a major genetic resource for the study of Huntington's disease. More than 100 living affected individuals in the family have all inherited the same defective gene from a common ancestor. Approximately 900 others are at 50% risk for inheriting the disease. Age of onset in the Venezuela disease pedigree varies widely, from several juvenile cases to others with initial manifestation of the disease late in life. Thus variation in age of onset within the Venezuela pedigree occurs without heterogeneity of the primary defect.

Factors that might affect age of onset have been reviewed (Conneally, Gusella and Wexler, 1986). Finch (1980) proposed that genetic modifiers in the form of ageing genes might be responsible for the variation in age of onset and suggested that the increase in Huntington's disease in midlife is due to an interaction of the defect with normal age related changes within the basal ganglia. According to this hypothesis, individuals with a 'superior' set of ageing genes would show symptoms of Huntington's disease later than those who carried an 'inferior' set of genes. A test of the theory would be to compare age of death among affected and unaffected members of the same family. When Farrer and Conneally (1985) evaluated 214 families with normal and affected sibs, their results were consistent with the hypothesis that 'ageing genes' may account for much of the variability in age of onset in Huntington's disease.

A second explanation for variable age of onset which has received much attention is the maternal protection model (Boehnke, Conneally and Lange, 1983; Myers *et al.*, 1983). The majority of individuals who suffer juvenile onset of Huntington's disease (symptoms appear before the age of 20) inherit the disease from their father. This phenomenon, which was first reported by Merritt *et al.* (1969) was later studied more thoroughly by Farrer and Conneally (1985), who found that 23 of 28 juvenile-onset offspring had fathers with Huntington's disease and that 1.6 times more late onset offspring were born to affected mothers than to affected fathers. Boehnke, Conneally and Lange (1983) and Myers *et al.* (1983) have postulated a maternally transmitted factor, likely residing in the cytoplasm, which delays onset in offspring from affected mothers. The factor could be encoded by a mitochondrial gene and might be expected to have a reproductive advantage in Huntington's disease since late onset cases would be more likely to procreate. An alternative explanation for the lack of juvenile onset progeny from mothers with Huntington's disease is that the intrauterine environment might alter development of the fetus in such a way as to delay the onset of the disease. This possibility will not be amenable to investigation until a true animal model of Huntington's disease is available.

Neuropathology

Postmortem examination of Huntington's disease patients indicates that development, as well as connectivity, of the cells of the central nervous system is normal, but that the mutant gene promotes the premature death of selective neurons within the brain. Other tissues outside the central nervous system are either unaffected or effectively normal because they are able to regenerate. The area of the brain apparently affected first and ultimately most severely is the striatum (caudate and putamen).

Important strides towards understanding the pattern of cell death in Huntington's disease have resulted from the identification by staining techniques and Golgi impregnation procedures of five types of neurons in the normal human striatum (DiFiglia, Pasik and Pasik, 1976, 1980; Graveland, Williams and DiFiglia, 1985; Graybiel, 1986). In addition to their distinctive morphologies, these neurons can also be identified based upon their content of the more than 30 possible neurotransmitters now known to be present in this region of the brain (DiFiglia, Pasik and Pasik, 1980; Ferrante *et al.*, 1985; Kowall, Ferrante and Martin, 1987). The five recognizable cell types differ dramatically in their susceptibility to Huntington's disease, with the earliest evidence for effects of the defect occurring in altered shape of medium-sized spiny neurons (Graveland, Williams and DiFiglia, 1985).

Neuronal loss is not restricted to discrete neuroanatomical regions of the brain. Although the striatum is the most severely affected, cell death in the globus pallidus, cerebral cortex, thalamic and subthalamic nuclei, and in some cases brainstem and spinal cord, is also apparent (Bruyn, Bots and Dom, 1979). In advanced cases, 20–30% loss in total brain weight is not uncommon. A study of 163 people who died of Huntington's disease was conducted to correlate clinical manifestations of the disease with the degree of neuropathological change in the striatum (Vonsattel *et al.*, 1985). In a few cases, unambiguous clinical symptoms were documented in patients with no gross evidence of neuropathological change, suggesting that cell death may lag behind impairment of neuronal function.

Detailed investigations of neurotransmitter levels in postmortem Huntington's disease brains have been intensively pursued, partly in hopes of developing a corrective therapy (Martin and Gusella, 1986). *Figure 12.1* summarizes the neuronal circuitry of the basal ganglia. The levels of a number of neurotransmitters as well as biosynthetic enzymes and receptor-binding sites are abnormal in

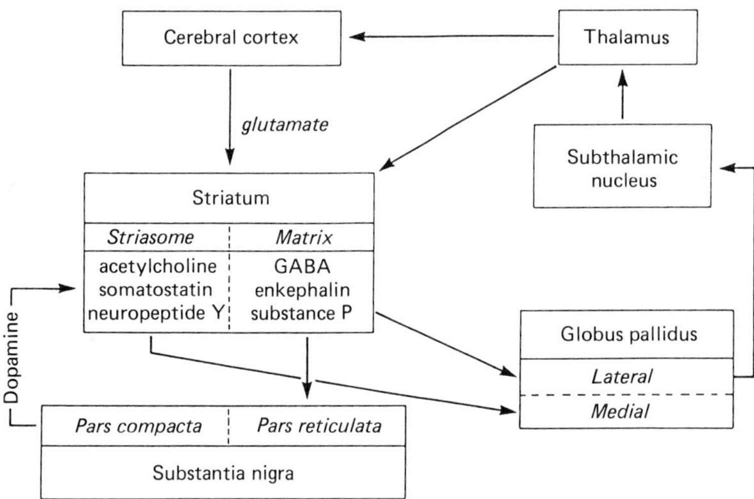

Figure 12.1 Principal circuitry of the basal ganglia. The striatum receives innervation from the cerebral cortex (glutaminergic), substantia nigra (pars compacta; dopaminergic), as well as from the brainstem (multiple neurotransmitters). Within the striatum, local interneurons contain acetylcholine, somatostatin, and neuropeptide Y while efferent neurons contain GABA, enkephalin, and substance P and project to the substantia nigra (pars reticulata), and globus pallidus. Glutamate, acetylcholine, and substance P have excitatory effects upon their target neuronal pathways, while dopamine and GABA exert inhibitory effects

Huntington's disease (Martin and Gusella, 1986; Bird, 1980). The levels of gamma-aminobutyric acid (GABA), enkephalin, substance P, and dynorphin are all decreased in the globus pallidus and substantia nigra of postmortem Huntington's disease brains. Acetylcholine levels are decreased in striatal tissue while somatostatin and neuropeptide Y levels are relatively increased. A major abnormality at the cellular level is the loss of GABA-containing neurons. The activity of glutamic acid decarboxylase is reduced by 85% in the striatum (Kuhl *et al.*, 1982) and 60% in the substantia nigra and globus pallidus (Bird *et al.*, 1973). The number of receptors for GABA is also reduced (Bird, 1980; Graybiel, 1986).

Characterization of specific cell types within the striatum in terms of neuropeptide and neurotransmitter content has recently become possible through the combination of Golgi impregnation, immunocytochemistry, and electron microscopy (Graybiel, 1986). The most abundant cells in the striatum are the medium-sized spiny neurons which appear to contain GABA, enkephalins, dynorphin, and substance P. Aspiny neurons, which are relatively spared in Huntington's disease, contain somatostatin and neuropeptide Y (DiFiglia and Aronin, 1982; DiFiglia, Aronin and Martin, 1982; Graybiel, 1986). Together with previous neurochemical investigations, these studies suggest that medium-sized GABAergic spiny neurons are the first cells to malfunction and die in Huntington's disease.

In view of studies in monkeys which demonstrate that only specific lesioning of the subthalamic nucleus is effective in eliciting chorea, Penney and Young (1986) have proposed that the chorea in Huntington's disease is initially an indirect result

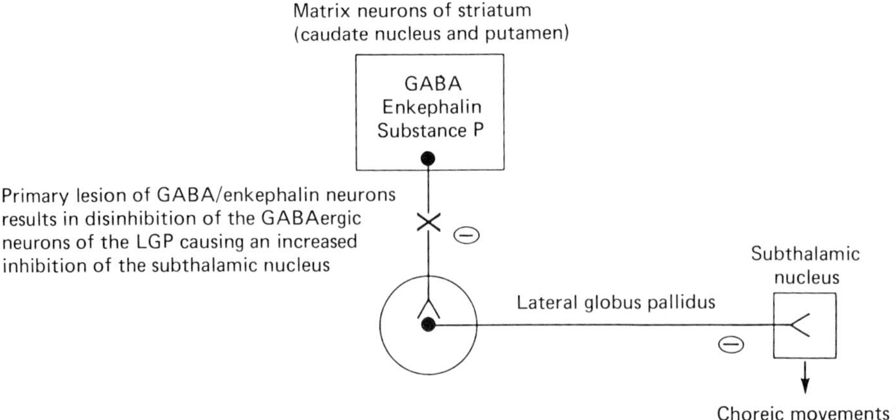

Figure 12.2 Hypothetical lesion promoting choreic movement disorder. The model, put forward by Penney and Young (1986), attempts to relate the prominent loss of GABA-containing neurons in the basal ganglia of Huntington's disease patients and the onset of chorea. As the normally inhibitory GABAergic neurons are selectively destroyed in the striatal matrix, the disinhibited efferent neurons of the lateral globus pallidus exert an increased inhibitory effect on the subthalamic nucleus. This could mimic a specific lesion in the latter region, which is known to elicit chorea in animal studies (Crossman, 1987)

of the loss of GABAergic neurons in the striatum. As outlined in *Figure 12.2*, selective loss of striatal GABAergic neurons would be expected to produce disinhibition of efferent inhibitory GABAergic neurons in the lateral globus pallidus. Since the latter innervate the subthalamic nucleus, increasing the inhibitory input to this area might mimic the effect of a specific lesion, producing chorea. Other aspects of the movement disorder in Huntington's disease (for example, dystonic movements, rigidity, and bradykinesia) could then result from secondary degeneration of neuronal elements in the basal ganglia leading to more global dysfunction as the disorder progresses.

Carter (1986) has recently put forward a proposal to explain the selective cell death, late onset and slow progression of Huntington's disease, based on the types of neurons which are primarily affected and on evidence for altered glucose metabolism in specific regions of Huntington's disease brains. Carter (1986) views the pathological changes in the disease as the results of a 'system degeneration' with the extent of the lesion determined by the presence of the disease gene, by a variable degree of vulnerability of different neurotransmitter systems, and by the functional characteristics of these systems. In this hypothesis, the Huntington's disease gene defect would alter energy metabolism in all neurons of affected individuals, with the effect on specific neurons resulting from their particular vulnerability to this alteration. Carter argues that the types of neurons which appear to be selectively affected in Huntington's disease are precisely those which would be most vulnerable to a defect in glucose metabolism. GABA, glutamate, and acetylcholine are all neurotransmitters which can be synthesized from glucose. To preserve glucose economy, cerebral metabolism salvages the carbon skeletons of these molecules for return to the Kreb's cycle (GABA shunt). This is not true for the catecholamines or 5-hydroxytryptophan, which are synthesized from dietary amino acids and whose associated neurons are apparently not affected in Huntington's disease.

In addition, metabolically active neurons would be more vulnerable to a gene defect which alters general brain energy metabolism. Unusually active cells would include small neurons with a relatively large surface area-to-volume ratio, neurons responsive to glutamate because of demands associated with excessive excitation, and neurons which are in continuous use such as those associated with voluntary movement. Carter proposes, therefore, that the concentration of lesions within the basal ganglia may simply reflect the fact that all of the proposed criteria conferring vulnerability are fulfilled by neurons in this brain region. Positron emission tomography has shown that glucose utilization is decreased in the caudate and putamen in Huntington's disease patients while total cerebral glucose consumption does not appear to be altered (Kuhl *et al.*, 1982, 1984). The argument put forth by Carter is that abnormal glucose metabolism is a logical explanation for the neuropathology of Huntington's disease, with the eventual cause of neuronal death in the disease being akin to hypoglycaemia or energy failure. While this hypothesis is speculative, it does attempt to explain the abundant neuroanatomical data which have recently begun to unravel the detailed cellular pathology of Huntington's disease. It may soon become testable as molecular genetic strategies guide the way to the eventual isolation and characterization of the defective DNA sequences which cause Huntington's disease.

THE MOLECULAR GENETIC APPROACH TO HUNTINGTON'S DISEASE

The application of recombinant DNA technology has had a dramatic impact on the study of many inherited human diseases for which traditional approaches have failed. Any disease with a well defined pattern of inheritance but whose primary defect has not been identified by standard biochemical and cell biological techniques, has become an excellent target for a new strategy being applied by molecular geneticists. The approach reverses the traditional route based upon defining a biochemical abnormality before direct isolation and characterization of the disease gene is possible. Instead, this strategy, which has been termed 'reverse genetics', relies on isolation of the disease gene based on its chromosomal location in order to use the defective DNA sequence to predict the protein defect. The genetic loci for Duchenne muscular dystrophy (Murray *et al.*, 1982; Davies *et al.*, 1983), Huntington's disease (Gusella *et al.*, 1983; Folstein *et al.*, 1985; Harper *et al.*, 1985; Haines *et al.*, 1986), cystic fibrosis (Tsui *et al.*, 1985; Knowlton *et al.*, 1985; White *et al.*, 1985; Wainwright *et al.*, 1985), adult polycystic kidney disease (Reeders *et al.*, 1985), familial Alzheimer's disease (St George-Hyslop *et al.*, 1987), and von Recklinghausen neurofibromatosis (Seizinger *et al.*, 1987), have all been linked to DNA markers making identification of the respective disease genes an imminent prospect. A portion of the Duchenne muscular dystrophy gene has already been cloned (Monaco *et al.*, 1986) and efforts to identify the related protein are underway, completing the cycle of the reverse genetics strategy where the protein which is altered in a genetic disease is identified by first searching out the causal DNA mutation. With the relevant DNA sequence in hand, it will be possible to study the normal function of the gene and the nature of the disease defect in any number of experimental systems. Understanding disease mechanisms will offer the hope of effective therapies while increasing our understanding of the mechanisms controlling normal and abnormal gene expression in the nervous system.

Huntington's disease was recognized several years ago as an ideal candidate to test the still theoretical approaches for chromosomal mapping using restriction fragment length polymorphisms (RFLPs). In Huntington's disease, dominant transmission, a low mutation rate, and complete penetrance makes tracing the inheritance of the defect relatively straightforward, although the late age of onset limits the number of informative generations in any given pedigree. To guarantee the success of the endeavour, large, well-characterized Huntington's disease pedigrees were sought and permanent cell lines were established for informative individuals. Foremost among the families investigated was the extended Venezuela disease kindred (Gusella *et al.*, 1983). With the pedigree resources guaranteeing eventual success, a DNA marker linked to Huntington's disease was discovered early in the effort. In 1983, the DNA probe G8 defining an anonymous DNA locus termed D4S10 was found to be linked to the disease gene which was thereby assigned to chromosome 4 (Gusella *et al.*, 1983). This represented the first chromosomal assignment of a previously unmapped autosomal defect using linkage to DNA markers and has subsequently been repeated for several other important disorders.

Linkage to the Huntington's disease locus

The discovery of linkage between G8 and Huntington's disease was based on the detection of two *HindIII* RFLPs by a randomly chosen single copy clone of human DNA (Gusella *et al.*, 1983). Four combinations of alleles, or haplotypes, are possible from the two varying *HindIII* sites, resulting in over 60% of individuals in the population being heterozygous (having two distinguishable alleles) for the locus. Heterozygosity in affected individuals is a prerequisite for tracing transmission of the Huntington's disease gene relative to the linked marker and determining whether a particular allele is inherited in a non-random manner along with the disease gene. To maximize the utility of the linked marker, additional RFLPs were sought, thereby increasing the overall level of heterozygosity. The D4S10 locus, originally defined by the clone G8, now consists of more than 50 kb of contiguous cloned sequence with single copy probes detecting 12 different RFLPs. The overall level of heterozygosity for D4S10 is greater than 0.95, making it informative in the vast majority of matings. *Figure 12.3* shows a map of the D4S10 locus with the three most informative RFLPs and the eight haplotypes they define. With these three polymorphisms alone, about 85% of members of the general population are heterozygous for the locus. It is important to note that different D4S10 alleles cosegregate with Huntington's disease in the various families tested. Thus, though D4S10 is tightly linked to the disease gene, predictive testing with this marker requires family screening sufficient to unequivocally determine which of the several D4S10 haplotypes is segregating with the defect in that particular pedigree.

The ultimate goal in the molecular genetic analysis of Huntington's disease is to clone and characterize the defect based on its map position. Evidence for linkage of D4S10 to Huntington's disease was first obtained from studying two large disease kindreds, an American family of Germanic ancestry and the very large Venezuela disease pedigree (Gusella *et al.*, 1983, 1986). In this initial analysis, no obligate recombination events were detected between the marker and the disease locus. Since that time we have analysed numerous additional large disease pedigrees for three reasons. First, it was necessary to optimize our estimate for the rate of

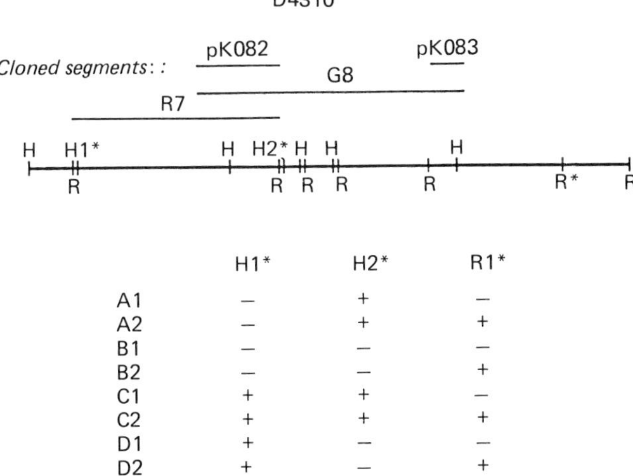

	H1*	H2*	R1*
A1	–	+	–
A2	–	+	+
B1	–	–	–
B2	–	–	+
C1	+	+	–
C2	+	+	+
D1	+	–	–
D2	+	–	+

Figure 12.3 Restriction fragment length polymorphism (RFLPs) at the D4S10 locus. The map shows the *HindIII* (*H*) and *Eco RI* (*R*) sites at the D4S10 locus. Asterisks indicate the polymorphic sites. The probes used to detect the various polymorphisms are indicated above the restriction map. G8 and R7 are overlapping genomic fragments both cloned in the bacteriophage Charon 4A. The G8 subclones, pK082 and pK083, are *Eco RI* fragments cloned in pBR328. Together, the probes can detect eight combinations, or haplotypes, of the three restriction site polymorphisms, thus making the locus informative for approximately 85% of the general population

recombination between the marker and disease loci as a measure of the distance separating them on the chromosome. Second, recombination events identified in this analysis could act as landmarks, denoting positions closer to the disease gene than D4S10. Finally, it was critical to test the possibility of non-allelic heterogeneity if D4S10 was to provide accurate predictive testing in Huntington's disease.

Figure 12.4 summarizes the linkage data between D4S10 and the Huntington's disease gene from three major studies. The data are derived from the two families mentioned above (Gusella *et al.*, 1983), several families from Wales (Harper *et al.*, 1985), one American black family, and a second American family of Germanic ancestry (Folstein *et al.*, 1985). From these pedigrees, it is clear that D4S10 displays about 4% recombination with Huntington's disease. The approximate 95% confidence interval is 1.5–6.5%. Assuming a uniform recombination rate in the human genome, this would correspond to a separation of 4×10^6 base pairs (bp). There is considerable evidence from several sources that recombination is not equal in all regions of the genome (Laurie and Hulten, 1985a, 1985b; Tanzi *et al.*, 1987), and this physical separation might well represent a significant overestimate. Two of the families used for these studies are particularly interesting because they represent opposite ends of the spectrum of symptoms seen in Huntington's disease. In the American black family, the disease has a relatively early age of onset (late 20s) with several juvenile cases and predominately motor dysfunction, while in the second American family onset is relatively late (early 60s) and accompanied by prominent psychiatric symptoms. All of the families showed positive linkage to D4S10 suggesting that they possess mutations at the same locus and that non-allelic heterogeneity does not occur in this disorder. A number of recombination events

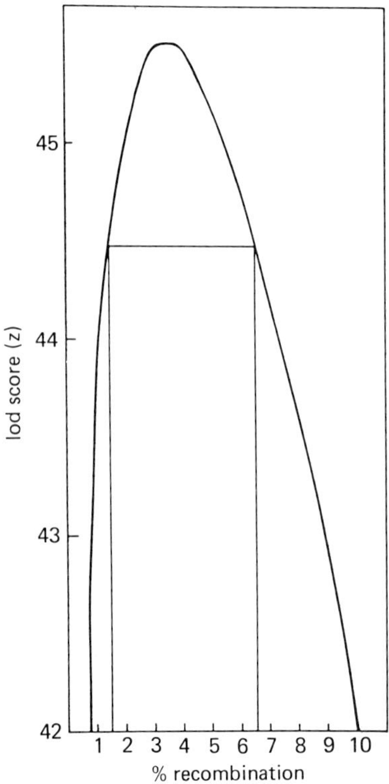

Figure 12.4 Linkage of the G8 marker (D4S10) to Huntington's disease. Linkage data for the families described in the text were analysed by the computer program LIPED (Ott, 1974, 1976). The curve describes the likelihood of linkage between D4S10 and Huntington's disease at different recombination frequencies to non-linkage. The curve peaks at 4% recombination, which is the best estimate for the genetic distance separating the two loci. The horizontal line drawn one order of magnitude below the curve peak defines the 95% confidence interval of 1.5–6.5%

were detected in this analysis and should prove useful for rapidly determining whether new markers segregate with D4S10, or with Huntington's disease in the 'crossover' affected individuals, facilitating the assignment of new loci closer to Huntington's disease.

Assessment of heterogeneity

If a genetic marker test is to be accurate for the prediction of Huntington's disease, it is essential to evaluate the degree to which non-allelic mutations give rise to the disorder. For testing to be useful, one must either know that heterogeneity does not exist, or be able to show that the test marker segregates with the disease in the family being studied, requiring a large pedigree. We have now tested 59 families with Huntington's disease for linkage to G8 (Haines *et al.*, 1986); 53 gave positive lod scores, while six small families gave insignificant negative scores. Statistical analysis of the data shows no evidence for non-allelic heterogeneity, and no difference in recombination frequency between the marker and Huntington's disease in the two sexes. The data suggest that more than 95% of Huntington's disease cases are caused by a mutation on chromosome 4, although testing of further families is necessary to increase the probability of complete homogeneity.

Characterization of homozygotes for Huntington's disease

In virtually all dominant human disorders for which homozygotes have been described, symptoms are significantly more severe in the homozygote than in the heterozygote (Pauli, 1983). These diseases are more accurately described as displaying 'incomplete dominance'. Taking advantage of both the close linkage of D4S10 to Huntington's disease and the extensive Venezuelan disease kindred, we have identified individuals who are very likely homozygous for the defective allele (Wexler *et al.*, 1987). These individuals were assessed for a range of phenotypic characteristics including age of onset, symptoms, and progression of the illness relative to typical disease heterozygotes from the Venezuelan disease pedigree. Our hope was that a difference in homozygotes compared to heterozygotes would lead to an assay for an altered protein associated with the disease phenotype. Surprisingly, no differences were detected. The study leads to two conclusions concerning the motor and cognitive dysfunction and psychiatric symptoms of Huntington's disease: first, they are not exaggerated by a double dose of the defective gene; and second, they are unaffected by the presence of the normal allele.

Mutations displaying complete dominance in *Drosophila* are thought to be associated with alterations which cause a 'gain in function' (White and Akam, 1985; Bender *et al.*, 1983). The imposition of a new function or activity upon a structural or regulatory protein would be consistent with complete dominance, whereas the elimination of normal function would be expected to show a dosage effect or even be lethal. Such a 'gain of function' might translate as overproduction of a normal protein or inappropriate production of a protein usually limited to a different tissue or to a different time of development. A mutation which leads to reduced expression would be more likely to show a dosage effect. Clearly, Huntington's disease cannot involve elimination of an essential enzyme activity since homozygosity is not lethal, but it could theoretically reduce an activity which must be strictly maintained above a threshold exceeding 50% of normal levels.

Precise physical mapping of D4S10

Chromosome 4 comprises about 6.5% of total genomic DNA. The search for additional DNA fragments near the Huntington's disease gene and for the disease gene itself would be facilitated by the precise localization of D4S10 to a small region of the chromosome. Furthermore, the generation of a cell line which could be used to identify new markers in the region around Huntington's disease would accelerate the search for the disease locus. To achieve these goals, we have used cells from patients with Wolf–Hirschhorn syndrome, a congenital anomaly involving heterozygous deletion of part of the short arm of chromosome 4 (Gusella *et al.*, 1985). Cases of Wolf–Hirschhorn syndrome are profoundly mentally retarded and have a characteristic craniofacial appearance including microcephaly, arched eyebrows, ocular hypertelorism, a broad malformed nose, lobeless ears, and micrognathia. The loss of chromosome 4 material can be the result of a *de novo* deletion or due to the inheritance of an unbalanced translocation, but most commonly involves the 4p16 band. D4S10 has been mapped to this terminal band of the chromosome both by detecting its deletion from Wolf–Hirschhorn syndrome patients (Gusella *et al.*, 1985), and by *in situ* hybridization of G8 to metaphase chromosome spreads (Wang *et al.*, 1985; Magenis *et al.*, 1986; Landegent *et al.*, 1986; Zabel *et al.*, 1986; MacDonald *et al.*, 1987).

Recently, we have constructed somatic cell hybrid lines containing derivative chromosomes 4 from Wolf–Hirschhorn syndrome patients as well as other individuals who contain translocation or deletion chromosomes (MacDonald *et al.*, 1987). Hybridization of the G8 probe to DNA from such cell lines has provided more precise physical mapping of D4S10, which has now been localized to the terminal 4p16.3 sub-band. *Figure 12.5* shows a typical mapping panel which we are now using to assign new markers to the region of D4S10. In this experiment, lanes of *HindIII* digested DNA from the various hybrid cell lines are electrophoretically separated along with lanes of control DNA, blotted onto nylon membrane, and hybridized to radioactive probe. The pattern of hybridization after autoradiography reveals the presence or absence of the probe in each of the cell lines. Probes with hybridization patterns identical to D4S10 are presumably located near the tip of the short arm of chromosome 4. In a concerted effort to identify markers closer to Huntington's disease, we have regionally mapped over 200 anonymous pieces of chromosome 4 DNA (Gilliam *et al.*, 1987). Only one of these has fallen in the same region as D4S10 (*see Figure 12.5*), suggesting that the size of this segment could be quite small. From the cytogenetic length of the chromosome, the region of 4p16.3 containing D4S10 should represent a maximum of 6×10^6 base pairs (bp). The relative lack of markers found near D4S10 indicates, however, that the marker may be located as little as 2×10^6 bp from the telomere of the chromosome.

The search for a flanking marker

The genetic linkage of D4S10 to the Huntington's disease gene localizes the disease locus to a region of 1.5–6.5 cM on either side of the DNA marker. Therefore, using this single marker, the location of Huntington's disease can be narrowed to a region of 10×10^6–20×10^6 base pairs (bp) of DNA. Although this represents less than 0.7% of the human genome or about 10% of chromosome 4, it is still too large an area to effectively screen directly for the disease gene since it could contain upwards of 500 genes. To restrict the area which must ultimately be searched for the presence of the disease gene, we have pursued the isolation of a 'flanking' marker, providing the capacity to identify recombination events on the other side of the Huntington's disease locus to D4S10. Crossovers detected by each marker would then serve to define the 'minimum genetic region' that must contain the defect. The potential size of this region is related to the total number of affected individuals who are informative for segregation of the particular marker allele. The family resources available set an upper limit of about 250 such individuals so that it is theoretically possible to delineate a region of 0.4% recombination on either side of the disease gene. The minimum genetic region could define an area of roughly one centimorgan (1% recombination; 1 cM) which contains the disease locus.

To date, all of the markers characterized have been located centromeric to D4S10 and have proved to be further from Huntington's disease than the original marker. Thus the disease locus is located in the small physical region bordered by D4S10 and the telomere. As noted above, this region could contain from 2×10^6 to 6×10^6 base pairs (bp) in size making it amenable to directed cloning strategies for identifying new DNA markers and ultimately cloning the disease gene. Since the distance between Huntington's disease and the telomere could be quite small, the

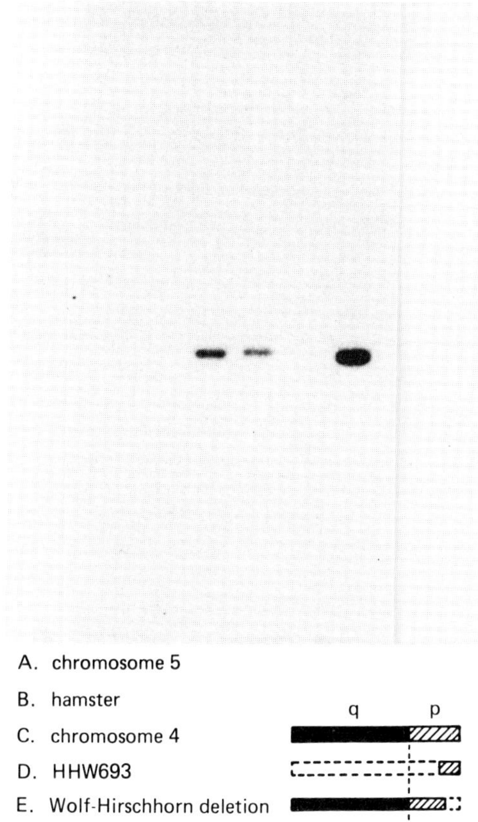

Figure 12.5 Regional localization of chromosome 4 probes. A mapping blot is shown for the single chromosome 4 marker known to map in the same chromosomal region at the tip of the short arm as G8. Each lane contains 5 μg of *HindIII*-digested genomic DNA that has been fractionated by agarose gel electrophoresis, transferred to nylon filter and hybridized to labelled probe. Hybrid cell lines containing chromosome 4 or portions thereof along with control lines with human chromosome 5 or no human chromosome are also represented. The drawing below the mapping blot illustrates the portion of chromosome 4 contained in the respective cell lines. The dark lines correspond to the long arm and the hatched regions to the short arm. The marker is positive for hybridization to chromosome 4 (lane C), to a cell line (HHW693) containing the terminal half of the short arm (lane D) (Wasmuth *et al.*, 1986), and to a cell line containing an uncharacterized amount of DNA from the tip region of the short arm (8C10) (lane F) (MacDonald *et al.*, unpublished observations). The marker is negative for hybridization to the control lanes (A,B) and to a Wolf–Hirschhorn deletion chromosome (lane E) which is missing the terminal 4p16.3 sub-band (MacDonald *et al.*, 1987)

task of finding a flanking marker may be extremely difficult. In this case, the telomere itself could provide the bordering landmark which, together with genetically detected crossovers below the defect, would define a minimum segment of chromosome 4 containing the disease locus.

Molecular strategies to map and analyse the minimum genetic region

Long-range restriction mapping

The recent development of pulsed field gradient gel systems (Schwartz and Cantor, 1984; Carle, Frank and Olsen, 1986), capable of separating fragments ranging from 10 kb to 2500 kb, offers a physical mapping approach to simplify genetic mapping when applied to chromosomal distances of up to several centimorgans in mammalian DNA. Such long-range restriction mapping has recently been applied to the chromosomal region surrounding the Duchenne muscular dystrophy locus (Burmeister and Lehrach, 1986; van Ommen *et al.*, 1986; Kenwrick *et al.*, 1987). An approximate 2000 kb stretch of DNA has been mapped for the presence of rare restriction cutting sites using the various cloned DNA markers from the Duchenne muscular dystrophy region as hybridization probes. This approach offers several distinct features which complement the genetic strategy. Genetic markers can be assigned to discrete physical regions along the chromosome based on their hybridization to homologous large restriction fragments. When a number of markers from a particular region are available as hybridization probes, it should be possible to align the entire area as overlapping restriction fragments, thereby establishing the exact physical distance between any two genetic markers. This coalescing of genetic and physical maps would make directed cloning of the minimal genetic region a possibility. A second advantage is that pulsed field gradient analysis offers a powerful approach to identify rapidly mutations caused by deletion, insertion, or translocation. One model for the primary mutation in Huntington's disease would be that a DNA insertion, perhaps of a retrovirus, causes enhanced expression of an adjacent locus. Such a model could account for both the low mutation rate of Huntington's disease, and the dominant mode of transmission.

Cloning fragments enriched by pulsed field gradient electrophoresis

In addition to long-range mapping, pulsed field gradient electrophoresis can be used preparatively to enrich for large DNA restriction fragments identified by hybridization to probes such as G8 (Michiels, Burmeister and Lehrach, 1987). In this way the genomic region extending in both directions from D4S10 to the bordering rare cutter sites could be cloned, systematically examined for disease-related alterations and used to generate new markers for genetic mapping and linkage disequilibrium studies. Depending on the vector used, cloning schemes can be designed to construct a complete library of the fragment, or, for a number of rare cutting enzymes, to clone only the terminal segments of the large restriction fragment. An ideal combination of physical and genetic mapping would be to map flanking markers to one large restriction fragment. This would place an upper limit on the physical size of the minimum genetic region and provide a strategy to clone the region containing the disease gene.

We are presently engaged in enrichment cloning strategies around the D4S10 locus. In practice, the enrichment for a homologous restriction fragment will depend on several factors:

(1) the degree to which the relevant CpG-rich restriction sites digest in the genomic DNA sample used (Bird, 1986);
(2) the degree of contamination by breakdown products from larger fragments;
(3) the size of the homologous fragment; and
(4) the enrichment that can be provided by using a somatic cell hybrid line with a restricted human chromosome complement.

We are fortunate to have a cell line, HHW693 (Wasmuth *et al.*, 1986), whose only human chromosome component is the terminal half of 4p and a small amount of chromosome 5, which together constitute less than 2% of the total human genome. Using this hybrid, it should be possible to achieve enrichments upwards of 1000-fold for particular large restriction fragments.

Jumping and linking libraries

A strategy which makes it possible to move unidirectionally along the chromosome involves 'jumping' from one to several hundred kilobases along a given chromosome with cloned sequences spaced at a distance equal to the jump intervals. Beginning at the D4S10 locus, it would be theoretically feasible to isolate five to ten new chromosomal segments spaced at intervals which together would span the Huntington's disease region. At this point the combination of physical mapping, chromosome walking, and enrichment cloning strategies should permit the cloning and characterization of virtually any subregion in the vicinity of the disease gene. For the jumping technique (Collins and Weissman, 1984; Poustka *et al.*, 1986), large DNA fragments are circularized in the presence of a selectable 'tag' plasmid, after which most of the internal sequences are cleaved away leaving the plasmid joined to the end segments of the original large DNA fragment. These recombinant molecules are joined to bacteriophage arms and packaged to form a 'jumping library'. Hybridization of the probe to a clone from this library gives a new chromosomal segment separated from the homologous probe by a distance equal to the original restriction fragment. Libraries derived from human DNAs cleaved either to completion with rare cutting enzymes (Poustka *et al.*, 1986, 1987) or partially digested with common cutting enzymes (Collins *et al.*, 1987) have been constructed and used successfully to jump to new sites on the chromosome. These libraries have been screened with clones from D4S10 and homologous clones are currently being characterized.

Jumping libraries can be used in conjunction with linking libraries to move in a stepwise manner in a given direction along the chromosome (Poustka *et al.*, 1986, 1987). Linking libraries are also especially well suited for constructing physical maps. This type of library contains a discrete subset of genomic sequences, for example, all clones which contain a particular rare cutting site such as that of *NotI*. A *NotI* linking library has been prepared (Frischauf *et al.*, 1987) from the hybrid cell line described previously which contains parts of the short arms of human chromosomes 5 and 4 (including the Huntington's disease region) as its only human DNA (Wasmuth *et al.*, 1986). Human clones have been isolated and are being screened by a panel of somatic cell hybrids to identify a unique set of probes derived from all or most of the *NotI* sites from the region at the tip of 4p. Several such probes are being mapped by pulsed field gradient electrophoresis and characterized as genetic markers to chart their location relative to the disease gene.

Identification of the Huntington's disease gene

The parallel development of hybrid cell lines specific for subregions of chromosome 4, especially band 4p16 (MacDonald *et al.*, 1987), and more powerful analytical techniques for long-range DNA mapping promise to overcome the dearth of markers near the disease locus at the tip of the short arm. Genetic and physical analysis of these markers should, in the near future, define the boundaries of the disease gene within genetic limits set by the availability of family resources. The experimental goal then changes from one of identifying closer markers to identifying the disease gene itself within a genetically defined and presumably cloned region. This region will contain an as yet unpredictable number of unrelated genes, polymorphisms, or even structural alterations, all of which will serve to camouflage the disease defect. Identifying the disease gene or defect within this context will present the next challenge to Huntington's disease researchers.

As described earlier, assuming we can monitor 250 meiotic events it should be possible to identify flanking markers which recombine with the defect in only 0.4% of cases. This would define a segment of about 1 cM genetic distance. Since increased recombination at the telomere should improve genetic resolution, the actual size of the segment may be much smaller than 1 million base pairs of DNA. A further way of decreasing the size of the region expected to contain the Huntington's disease gene is to test for linkage disequilibrium which may be observed very close to the gene. Applicable to mutations which occurred only once or at least very rarely in human history, linkage disequilibrium reflects the preservation of the immediate sequence environment of the original mutation, relatively undisturbed by recombination events. Polymorphisms in regions further from the disease locus will move towards equilibrium due to random meiotic recombination. In practice, a polymorphic variant, which occurs only infrequently in the general population but which appears in every Huntington's disease affected individual, would be the ideal demonstration of linkage disequilibrium with the disease gene. Depending on the historical timing of the original mutation, disequilibrium might permit localization of the gene to less than 100 000 base pairs, a region which approximates single gene resolution in the mammalian genome.

In the event that linkage disequilibrium studies do not lead us to single gene resolution, we will be left to identify the disease gene among a number of candidates. Identification of protein coding sequences within the minimal genetic region is the first priority. Such DNA can be analysed directly for disease-related mutations, or used as hybridization probes to determine which tissues express the gene, to test for differences in expression in normal versus Huntington's disease homozygous individuals, and to isolate cDNA clones for investigating RNA processing errors. It is hoped that either a brain specific pattern of expression or a distinctive RNA transcript in disease homozygotes will distinguish the disease gene from others in the area, but this is by no means certain. In the event that we are not able to narrow the field of candidates by disequilibrium studies or by analysis of the individual genes and their RNA transcripts, more long-term approaches will be warranted. We have begun mapping the Huntington's disease syntenic region in mice where more powerful genetic analysis is possible. Since coding sequences are usually conserved across species, it may be possible to map the syntenic region at greater resolution in mice, then transfer this information back to humans. Introduction of candidate genes into mouse embryos and assessment of neurological status in the resulting transgenic animals may ultimately be required to positively identify the Huntington's disease gene.

Under favourable circumstances, identification of the Huntington's disease gene will be fairly straightforward. For example, structural changes in the DNA or sequencing studies may reveal an obvious candidate. Even under less favourable circumstances, the combination of molecular biology and genetics provides such a myriad of experimental strategies that the eventual identification of the Huntington's disease gene can be predicted with reasonable confidence.

References

BENDER, W., AKAM, M., KARCH, F. *et al.* (1983) Molecular genetics of the biothorax complex in *Drosophila melanogaster*. *Science,* **221,** 23–29

BIRD, A. P. (1986) CpG-rich islands and the function of DNA methylation. *Nature (London),* **321,** 209–214

BIRD, E. D., MACKAY, A. V. P., RAYNER, C. N. and IVERSEN, L. L. (1973) Reduced glutamic acid decarboxylase activity of post-mortem brain in Huntington's chorea. *Lancet,* **i,** 1090–1020

BIRD, E. D., CARO, A. J. and PILLING, J. B. (1974) A sex related factor in the inheritance of Huntington's chorea. *Annals of Human Genetics,* **37,** 255–260

BIRD, E. D. (1980) Chemical pathology of Huntington's disease. *Annual Review of Pharmacology and Toxicology,* **20,** 533–551

BITTENBENDER, J. B. and QUADFASEL, F. A. (1962) Rigid akinetic forms of Huntington's chorea. *Archives of Neurology,* **7,** 275–288

BOEHNKE, M., CONNEALLY, P. M. and LANGE, K. (1983) Two models for a maternal factor in the inheritance of Huntington's disease. *American Journal of Human Genetics,* **35,** 845–860

BROTHERS, C. R. D. (1949) The history and incidence of Huntington's chorea in Tasmania. *Proceedings of the Royal Australasian College of Physicians,* **4,** 48–50

BROTHERS, C. R. D. (1964) Huntington's chorea in Victoria and Tasmania. *Journal of Neurological Sciences,* **1,** 405–420

BRUYN, G. W., BOTS, G. T. A. M. and DOM, R. (1979) Huntington's chorea: current neuropathological status. *Advances in Neurology,* **23,** 83–93

BURMEISTER, M. and LEHRACH, H. (1986) Long-range restriction map around the Duchenne muscular dystrophy gene. *Nature (London),* **324,** 582–585

CARLE, G. F., FRANK, M. and OLSEN, M. V. (1986) Electrophoretic separation of large DNA molecules by periodic inversion of the electric field. *Science,* **232,** 65–68

CARTER, C. J. (1986) The biochemical pathology of Huntington's disease. *Reviews in the Neurosciences,* **1,** 1–34

COLLINS, F. S. and WEISSMAN, S. M. (1984) Directional cloning of DNA fragments at a large distance from an initial probe: a circularization method. *Proceedings of the National Academy of Sciences USA,* **81,** 6812–6816

COLLINS, F. S., DRUMM, M. S., COLE, J. L. *et al.* (1987) Construction of a general human chromosome jumping library with application to cystic fibrosis. *Science,* **235,** 1046–1049

CONNEALLY, P. M., GUSELLA, J. F. and WEXLER, N. S. (1986) Huntington disease genetics, presymptomatic and prenatal diagnosis. *Birth Defects Symposium, Albany, New York, 29–30 September.* New York: Alan Liss, Inc, in press

CROSSMAN, A. R. (1987) Primate models of dyskinesia: The experimental approach to the study of basal ganglia-related involuntary movement disorder. *Neuroscience,* **21,** 1–40

DAVIES, K. E., PEARSON, P. L., HARPER, P. S. *et al.* (1983) Linkage analysis of two cloned DNA sequences flanking the Duchenne muscular dystrophy locus on the short arm of the X chromosome. *Nucleic Acids Research,* **11,** 2303–2312

DiFIGLIA, M., PASIK, P. and PASIK, T. (1976) A Golgi study of neuronal types in the neostriatum of monkeys. *Brain Research,* **114,** 245–256

DiFIGLIA, M., PASIK, T. and PASIK, P. (1980) Ultrastructure of Golgi-impregnated and goldtoned spiny and aspiny neurons in the monkey neostriatum. *Journal of Neurocytology,* **9,** 471–492

DiFIGLIA, M., ARONIN, N. and MARTIN, J. B. (1982) Light and electron microscopic localization of immunoreactive leu-enkephalin in the monkey basal ganglia. *Journal of Neuroscience,* **2,** 303–320

DiFIGLIA, M. and ARONIN, N. (1982) Ultrastructural features of immunoreactive somatostatin neurons in the rat caudate nucleus. *Journal of Neuroscience,* **2,** 267–274

FARRER, L. A. and CONNEALLY, P. M. (1985) A genetic model for age at onset in Huntington's disease. *American Journal of Human Genetics,* **37,** 350–357

FERRANTE, R. J., KOWALL, N. W., BEAL, M. F., RICHARDSON, E. P. JR, BIRD, E. D. and MARTIN, J. B. (1985) Selective sparing of a class of striatal neurons in Huntington's disease. *Science,* **230,** 561–563

FOLSTEIN, S., PHILLIPS, J., MYERS, D. *et al.* (1985) Huntington's disease: two families with differing clinical features show linkage to the G8 probe. *Science,* **229,** 776–779

FINCH, C. E. (1980) The relationship of aging genes in the basal ganglia to manifestations of Huntington's chorea. *Annals of Neurology,* **7,** 406–410

FRISCHAUF, A. M., POHL, T., SMITH, B. *et al.* (1987) Cloning of *NotI* site containing sequences (linking clones) from the regions 4p15.1–4pter and 5p up to 5p15.1 and further characterization of the clones expected to map close to the Huntington's disease locus. *Cytogenetics and Cell Genetics,* in press

GILLIAM, T. C., HEALY, S. T. MacDONALD, M. E. *et al.* (1987) Isolation of polymorphic DNA fragments from human chromosome 4. *Nucleic Acids Research,* **15,** 1445–1458

GRAVELAND, G. A., WILLIAMS, R. S. and DiFIGLIA, M. (1985) A Golgi study of the human neostriatum: neurons and afferent fibers. *Journal of Comparative Neurology,* **234,** 317–333

GRAYBIEL, A. M. (1986) Neuropeptides in the basal ganglia. In: *Neuropeptides in Neurologic and Psychiatric Disease,* edited by J. B. Martin and J. Barchas, pp. 135–161. New York: Raven Press

GUSELLA, J. F., WEXLER, N. S., CONNEALLY, P. M. *et al.* (1983) A polymorphic DNA marker genetically linked to Huntington's disease. *Nature (London),* **306,** 234–238

GUSELLA, J. F., TANZI, R. E., BADER, P. I. *et al.* (1985) Deletion of the Huntington's disease linked G8 (D4S10) locus in Wolf–Hirschhorn syndrome. *Nature (London),* **318,** 75–79

GUSELLA, J. F., GILLIAM, T. C., TANZI, R. E. *et al.* (1986) Molecular genetics of Huntington's disease. *Cold Spring Harbor Symposium of Quantitative Biology,* **LI,** 359–364

HAINES, J., TANZI, R., WEXLER, N. *et al.* (1986) No evidence of linkage heterogeneity between Huntington disease (HD) and G8 (D4S10). *American Journal of Human Genetics,* **39,** A156

HARPER, P. S., YOUNGMAN, S., ANDERSON, M. A. *et al.* (1985) Genetic linkage between Huntington's disease and the DNA polymorphism G8 in South Wales families. *Journal of Medical Genetics,* **22,** 447–450

HAYDEN, M. R. (1981) *Huntington's Chorea.* New York: Springer-Verlag

HAYDEN, M. R. and BEIGHTON, P. H. (1982) Genetic aspects of Huntington's chorea: results of a national survey. *American Journal of Human Genetics,* **11,** 135–141

HAYDEN, M. R., MacGREGOR, J. M., SAFFER, D. S. and BEIGHTON, P. H. (1982) The high frequency of juvenile Huntington's chorea in South Africa. *Journal of Medical Genetics,* **19,** 94–97

HUNTINGTON, G. (1872) On chorea. *Medical and Surgical Reporter,* **26,** 317–321

KENWRICK, S., PATTERSON, M., SPEER, A., FISCHBECK, K. and DAVIES, K. (1987) Molecular analysis of the Duchenne muscular dystrophy region using pulsed field gel electrophoresis. *Cell,* **48,** 351–357

KNOWLTON, R. G., COHEN-HAGENAUER, O., VAN CONG, N. *et al.* (1985) A polymorphic DNA marker linked to cystic fibrosis is located on chromosome 7. *Nature (London),* **318,** 380–382

KOWALL, N. W., FERRANTE, R. J., MARTIN, J. B. (1988) Patterns of cell loss in Huntington's disease. *Trends in Neurosciences,* in press

KUHL, D. R., PHELPS, M. E., MARKHAM, C. H. *et al.* (1982) Cerebral metabolism and atrophy in Huntington's disease determined by FDG and computed tomographic scan. *Annals of Neurology,* **12,** 425–434

KUHL, D. E., METTER, E. J., RIEGE, W. H. and MARKHAM, C. H. (1984) Patterns of cerebral glucose utilization in Parkinson's disease and Huntington's disease. *Annals of Neurology,* **15 (Suppl.),** S119–S125

LANDEGENT, J. E., JANSEN, I. N., DE WAL, N. *et al.* (1986) Fine mapping of the Huntington's disease-linked D4S10 locus by non-radioactive *in situ* hybridization. *Human Genetics,* **73,** 354–357

LAURIE, D. A. and HULTEN, M. A. (1985a) Further studies of bivalent chiasma frequency in human males with normal karyotypes. *Annals of Human Genetics,* **49,** 189–201

LAURIE, D. A. and HULTEN, M. A. (1985b) Further studies on chiasma distribution and interference in the human male. *Annals of Human Genetics,* **49,** 203–214

MacDONALD, M. E., ANDERSON, M. A., GILLIAM, T. C. *et al.* (1987) A somatic cell hybrid panel for localizing DNA segments near the Huntington's disease gene. *Genomics,* **1,** 29–34

MAGENIS, R. E., GUSELLA, J., WELIKY, K. *et al.* (1986) Huntington's disease linked restriction fragment length polymorphism localized within band p16.1 of chromosome 4 by *in situ* hybridization. *American Journal of Human Genetics,* **39,** 383–391

MARTIN, J. B. and GUSELLA, J. F. (1986) Huntington's disease: pathogenesis and management. *New England Journal of Medicine,* **315,** 1267–1276

MERRITT, A. D., CONNEALLY, P. M., RAHMAN, N. F. and DREW, A. L. (1969) Juvenile Huntington's chorea. In: *Progress in Neurogenetics,* edited by A. Barbeau and T. R. Brunette, pp. 645–650. Amsterdam: Excerpta Medica Foundation

MICHIELS, F., BURMEISTER, M. and LEHRACH, H. (1987) Derivation of clones close to *met* by preparative field inversion gel electrophoresis. *Science,* **236,** 1305–1308

MONACO, A. P., NEVE, R. L., COLLETTI-FEENER, C., BERTELSON, C. J., KURNIT, D. M. and KUNKEL, L. M. (1986) Isolation of candidate cDNAs for portions of the Duchenne muscular dystrophy gene. *Nature (London),* **323,** 646–650

MURRAY, J. M., DAVIES, K. E., HARPER, P. S., MEREDITH, L., MUELLER, C. R. and WILLIAMSON, R. (1982) Linkage relationship of a cloned DNA sequence on the short arm of the X chromosome to Duchenne muscular dystrophy. *Nature (London),* **300,** 69–71

MYERS, R. H. and MARTIN, J. B. (1982) Huntington's disease. *Seminars in Neurology,* **2,** 365–372

MYERS, R. H., GOLDMAN, D., BIRD, E. D., SAX, D. S., MERRIL, C. R., SHOENFELD, M. and WOLF, P. A. (1983) Maternal transmission of Huntington's disease. *Lancet,* **ii,** 208–210

NEGRETTE, A. (1963) *Corea de Huntington.* Thesis Universidad Central de Venezuela, Venezuela

OTT, J. (1974) Estimation of the recombination fraction in human pedigrees: efficient computation of the likelihood for human linkage studies. *American Journal of Human Genetics,* **26,** 588–597

OTT, J. (1976) A computer program for linkage analysis of general human pedigrees. *American Journal of Human Genetics,* **28,** 528–529

PAULI, R. M. (1983) Dominance and homozygosity in man. *American Journal of Medical Genetics,* **16,** 455–458

PENNEY, J. B. and YOUNG, A. B. (1986) Striatal inhomogeneities and basal ganglia function. *Movement Disorders,* **1,** 3–15

POUSTKA, A., POHL, T. M., BARLOW, D. P., FRISCHAUF, A. M. and LEHRACH, H. (1987) Construction and use of human chromosome jumping libraries from *NotI*-digested DNA. *Nature (London),* **325,** 353–355

POUSTKA, A. M., POHL, T., BARLOW, D. *et al.* (1986) Molecular approaches to mammalian genetics. *Cold Spring Harbor Symposium of Quantitative Biology,* **LI,** 131–140

REEDERS, S. T., BREUNING, M. H., DAVIES, K. E. *et al.* (1985) A highly polymorphic DNA marker linked to adult polycystic kidney disease on chromosome 16. *Nature (London),* **317,** 542–544

SEIZINGER, B. R., ROULEAU, G. A., OZELIUS, L. J. *et al.* (1987) Genetic linkage of von Recklinghausen neurofibromatosis to the nerve growth factor receptor gene. *Cell,* **49,** 589–594

SCHOENFELD, M., MYERS, R. H., CUPPLES, L. A., BERKMAN, B., SAX, D. S. and CLARK, E. (1984) Increased rate of suicide among patients with Huntington's disease. *Journal of Neurology, Neurosurgery, and Psychiatry,* **47,** 1283–1287

SCHWARTZ, D. C. and CANTOR, C. R. (1984) Separation of yeast chromosome-sized DNAs by pulsed field gradient gel electrophoresis. *Cell,* **37,** 67–75

ST GEORGE-HYSLOP, P. H., TANZI, R. E., POLINSKY, R. J. *et al.* (1987) The genetic defect causing familial Alzheimer's disease maps on chromosome 21. *Science,* **235,** 885–890

TANZI, R. E., HAINES, J. L., STEWART, G. D. *et al.* (1987) Genetic linkage map for the long arm of human chromosome 21. *Proceedings of the National Academy of Sciences USA,* submitted

TSUI, L.-C., BUCHWALD, M., BARKER, D. *et al.* (1985) Cystic fibrosis locus defined by a genetically linked polymorphic DNA marker. *Science,* **230,** 1054–1057

VAN OMMEN, G. J. B., HOFKER, J. M. H., MONACO, A. P. *et al.* (1986) A physical map of 4 million base pairs around the Duchenne muscular dystrophy gene on the human X chromosome. *Cell,* **47,** 499–504

VONSATTEL, J. P., MYERS, R. H., STEVENS, T. J. *et al.* (1985) Selective sparing of a class of striatal neurons in Huntington's disease. *Journal of Neuropathology and Experimental Neurology,* **44,** 559–577

WANG, H. S., GREENBERG, C. R., KALOUSEK, D., GUSELLA, J., HOUSMAN, D. and HAYDEN, M. R. (1985) Subregional assignment of the linked marker D4S10 (G8) for Huntington disease by *in situ* hybridization. *American Journal of Human Genetics,* **39,** 392–396

WAINWRIGHT, B., SCAMBLER, P. J., SCHMIDTKE, J. *et al.* (1985) Localization of cystic fibrosis locus to 7cen-q22. *Nature (London),* **318,** 384–386

WASMUTH, J. J., CARLOCK, L. R., SMITH, B. and IMMKEN, L. L. (1986) A cell hybrid and recombinant DNA library which facilitate identification of polymorphic loci in the vicinity of the Huntington's disease gene. *American Journal of Human Genetics,* **39,** 397–403

WENT, L. N., VEGTER-VAN DER VLIS, M., VOLKERS, W. and COLLEWIJN, H. (1975) Huntington's chorea. In: *Early Diagnosis and Prevention of Genetic Diseases,* edited by L. N. Went *et al.,* pp. 13–25. Leiden: Leiden University Press

WENT, L. N., VAN DER VLIS, M., BRUYN, G. W. and VOLKERS, W. S. (1983) Huntington's chorea in the Netherlands. *Annals of Human Genetics,* **47,** 205–214

WEXLER, N. S., YOUNG, A. B., TANZI, R. E. *et al.* (1987) Homozygotes for Huntington's disease. *Nature (London),* **326,** 194–197

WHITE, R. A. H. and AKAM, M. E. (1985) Contrabithorax mutations cause inappropriate expression of ultrabithorax products in *Drosophila. Nature (London),* **318,** 567–569

WHITE, R. L., WOODWARD, S., LEPPERT, M. *et al.* (1985) A closely linked genetic marker for cystic fibrosis. *Nature (London),* **318,** 382–384

ZABEL, B. U., NAYLOR, S. L., SAKAGUCHI, A. Y. and GUSELLA, J. F. (1986) Mapping of the DNA locus D4S10 and the linked Huntington's disease gene to 4p16-p15. *Cytogenetics and Cell Genetics,* **42,** 187–190

13
Molecular genetics and muscular dystrophy

Peter S. Harper

INTRODUCTION

The muscular dystrophies form a group of progressive, genetically determined primary muscle disorders that are a major source of disability and early death in childhood and early adult life. Their clinical features and classification (*Table 13.1*), based on pathological and electrophysiological abnormalities, together with patterns of inheritance, have been well studied, but in no case has the primary biochemical defect been established, despite much careful experimental work over several decades. Muscular dystrophies thus form a particularly challenging target for the use of the 'reverse genetics' approach to map, identify and characterize the various genes responsible, to analyse the molecular pathology at the gene level responsible for the disease states, and to identify the proteins for which these genes code.

Table 13.1 Major forms of human muscular dystrophy

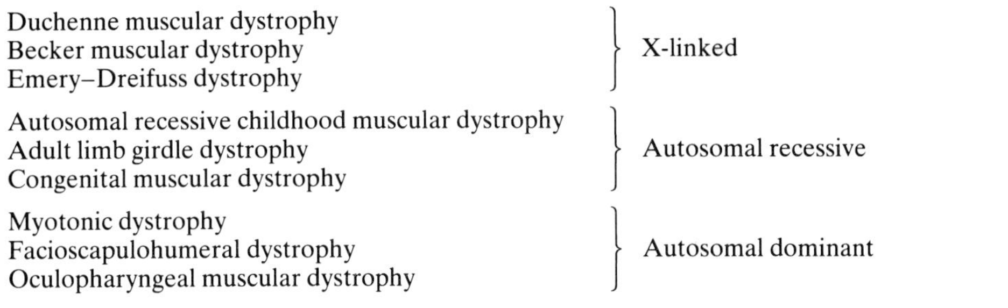

Disorder	Inheritance
Duchenne muscular dystrophy Becker muscular dystrophy Emery–Dreifuss dystrophy	X-linked
Autosomal recessive childhood muscular dystrophy Adult limb girdle dystrophy Congenital muscular dystrophy	Autosomal recessive
Myotonic dystrophy Facioscapulohumeral dystrophy Oculopharyngeal muscular dystrophy	Autosomal dominant

As a result of research over the past few years, we already have a considerable body of information leading towards these goals, particularly in the case of the X-linked muscular dystrophies, which provided the first genetic disorder to be studied in this way, and the first disease gene to be fully characterized in the absence of any information as to the nature of its possible product. This work has had profound implications for research into other neuromuscular disorders, and has indeed transformed the outlook for those many mendelian disorders where lack of understanding at the protein level has, until now, held back the prospects for clarifying their nature.

DUCHENNE AND BECKER MUSCULAR DYSTROPHIES

These two forms of X-linked muscular dystrophy will be considered together, since it is now established that they form a single entity as judged by a single genetic locus and comparable molecular pathology, even though they normally run separately in families and show marked clinical differences that are reflected in problems of management and genetic counselling. *Table 13.2* summarizes respectively these

Table 13.2 Becker and Duchenne muscular dystrophies: similarities and differences

Features	*Becker muscular dystrophy*	*Duchenne muscular dystrophy*
Onset	Variable; commonly late childhood to early adult life	2–4 years
Age in wheelchair	Variable > 95% after 12 years	> 95% before 12 years
Age at death	Rare before 30 years	Commonly 15–25 years
Mental retardation	May occur	Frequent
Cardiac involvement	Uncommon	Usual
Calf hypertrophy	Usual; often very marked	Usual
Creatine kinase (CK) levels	Greatly raised in childhood	Greatly raised

similarities and differences. The development of our understanding of the disorder in molecular terms can be considered in terms of four phases, each with considerable overlap:

(1) genetic studies prior to the development of molecular techniques;
(2) genetic linkage studies based on DNA restriction fragment length polymorphisms (RFLPs);
(3) studies of X autosome translocations in affected girls;
(4) genetic deletions, visible and invisible;
(5) characterization of the Duchenne muscular dystrophy gene based on these combined approaches.

Early genetic studies in Duchenne muscular dystrophy

The X-linked nature of Duchenne muscular dystrophy was apparent from the earliest clinical studies (Gowers, 1879), as was the case for the milder disorder recognized by Becker and Kiener (1955) and now known as Becker muscular dystrophy. *Table 13.3* summarizes some of the relevant facts resulting from these early studies, in particular those relating to carrier females. Estimates of mutation rate are approximate owing to the dependence on carrier detection tests falling

Table 13.3 Duchenne and Becker muscular dystrophy: basic genetic characteristics

Features	*Duchenne muscular dystrophy*	*Becker muscular dystrophy*
Incidence (male births)	≃ 1 in 3000	≃ 1 in 30 000
Genetic fitness	0	≃ 70%
Carrier detection (CK)	Two-thirds outside normal range	Under half outside normal range
Manifesting carriers	Around 10%	Rare
Early linkage data (Xg and colour blindness)	No linkage	Linkage with colour blindness suggested

short of complete accuracy, while in Becker muscular dystrophy there is difficulty in distinguishing the isolated case from one of autosomal recessive limb girdle dystrophy.

From the point of view of gene mapping it can be seen that the only firm information given by these early studies was the exclusion of measurable linkage with the Xg blood group at the distal end of the short arm of the X chromosome (Blyth *et al.*, 1965). The finding of suggestive, though not conclusive, evidence that Becker muscular dystrophy was linked to colour blindness (Skinner, Smith and Emery, 1974), initially strengthened by weak positive evidence for linkage to glucose 6 phosphate dehydrogenase (G6PD) and Becker muscular dystrophy (Zatz *et al.*, 1974) led to the concept of two separate loci for X-linked dystrophies, that for Becker muscular dystrophy being distally on the long arm near colour blindness and G6PD, while that for Duchenne muscular dystrophy was at some other undefined locus. As will be seen, this was one of the first hypotheses that could be tested by the new DNA markers.

A further area of debate arising from the early genetic studies, still not fully resolved, has been over the proportion of cases of Duchenne muscular dystrophy resulting from new mutation. As a genetically lethal X-linked disorder, Duchenne muscular dystrophy should follow the rule proposed by Haldane (1935), with one-third of cases representing mutations, equivalent to the one-third of the disease genes being lost in each generation by occurring in a male. Uncertainties in tests of carrier detection not only make estimates of the absolute mutation rate difficult, but also create problems in determining whether individual isolated cases are inherited or result from new mutations. There have been suggestions that most isolated cases are born to carrier mothers (Roses *et al.*, 1977), though most studies have supported the expected proportion (Yasuda and Kondo, 1980; Caskey *et al.*, 1980). Evidence from molecular studies is now becoming available that can in some cases identify the precise origin of the mutation and should resolve this question.

Apart from genetic studies, a large amount of work on the ultrastructure and biochemistry of muscle and other tissues in Duchenne muscular dystrophy has been carried out over the past 30 years aimed at detecting a primary defect. Details of this can be found in previous reviews (Engel, 1986), but since the data are essentially negative or inconclusive they will not be discussed here; undoubtedly they will in many cases be re-evaluated when the gene product is finally identified from the molecular studies described below.

Molecular approaches to Duchenne muscular dystrophy

The localization and identification of the Duchenne muscular dystrophy gene has been the subject of work over the past 7 years that has been remarkable both in its intensity and in its ingenuity in the application of new techniques and strategies. *Table 13.4* summarizes some of these; their convergence and combined application to the problem has been the main reason for the successful outcome, and for Duchenne muscular dystrophy being regarded as a model for molecular research into other genetic disorders. A further reason for success has been the close degree of collaboration between different groups (Zatz *et al.*, 1974; Kunkel *et al.*, 1986), and between molecular and clinical geneticists, allowing the resources provided by family studies and by unusual forms of the disorder to be exploited to the full. The principal approaches will be described in turn.

Table 13.4 Resources for molecular studies in muscular dystrophies

Primary resources
Duchenne muscular dystrophy girls with X-autosome translocation
Duchenne muscular dystrophy boys with visible deletions
Other patients with X chromosome rearrangements
Secondary resources
DNA polymorphisms (RFLPs)
DNA libraries from specific chromosomal regions
Cloned sequences from junction fragments
Hybrid cell lines

Genetic linkage at the DNA level

Early attempts at mapping most X-linked disorders were severely limited by the lack of suitable polymorphic markers. The only one with a high degree of polymorphism, the Xg blood group, proved to be located terminally on the short arm, while colour blindness and G6PD deficiency, located close together near the end of the long arm, were relatively uninformative in most population groups studied. Thus the recognition that DNA sequences showing restriction fragment length polymorphisms (RFLPs) were relatively abundant in the human genome and could act as genetic markers (Botstein *et al.*, 1980), was a turning point in the mapping of the X chromosome. Their codominant inheritance and independence of X inactivation was especially valuable for the X chromosome.

The study of Duchenne muscular dystrophy provided the first X chromosome RFLPs to be discovered. Prior to this the source had generally been total human DNA (Maniatis *et al.*, 1978), but the development of techniques of chromosome sorting based on flow cytometry gave the possibility of obtaining DNA relatively specific for the X chromosome (Kunkel *et al.*, 1982). In 1981 Davies and colleagues used such a preparation to prepare the first human DNA library specific for the X chromosome and isolated clones from this that showed RFLPs. These could now be tested for genetic linkage to Duchenne muscular dystrophy and other disorders.

The first such sequence to be isolated was in fact located on the X chromosome long arm, and in view of the suggestions already arising at that time of a short arm localization for Duchenne muscular dystrophy, it was the second polymorphic sequence (RC8) that was chosen to study against families with Duchenne muscular dystrophy (Murray *et al.*, 1982). Studies of DNA from hybrid cell lines had shown that this sequence was on the short arm, and though it was not highly polymorphic, clear evidence of linkage was obtained with Duchenne muscular dystrophy, the first occasion on which a DNA polymorphism had been used to localize a disease gene. Shortly afterwards a second marker, (L1.28), became available, isolated from a total human DNA library along with a series of other sequences, but subsequently located on the X chromosome close to the centromere (Hofker *et al.*, 1985).

L1.28 proved to be both a more informative marker than RC8 and one that was easier to work with, with the result that conclusive data were rapidly obtained (Davies *et al.*, 1983) even though the genetic distance (around 15 centimorgans (cM) from Duchenne muscular dystrophy) proved to be similar to that between the disease and RC8. Of particular importance was the finding that there was frequent recombination between the two markers, showing that the Duchenne locus must lie between them, a conclusion supported by their physical localization. The Duchenne muscular dystrophy gene was thus firmly bracketed by two DNA polymorphisms, opening the way for rapid further developments.

It is worth noting how many major conclusions could be drawn at this relatively early stage, with two markers flanking the Duchenne gene at the considerable distance of 15 cM. Firstly a search for heterogeneity within Duchenne muscular dystrophy could be made, something that had been suggested previously and generally would not have been unexpected. Division of families into those with and without mental retardation, admittedly a crude and somewhat arbitrary division, showed that both groups had the same linkage relationships with the two marker genes, with no suggestion of heterogeneity (O'Brien *et al.*, 1983).

An extension of heterogeneity testing was given by the analysis of Becker muscular dystrophy families undertaken by Kingston and colleagues (Kingston *et al.*, 1983, 1984). It came as a considerable surprise in view of previous linkage studies when the data showed, first with L1.28 and then with RC8, that Becker muscular dystrophy was in the same region as the Duchenne version, rather than on the long arm. This conclusion was supported by studies with other markers (Brown *et al.*, 1985), making the presumption of allelism with Duchenne muscular dystrophy, a strong one that has since been confirmed.

The feasibility of clinical application was also demonstrated at this time (Harper *et al.*, 1983), particularly for those families informative for both the flanking markers, or where pre-existing carrier information was available from creatine kinase (CK) analysis. The proportion of maximally informative families at this stage was too low to allow prenatal diagnosis to be considered as an option.

With the Duchenne muscular dystrophy gene flanked, efforts could now be concentrated on isolating further sequences in the region which might provide more accurate prediction in a higher proportion of families as well as coming closer to the Duchenne gene itself. During 1984 and 1985 a number of such sequences was found and evaluated, resulting in a detailed map around the Duchenne gene (Brown *et al.*, 1985) and a general framework of the X chromosome map itself (Drayna and White, 1985). These markers were principally isolated from flow sorted X chromosome libraries but one (OTC) was derived from a gene specific cDNA clone for the enzyme ornithine transcarbamylase (Davies *et al.*, 1985). For the purpose of diagnostic application they formed two blocks, one distal (RC8, D2 and C7), the other proximal to the DMD locus (754, OTC and L1.28), greatly increasing the number of families where at least one of each block were informative, so providing flanking markers (Boyd and Buckle, 1986).

Despite the rapid increase in the numbers of linked markers, none was showing extremely close linkage to Duchenne muscular dystrophy. Notably probe 754, which mapped on cell hybrid studies (Francke *et al.*, 1985) within a small deletion characterized by Duchenne muscular dystrophy and other disorders, showed around 10% recombination. The explanation for this was not clear at this stage, but in addition to reducing the accuracy of diagnostic prediction it also posed problems for the principal target of identifying the Duchenne gene itself using linked markers as a starting point. Thus other strategies, already in progress, began to appear more feasible in achieving this goal.

Duchenne muscular dystrophy girls and X-autosome translocations

The hypothesis that balanced X-autosome translocations can give valuable information on the localization of genes on the X chromosome is a further example of a principle first validated for Duchenne muscular dystrophies but with general implication for other X-linked disorders. At least 16 X-autosomal translocations in girls associated with muscular dystrophy have been described since 1977, and the findings have been reviewed by Boyd and colleagues (Boyd and Buckle, 1986; Boyd *et al.*, 1986), who have also compared the X chromosome breakpoints in the various cases using cytogenetics and molecular techniques. Essentially, these studies show that all cases with a Duchenne or Becker muscular dystrophy phenotype show a breakpoint in the p21 region of the X chromosome, while the autosomes involved are variable. However the site of the X breakpoint is not absolutely identical in different patients either cytogenetically or when tested with the various probes now available from within the Duchenne gene.

The clinical features of these affected girls have in all cases been compatible with X-linked muscular dystrophy, but with widely differing severity. The best studied of all, the X:21 translocation originally reported by Verellen and coworkers (Verellen *et al.*, 1977; Verellen-Dumoulin *et al.*, 1984), and subsequently used to obtain sequences from the breakpoint, showed a course that would normally be classified as Becker muscular dystrophy, though most of the others have been typical of the Duchenne type. All of the cases reported have been isolated, with no cytogenetic abnormality or evidence of Duchenne muscular dystrophy carrier state in relatives.

The constancy of the breakpoint in the p21 area in these cases, together with the genetic linkage evidence also placing the Duchenne muscular dystrophy gene in this region, led to the hypothesis that the translocation had a direct effect in producing the disorder (Lindenbaum *et al.*, 1979), either by disrupting the gene or by its interfering with its function in some other way. This now has been confirmed by Worton and colleagues (Worton *et al.*, 1984; Ray *et al.*, 1985) who have performed detailed molecular studies on the X:21 translocation already mentioned. They utilized the fact that in this case the autosomal breakpoint was through the ribsomal genes situated on the short arm of chromosome 21. These show a repeating pattern that could be identified and followed until the different sequences were reached at each end specific to the X chromosome and to the rest of chromosome 21 respectively. The hypothesis that the X junction should represent a point within the Duchenne muscular dystrophy gene was confirmed when cloning of this fragment and testing it against Duchenne patients showed evidence for gene deletion in a proportion of cases. To place this finding in its context requires description of the work already undertaken by Kunkel and colleagues on deleted sequences, using yet another approach, the study of a cytogenetically visible deletion.

Deletions and Duchenne muscular dystrophy

Although cytogenetic analysis of boys with Duchenne muscular dystrophy had previously been constantly negative, the study of some unusual and complex cases showing chromosomal defects provided a critical opportunity for the identification of the Duchenne gene. Notable among these was the patient 'BB' who, in addition to Duchenne muscular dystrophy, had mental retardation, together with chronic granulomatous disease, the Mcleod red cell phenotype and retinitis pigmentosa (Francke *et al.,* 1985). Cytogenetic study showed a visible deletion in the p21 region, and the suggestion was made that the patient had lost a number of specific genes in this region, including those for the disorders mentioned. Subsequently further patients were recognized with Duchenne muscular dystrophy, adrenal hypoplasia and glycerol kinase deficiency (Guggenheim *et al.,* 1980), again generally accompanied by mental retardation, and mostly showing a slightly more distally placed visible deletion. Such deletion syndromes have now become a valuable source of evidence (Francke *et al.,* 1987) regarding groups of adjacent genes, a further lesson of general importance arising from the specific study of Duchenne muscular dystrophy.

When the then available linked probes were tested against DNA from 'BB', only probe 754 was deleted, confirming the relatively large distance of most of them from the Duchenne gene. The approach of Kunkel *et al.* (1985), was to hybridize DNA from 'BB' with normal DNA under carefully controlled conditions and with the DNA from 'BB' in large excess, with the result that only those normal sequences not having a counterpart in 'BB' would be unhybridized and available for cloning.

Among the sequences obtained in this way, one (pERT 87) showed lack of hybridization with a proportion of DNA samples from Duchenne muscular dystrophy boys who themselves had shown no cytogenetic abnormality (Monaco *et al.*, 1985). The presumption was that this sequence lay within the Duchenne gene, a view strengthened by the finding that a proportion of these cases also showed deletion with the junction fragment 'XJ' probe isolated by Worton *et al.* (1984) from the X:21 translocation patient. This recognition of gene deletions in 'simple' cases of Duchenne muscular dystrophy provided the first direct evidence for molecular pathology in this disorder.

The information available on gene deletions in Duchenne muscular dystrophy has subsequently increased considerably, particularly as probes within the gene have become available and can already be used to analyse heterogeneity of mutational events in the same way that earlier, relatively weak linkage data could be used to test for multilocus heterogeneity. A large collaborative study (Kunkel *et al.*, 1986) showed that the deletions were of varying length, that some but not all were shared between pERT87 and XJ, and that there was no obvious difference in phenotype between deletion and non-deletion cases (leaving aside the few with visible cytogenetic deletions); deletions have also been recorded in some Becker muscular dystrophy patients, confirming the allelism of the disorders.

The proportion of cases showing deletion is steadily increasing as new probes are available that cover different portions of the Duchenne muscular dystrophy region; while currently around 12%, the application of pulsed field techniques (*see below,* p. 191) and the recent study of cDNA probes for exons of the gene are beginning to suggest that as many as half of all cases could result from deletion.

The recognition of deletions is of considerable diagnostic and predictive importance for Duchenne muscular dystrophy families (Harper and Thomas, 1986). It allows an 'all-or-none' test to be applied prenatally in pregnancies at risk among relatives, which gives considerably greater confidence than the probabalistic estimates of risk that result from application of linkage data. Even more important, such a test is independent of family structure, provided there is at least one living affected member who can be tested for deletion. This adds weight to the importance of ensuring that DNA is isolated and stored on all patients before they die.

Study of the pERT87 and XJ sequences provided another valuable advance, in addition to the recognition of deletions. Restriction fragment length polymorphisms (RFLPs) were detected for both clones and their number increased when adjacent sequences were isolated. As expected, these RFLPs show close linkage with the disorder when tested in families not showing a deletion (Thompson *et al.*, 1986; Walker *et al.*, 1986). However, the linkage is far from absolute, estimates of around 5% recombination being found for both pERT87 and XJ. This is in striking contrast with the near complete absence of recombination found for intragenic probes at other loci, and initially appeared to support the suggestion of instability in this region that had originated from earlier linkage studies using other probes. The alternative explanation was that the gene is extremely large; evidence supporting this was already available from the cytogenetic data on X-autosome translocations, as well as from the fact that sequences at some distance from each other, such as pERT87 and XJ, were both capable of detecting deletions. Further studies have now made it clear that the Duchenne muscular dystrophy gene is indeed extremely large (Monaco and Kunkel, 1987), and is currently much the largest human gene known.

Isolation of the Duchenne muscular dystrophy gene

The work described so far pointed to a number of properties of the Duchenne gene which were both puzzling and unusual. It seemed almost certain that the pERT87 and XJ clones were within the gene itself, but the full extent and nature of the gene remained uncertain. Likewise it was clear that the gene covered a large chromosomal region, but it was not known whether there were a number of distinct adjacent subunits or whether there was a single very large gene and gene product.

Most of our subsequent knowledge has come from detailed analysis of the DNA around pERT87, with both direct gene walking and the analysis of the endpoints of deletions that have provided new 'jump' clones for analysis. Kunkel, Monaco and colleagues (Monaco *et al.*, 1986; Monaco and Kunkel, 1987) searched for sequences conserved across widely different species in order to find a likely coding sequence, and were able to find one which identified a large 16 kb mRNA transcript. This could in turn be used to screen a cDNA library, and recognize cDNA sequences corresponding to exons of the gene. A considerable number of these have now been recognized and mapped in relation to the genomic DNA. The picture that emerges is of numerous relatively small exons interspersed among large non-coding sequences. Corresponding information from analysis of the XJ region is progressively extending our map of the Duchenne muscular dystrophy gene; it is likely to be complete in the near future (Koenig *et al.*, 1987).

The DNA transcript appears confined to muscle, both fetal and adult (Monaco *et al.*, 1986). Current work is focusing on what protein it corresponds to, and this is likely to be known by the time this book is published. If expression is indeed confined to muscle, and if the protein is as large as would be predicted from the size of the gene and DNA transcript, the possibilities are limited. One 'candidate protein' under scrutiny has been nebulin, a high molecular weight muscle protein whose electron microscopic appearance and distribution suggest a key role in maintaining stability of the contractile elements (Horowits *et al.*, 1986). Preliminary reports have shown alterations in Duchenne muscular dystrophy muscle (Wood *et al.*, 1987), but whether these are primary or not remains undetermined. The key test will be whether the gene or genes coding for this protein are X-linked; if not, it can confidently be excluded. In the meantime it should prove possible to use predictions of structure derived from the DNA sequence protein data to compare with any known protein being considered, or to guide the search for an unknown one.

An additional technique that is contributing to our knowledge of the Duchenne muscular dystrophy gene is that of pulsed field gel electrophoresis, which allows separation of large DNA fragments, several million base pairs (bp) in size. This has already been used to map the Duchenne muscular dystrophy region (Monaco *et al.*, 1986; van Ommen *et al.*, 1986; Kenwrick *et al.*, 1987), and may also prove valuable in identifying deletions not detectable by currently available probes.

Practical aspects

A number of the clinical applications of the new molecular developments in Duchenne muscular dystrophy have been mentioned already, but it is worth restating and outlining them here. It must be emphasized that the situation is an extremely fluid one; new developments continue to occur on a month-by-month basis, and while the new techniques have so far been largely DNA-orientated, this may change completely once a specific protein has been confirmed. The following points should be considered by those clinicians involved with Duchenne muscular dystrophy families who ask themselves and others how these developments should influence their clinical practice. Several detailed schemes have been put forward (Harper and Thomas, 1986; Bakker *et al.*, 1986).

A genetic register for X-linked muscular dystrophies

This is something that needs no molecular expertise and which can only be organized by clinicians, whether neurologists or clinical geneticists. Effective application of preventive measures can only be achieved if a comprehensive, accurate and regularly updated register of all affected individuals and their relatives at risk exists. This should preferably be on a geographical basis; a population size of 1–4 million, such as exists in most of the UK health regions, is sufficient to be viable, but not so large as to be impersonal. Such a register will probably contain around 200 Duchenne muscular dystrophy families, with over 1000 females at significant risk. To maintain it will require a full-time worker to maintain contact with members and ensure regular updating of information and testing of younger girls as they grow up. The use of computers now makes the task much less arduous, but it is still a major undertaking (Read *et al.*, 1986). However, such a system is greatly preferable to recurrent 'emergency' testing of at-risk pregnancies, where family members have not been typed previously.

Banking of DNA samples

Banking of DNA samples from affected patients and key relatives is again an area that does not require sophisticated molecular expertise and that provides a foundation on which future analysis can rest. DNA from a single 20 ml blood sample is probably adequate for service aspects, though for research purposes setting up lymphoblastoid cell lines is of great value.

Thorough carrier testing using creatine kinase (CK) analysis

DNA testing has not removed the need for this and there are many families where it will give definitive results as to carrier status, particularly if abnormal. Comparison of DNA and CK results has shown little disagreement, the end result being a sharper discrimination of carrier status if both approaches are used together (Williams *et al.*, 1986; Hodgson *et al.*, 1987).

Deletion testing

This is increasingly important since the recognition of a deletion will allow specific prenatal diagnosis. Until probes covering the entire gene are available, the failure to detect a deletion should be regarded as provisional. One cautionary note should be sounded; it seems likely that some deletion cases arise from germinal mosaicism in the mother (Bakker *et al.*, 1987). Thus a mother of an isolated case of Duchenne muscular dystrophy due to deletion who herself shows no evidence of the deletion should not be considered free of risk for a further affected child; subsequent pregnancies should be monitored for a deletion.

Linkage prediction

Here much will depend on family structure, and before hopes are raised too high for either carrier or prenatal prediction this needs to be carefully studied. Although some degree of prediction may be feasible even for the isolated case, it will be more definitive if the mother is an obligatory carrier and if affected (or unaffected) males in the previous generation are available; it must be remembered that all daughters of a Becker muscular dystrophy patient are carriers and that no amount of DNA or CK testing will alter this. Since mutation causing the disorder can occur at any point along the gene there will be an error of up to 5% due to recombination using intragenic probes, unless it can also be shown that flanking markers show no recombination, in which case error is most unlikely (Bertelson *et al.*, 1986).

Counselling

Predictive studies must be accompanied by full counselling. This may sound self-evident, but it is essential that families understand the limitations of the available tests, the timing of prenatal procedures (especially the need to report a pregnancy immediately if chorion biopsy is to be considered) and, in the case of more distant relatives, that they appreciate the nature of the disorder. Wherever possible the carrier status of women at risk and their informativeness for prenatal testing should be established before a pregnancy is undertaken.

In summary, the application of molecular techniques to the prevention of X-linked muscular dystrophies is a demanding and often difficult exercise, but one which is rewarding for both the family and clinician if done well. It demands an exceptionally high degree of collaboration between clinical and laboratory workers, with the clinical geneticist being in most instances the person who is best suited to undertake the coordinating role.

OTHER MUSCULAR DYSTROPHIES

The approaches used so successfully for Duchenne muscular dystrophy are already being applied to other muscular dystrophies, and it is likely that the genes for all the major forms will be accurately identified in the near future. The current situation for the various forms is summarized briefly below.

Emery–Dreifuss muscular dystrophy

This is a clinically distinct X-linked disorder (Dreifuss and Hogan, 1961; Emery and Dreifuss, 1966; Hopkins, Jackson and Elsas, 1981), characterized by relatively benign muscle wasting affecting humeral and peroneal groups as well as limb girdle muscles, contractures, and involving the heart at an early stage, with progressive atrial paralysis. A number of large kindreds are known, and the disorder probably includes some families (and isolated male cases) previously identified as X-linked humeroperoneal and scapuloperoneal syndromes.

Genetic linkage studies have firmly located the gene to the distal long arm of the X chromosome (Boswinkel *et al.*, 1985; Thomas *et al.*, 1986; Yates *et al.*, 1986), with moderately close linkage to the factor 8 locus, as well as to the DNA probes St14 and DX13; there is also close linkage to the genes for colour blindness, which have been cloned and show useful polymorphism.

A possible animal model that may prove relevant to Emery Dreifuss dystrophy is the *mdx* mouse (Bulfield *et al.*, 1984). This X-linked myopathy was originally thought likely to be homologous to Duchenne muscular dystrophy but is now considered more comparable to Emery Dreifuss dystrophy. Unfortunately the mouse linkage map of the X chromosome can not be related simply to that of the human, so it is not yet clear to which (if either) human dystrophy this mutant corresponds.

X-linked myotubular myopathy

This uncommon and usually lethal congenital myopathy (Barth, van Wijngaarden and Bethlem, 1975) also shows preliminary evidence for a distal long arm localization, but this requires confirmation.

Myotonic dystrophy

This has been the subject of considerable genetic study (Harper, 1979, 1986), including linkage studies with protein markers that showed clear linkage with the secretor and Lutheran blood group loci (Renwick *et al.*, 1971; Harper *et al.*, 1972). This linkage group is now known to be on chromosome 19 (Whitehead *et al.*, 1982), which is becoming one of the best mapped autosomes (Brook *et al.*, 1986). Close markers include the enzyme peptidase D, as well as a number of DNA sequences, the best documented and probably the closest being the gene APOC2 (Shaw *et al.*, 1985), with a recombination fraction of around 4%. This probe is informative in most families and has been applied both in prediction for apparently healthy individuals at risk and also prenatally (Meredith *et al.*, 1986). Among the other DNA probes linked to myotonic dystrophy, LDR 152, derived from a flow sorted chromosome 19 library (Bartlett *et al.*, 1987), also shows close linkage, but the ordering in relation to APOC2 and the disease gene is uncertain. Molecular studies starting from this sequence have recently identified neighbouring 'HTF islands' (corresponding to unmethylated regions often associated with the 5′ ends of specific genes). Approaches such as this and pulsed field electrophoresis are beginning to produce a detailed molecular map of the proximal long arm of chromosome 19 in a way comparable to that already achieved for the Duchenne muscular dystrophy region of the X chromosome. A lack of chromosomal deletions or translocations producing myotonic dystrophy is one factor that has held back progress to this end.

Several candidate genes have become available for testing in relation to myotonic dystrophy on account of their chromosomal location and their relationship to functions known to be impaired in myotonic dystrophy. The first of these was the insulin receptor gene, found to be on chromosome 19 (Yang-Feng, Francke and Ullrich, 1985), which was of interest in view of the known abnormality in insulin sensitivity shown by most patients with myotonic dystrophy. This provided a good example of how rapidly a candidate gene can be assessed and rejected, since it was soon found that the insulin receptor gene mapped physically to a different chromosomal region when tested against a panel of hybrid cell lines containing different portions of 19 (the distal short arm as opposed to the proximal long arm) (Shaw *et al.*, 1986). Also, recombination could be demonstrated between the two, firmly excluding a causative relationship.

More recent candidate genes which map in the appropriate region of 19 include the muscle form of creatine kinase (Nigro *et al.*, 1987) and those for sodium–potassium (Na–K) ATPase (Shaw and Eiberg, 1987). The latter is of particular interest in view of the likely membrane defect responsible for the myotonia, even though previous studies of ATPase in myotonic dystrophy have been inconclusive.

Facioscapulohumeral dystrophy

This disorder, following autosomal dominant inheritance and in general the mildest of the muscular dystrophies, has been the subject of some comprehensive general genetic studies, including linkage to protein markers (Padberg, 1982). Although positive results were obtained with the marker Gm, subsequent DNA analysis using immunoglobulin gene probes and other markers in the adjacent area of chromosome 14 have proved negative (Lunt *et al.*, 1988). Further studies are in progress and an exclusion map has so far excluded the disorder from at least 30% of the genome (Lunt *et al.*, 1988), so definitive results are likely before long.

Autosomal recessive limb girdle dystrophy

Although this has long been classed as a separate disorder, the more closely it is studied the rarer it becomes (Yates and Emery, 1985). In the United Kingdom at least most cases prove to be Becker muscular dystrophy (isolated males), manifesting carriers for Duchenne muscular dystrophy (females), or spinal muscular atrophies (incompletely studied cases). The residuum is too scanty, and probably still too heterogeneous, to allow gene mapping studies. In some countries this form appears to be more common (Moser, 1966), so the best hope for identifying the gene is probably to study large inbred kindreds with multiple cases.

Conclusions

The techniques of molecular genetics have resulted in major increases in our understanding of the muscular dystrophies and have already brought about a radical change in the approach to research into this group of disorders. For both Duchenne and Becker muscular dystrophy we now have a detailed understanding of the gene and some of its pathology, and are poised to convert this into a comparable knowledge at the protein level. The other dystrophies, in particular myotonic dystrophy, are following a similar course and there is a clear expectation that the localization and characterization of the genes for the major forms of muscular dystrophy will be achieved in the near future.

These developments have already brought with them practical gains in terms of improved carrier detection and prenatal diagnosis for those forms where gene localization is relatively accurate. The challenge is now to see whether molecular techniques will be able to help in modifying the course of the disorder. Meanwhile, the techniques and concepts that have arisen through the molecular analysis of the muscular dystrophies are already bearing fruit in the field of genetic disorders generally, and inherited neurological disorders in particular.

References

BAKKER, E., BONTEN, E. J., DE LANGE, L. F. *et al.* (1986) DNA probe analysis for carrier detection and prenatal diagnosis of Duchenne muscular dystrophy. *Journal of Medical Genetics,* **23,** 573–580

BAKKER, E., VAN BROECKHOVEN, C., BONTEN, E. J. *et al.* (1977) Germline mosaicism and Duchenne muscular dystrophy mutations. *Nature,* **329,** 554–556

BARTH, P. G., VAN WIJNGAARDEN, G. K. and BETHLEM, J. (1975) X-linked myotubular myopathy with fatal neonatal asphyxia. *Neurology,* **25,** 531–536

BARTLETT, R. J., PERICAK-VANCE, M. A., YAMAOKA, L. *et al.* (1987) A new probe for the diagnosis of myotonic muscular dystrophy. *Science,* **235,** 1641–1650

BECKER, P. E. and KIENER, F. (1955) Eine neue X-chromosomale Muskeldystrophie. *Archiv für Psychiatrie und Nervenkrankheiten,* **193,** 427–448

BERTELSON, C. J., BARTLEY, J. A., MONACO, A. P., COLLETTI-FEENER, C., FISCHBECK, K. and KINKEL, L. M. (1986) Localisation of Xp21 meiotic exchange points in Duchenne muscular dystrophy families. *Journal of Medical Genetics,* **23,** 531–537

BLYTH, H., CARTER, C. O., DUBOWITZ, V. *et al.* (1965) Duchenne's muscular dystrophy and the Xg blood groups: a search for linkage. *Journal of Medical Genetics,* **2,** 157–160

BOSWINKEL, E., WALKER, A., HODGSON, S. *et al.* (1985) Linkage analysis using eight DNA polymorphisms along the length of the X chromosome locates the gene for Emery–Dreifuss muscular dystrophy to distal Xq. *Cytogenetics and Cell Genetics,* **40,** 586

BOTSTEIN, D., WHITE, R. L., SKOLNICK, M. and DAVIES, R. W. (1980) Construction of a genetic linkage map in man using restriction fragment length polymorphisms. *American Journal of Human Genetics,* **32,** 314–331

BOYD, Y. and BUCKLE, V. J. (1986) Cytogenetic heterogeneity of translocations associated with Duchenne muscular dystrophy. *Clinical Genetics,* **29,** 108–115

BOYD, Y., BUCKLE, V., HOLT, S., MUNRO, E., HUNTER, D. and CRAIG, I. (1986) Muscular dystrophy in girls with X: autosome translocations. *Journal of Medical Genetics,* **23,** 484–490

BROOK, J. D., SHAW, D. J., THOMAS, N. S. T., MEREDITH, A. L., COWELL, J. and HARPER, P. S. (1986) Mapping genetic markers on human chromosome 19 using subchromosomal fragments in somatic cell hybrids. *Cytogenetics and Cell Genetics,* **41,** 30–37

BROWN, C. S., THOMAS, N. S. T., SARFARAZI, M. *et al.* (1985) Genetic linkage relationships of seven DNA probes with Duchenne and Becker muscular dystrophies. *Human Genetics,* **71,** 62–74

BULFIELD, G., SILLER, W. G., WIGHT, P. A. L. and MOORE, K. J. (1984) X chromosome-linked muscular dystrophy (mdx) in the mouse. *Proceedings of the National Academy of Science (USA),* **81,** 1189–1192

CASKEY, C. T., NUSSBAUM, R. L., COHAN, L. C. and POLLACK, L. (1980) Sporadic occurrence of Duchenne muscular dystrophy: evidence for new mutation. *Clinical Genetics,* **18,** 329–341

DAVIES, K. E., BRIAND, P., IONASESCU, V. *et al.* (1985) Gene for OTC: characterization and linkage to Duchenne muscular dystrophy. *Nucleic Acids Research,* **13,** 155–165

DAVIES, K. E., PEARSON, P. L., HARPER, P. S., MURRAY, J. M., O'BRIEN, T., SARFARAZI, M. and WILLIAMSON, R. (1983) Linkage analysis of two cloned DNA sequences flanking the Duchenne muscular dystrophy locus on the short arm of the human X-chromosome. *Nucleic Acids Research,* **11,** 2303–2312

DAVIES, K. E., YOUNG, B. D., ELLES, R. G., HILL, M. E. and WILLIAMSON, R. (1981) Cloning of a representative genomic library of the human X-chromosome after sorting by flow cytometry. *Nature (London),* **293,** 374–376

DRAYNA, D. and WHITE, R. (1985) The genetic linkage map of the human X chromosome. *Science,* **230,** 753–758

DREIFUSS, F. H. and HOGAN, G. R. (1961) Survival in X-chromosomal muscular dystrophy. *Neurology,* **11,** 734–737

EMERY, A. E. H. and DREIFUSS, F. E. (1966) Unusual type of benign X-linked muscular dystrophy. *Journal of Neurology, Neurosurgery and Psychiatry,* **29,** 338–342

ENGEL, A. G. (1986) Duchenne dystrophy. In *Myology,* edited by A. G. Engel and B. Q. Banker, pp. 1185–1240. New York: McGraw-Hill

FRANCKE, U., HARPER, J. F., DARRA, B. T. *et al.* (1987) Congenital adrenal hypoplasia, myopathy, and glycerol kinase deficiency: Molecular genetic evidence for deletions. *American Journal of Human Genetics,* **40,** 212–227

FRANCKE, U., OCHS, H. D., DE MARTINVILLE, B. *et al.* (1985) Minor Xp21 chromosome deletion in a male associated with expression of Duchenne muscular dystrophy, chronic granulomatous disease, retinitis pigmentosa and McLeod syndrome. *American Journal of Human Genetics,* **37,** 250–267

GOWERS, W. G. (1879) *Pseudo-hypertrophic Muscular Paralysis, A Clinical Lecture.* London: J. and A. Churchill

GUGGENHEIM, M. A., McCABE, E. R. B., ROIG, M. *et al.* (1980) Glycerol kinase deficiency with neuromuscular, skeletal and adrenal abnormalities. *Annals of Neurology,* **7,** 441–449

HALDANE, J. B. S. (1935) The rate of spontaneous mutation of a human gene. *Journal of Genetics,* **31,** 317–326

HARPER, P. S. (1979) *Myotonic Dystrophy.* Philadelphia: Saunders

HARPER, P. S. (1986) Myotonic disorders. In: *Myology,* edited by A. G. Engel and B. Q. Banker, pp. 1267–1296. New York: McGraw-Hill

HARPER, P. S., O'BRIEN, T., MURRAY, J. M., DAVIES, K. M., PEARSON, P. and WILLIAMSON, R. (1983) The use of linked DNA polymorphisms for genotype prediction in families with Duchenne muscular dystrophy. *Journal of Medical Genetics,* **20,** 252–254

HARPER, P. S., RIVAS, M. L., BIAS, W. B. M., HUTCHINSON, J. R., DYKEN, P. R. and McKUSICK, V. A. (1972) Genetic linkage confirmed between the loci for myotonic dystrophy, ABH secretion and Lutheran blood group. *American Journal of Human Genetics,* **24,** 310–316

HARPER, P. S. and THOMAS, N. S. T. (1986) A molecular approach to genetic counselling in the X-linked muscular dystrophies. *American Journal of Medical Genetics,* **25,** 687–702

HODGSON, S., WALKER, A., COLE, C. *et al.* (1987) The application of linkage analysis to genetic counselling in families with Duchenne or Becker muscular dystrophy. *Journal of Medical Genetics,* **24,** 152–159

HOFKER, M. H., WAPENAAR, M. C., GOOR, N., BAKKER, B., VAN OMMEN, G. J. B. and PEARSON, P. L. (1985) Isolation of probes detecting restriction fragment length polymorphisms from X chromosome specific libraries: potential use for diagnosis of Duchenne muscular dystrophy. *Human Genetics,* **70,** 148–156

HOPKINS, L. C., JACKSON, J. A. and ELSAS, L. J. (1981) Emery–Dreifuss humeroperoneal muscular dystrophy: an X-linked myopathy with unusual contractures and bradycardia. *Annals of Neurology,* **10,** 230–237

HOROWITS, R., KEMPNER, E. S., BISHER, M. E. and PODOLSKY, R. J. (1986) A physiological role for titin and nebulin in skeletal muscle. *Nature (London),* **323,** 160–164

KENWRICK, S., PATTERSON, M., SPEER, A., FISCHBECK, K. and DAVIES, K. (1987) Molecular analysis of the Duchenne muscular dystrophy region using pulsed field gel electrophoresis. *Cell,* **48,** 351–357

KINGSTON, H. M., SARFARAZI, M., THOMAS, N. S. T. and HARPER, P. S. (1984) Localisation of the Becker muscular dystrophy gene on the short arm of the X chromosome by linkage to cloned DNA sequences. *Human Genetics,* **67,** 6–17

KINGSTON, H. M., THOMAS, N. S. T., PEARSON, P. L., SARFARAZI, M. and HARPER, P. S. (1983) Genetic linkage between Becker muscular dystrophy and a polymorphic DNA sequence on the short arm of the X chromosome. *Journal of Medical Genetics,* **20,** 255–258

KOENIG, M., HOFFMAN, E. P., BERTELSON, C. J., MONACO, A. P., FEENER, C., KUNKEL, L. M. (1987) Complete cloning of the Duchenne muscular dystrophy (DMD) cDNA and preliminary genomic organisation of the DMD gene in normal and affected individuals, *Cell,* **50,** 509–517

KUNKEL, L. M., HEJTMANCIK, J. F., CASKEY, C. T. *et al.* (1986) Analysis of deletions in DNA from patients with Becker and Duchenne muscular dystrophy. *Nature (London),* **322,** 73–77

KUNKEL, L. M., MONACO, A. P., MIDDLESWORTH, W., OCHS, H. D. and LATT, S. A. (1985) Specific cloning of DNA fragments absent from the DNA of a male patient with an X chromosome deletion. *Proceedings of the National Academy of Science USA,* **82,** 4778–4782

KUNKEL, L. M., TANTRAVAHI, U., EISENHARD, M. and LATT, S. A. (1982) Regional localisation on the human X of DNA sequences cloned from flow sorted chromosomes. *Nucleic Acids Research,* **10,** 1557–1561

LINDENBAUM, R. H., CLARKE, G., PATEL, C., MONCRIEFF, M. and HUGHES, J. T. (1979) Muscular dystrophy in an X:1 translocation female suggests that Duchenne locus is on X chromosome short arm. *Journal of Medical Genetics,* **16,** 389–392

LUNT, P. W., UPADHAYA, M., NOADES, J., SARFARAZI, M., HARPER, P. S. (1988) Fascioscapulohumeral muscular dystrophy. *Cytogenetics and Cell Genetics* (in press)

MANIATIS, T., HARDISON, R., LACY, E. *et al.* (1978) The isolation of structural genes from libraries of eukaryotic DNA. *Cell,* **15,** 687–693

MEREDITH, A. L., HUSON, S. M., LUNT, P. W. *et al.* (1986) Application of a closely linked polymorphism of restriction fragment length to counselling and prenatal testing in families with myotonic dystrophy. *British Medical Journal,* **293,** 1353–1356

MONACO, A. P., BERTELSON, C. J., MIDDLESWORTH, W. *et al.* (1985) Detection of deletions spanning the Duchenne muscular dystrophy locus using a tightly linked DNA segment. *Nature (London),* **316,** 842–845

MONACO, A. P. and KUNKEL, L. M. (1987) A giant locus for the Duchenne and Becker muscular dystrophy gene. *Trends in Genetics,* **3,** 33–37

MONACO, A. P., NEVE, R. L., COLLETTI-FEENER, C., BERTELSON, C. J., KURNIT, D. M. and KUNKEL, L. M. (1986) Isolation of candidate cDNAs for portions of the Duchenne muscular dystrophy gene. *Nature (London),* **323,** 646–650

MOSER, H. (1966) Progressive Muskeldystrophie. VII. Haufigdeit, klinik and Genetik der Typen I and II. *Schweizerische Medizinische Wochenschrift,* **96,** 169–174, 205–211

MURRAY, J. M., DAVIES, K. E., HARPER, P. S., MEREDITH, L., MUELLER, C. R. and WILLIAMSON, R. (1982) Linkage relationship of a cloned DNA sequence on the short arm of the X chromosome to Duchenne muscular dystrophy. *Nature (London),* **300,** 69–71

NIGRO, J. M., SCHWEINFEST, C. W., RAJKOVIC, A. *et al.* (1987) cDNA cloning and mapping of the human creatine kinase M gene to 19q13. *American Journal of Human Genetics,* **40,** 115–125

O'BRIEN, T. A., HARPER, P. S., DAVIES, K., MURRAY, J. M., SARFARAZI, M. and WILLIAMSON, R. (1983) Lack of heterogeneity in Duchenne muscular dystrophy. *Journal of Medical Genetics,* **20,** 249–251

PADBERG, G. (1982) *Facioscapulohumeral disease.* Thesis, University of Leiden

RAY, P. N., BELFALL, B., DUFF, C. *et al.* (1985) Cloning of the breakpoint of an X:21 translocation associated with Duchenne muscular dystrophy. *Nature (London),* **318,** 672–675

READ, A. P., KERZIN-STORRAR, L., MOUNTFORD, R. C., ELLES, R. G. and HARRIS, R. (1986) A register based system for gene tracking in Duchenne muscular dystrophy. *Journal of Medical Genetics,* **23,** 581–586

RENWICK, J. H., BUNDEY, S. E., FERGUSON-SMITH, M. A. and IZATT, M. M. (1971) Confirmation of the linkage of the loci for myotonic dystrophy and ABH secretion. *Journal of Medical Genetics,* **8,** 407

ROSES, A. D., ROSES, M. J., METCALF, B. S., HULL, K. L., NICHOLSON, G. A., HARTWIG, G. B. and ROE, C. R. (1977) Pedigree testing in Duchenne muscular dystrophy. *Annals of Neurology,* **2,** 270–278

SHAW, D. J. and EIBERG, H. (1987) Report on chromosomes 17, 18, 19. *Cytogenetics and Cell Genetics* (in press)

SHAW, D. J., MEREDITH, A. L., BROOK, J. D. *et al.* (1986) Linkage relationships of the insulin receptor gene with the complement component 3, LDL receptor, apolipoprotein C2 and myotonic dystrophy loci on chromosome 19. *Human Genetics,* **74,** 267–269

SHAW, D. J., MEREDITH, A. L., SARFARAZI, M., HUSON, S. M., BROOK, J. D., MYKLEBOST, O. and HARPER, P. S. (1985) The apolipoprotein CII gene: sub-chromosomal localisation and linkage to the myotonic dystrophy locus. *Human Genetics,* **70,** 271–273

SKINNER, R., SMITH, C. and EMERY, A. E. H. (1974) Linkage between the loci for benign (Becker-type) X-borne muscular dystrophy and deutan colour blindness. *Journal of Medical Genetics,* **11,** 317–320

THOMPSON, M. W., RAY, P. N., BELFALL, B., DUFF, C., LOGAN, C., OSS, I. and WORTON, R. G. (1986) Linkage analysis of polymorphisms within the DNA fragment XJ cloned from the breakpoint of an X:21 translocation associated with X-linked muscular dystrophy. *Journal of Medical Genetics,* **23,** 548–555

THOMAS, N. S. T., WILLIAMS, H., ELSAS, L. J., HOPKINS, L. C., SARFARAZI, M. and HARPER, P. S. (1986) Localisation of the gene for Emery–Dreifuss muscular dystrophy to the distal long arm of the X chromosome. *Journal of Medical Genetics,* **23,** 596–598

VAN OMMEN, G. J. B., VERKERK, J. M. H., HOFKER, M. H. *et al.* (1986) A physical map of 4 million bp around the Duchenne muscular dystrophy gene on the human X chromosome. *Cell,* **47,** 499–504

VERELLEN, C., DE MEYER, R., FREUND, M., LATERRE, C., SCHOLBERG, B. and FREDERIC, J. (1977) Progressive muscular dystrophy of the Duchenne type in a young girl associated with an aberration of chromosome X. In: *Proceedings of the Fifth International Congress of Birth Defects,* p. 42. Amsterdam: Excerpta Medica

VERELLEN-DUMOULIN, C. H., FREUND, M., DE MEYER, R. *et al.* (1984) Expression of an X-linked muscular dystrophy in a female due to translocation involving Xp21 and non-random inactivation of the normal X chromosome. *Human Genetics,* **67,** 115–119

WALKER, A., HART, K., COLE, C., HODGSON, S., JOHNSON, L., DUBOWITZ, V. *et al.* (1986) Linkage studies in Duchenne and Becker muscular dystrophies. *Journal of Medical Genetics,* **23,** 538–547

WHITEHEAD, A. S., SOLOMON, E., CHAMBERS, S., BODMER, W. F., POVEY, S. and FEY, G. (1982) Assignment of the structural gene for the third component of human complement to chromosome 19. *Proceedings of the National Academy of Science USA,* **79,** 5021

WILLIAMS, H., SARFARAZI, M., BROWN, C., THOMAS, N. and HARPER, P. S. (1986) The use of flanking markers in the prediction for Duchenne muscular dystrophy. *Archives of Disease in Childhood,* **61,** 218–222

WOOD, D. S., ZEVIANI, M., PRELLE, A. *et al.* (1987) Is nebulin the defective gene product in Duchenne muscular dystrophy? *New England Journal of Medicine,* **316,** 107–108

WORTON, R. G., DUFF, C., SYLVESTER, J. E., SCHMICKEL, R. D. and WILLARD, H. F. (1984) Duchenne muscular dystrophy involving translocation of the DMD gene next to ribsomal RNA genes. *Science,* **224,** 1447–1449

YANG-FENG, T. L., FRANCKE, U. and ULLRICH, A. (1985) Gene for human insulin receptor: localisation to site on chromosome 19 involved in pre-B-cell leukemia. *Science,* **228,** 728–731

YASUDA, N. and KONDO, K. (1980) No sex difference in mutation rates of Duchenne muscular dystrophy. *Journal of Medical Genetics,* **17,** 106–111

YATES, J. R. W., AFFARA, N. A., JAMIESON, D. M. *et al.* (1986) Emery–Dreifuss muscular dystrophy: localisation to Xq27.3→qter confirmed by linkage to the factor VIII gene. *Journal of Medical Genetics,* **23,** 587–590

YATES, J. R. W. and EMERY, A. E. H. (1985) A population study of adult onset limb-girdle muscular dystrophy. *Journal of Medical Genetics,* **22,** 250–257

ZATZ, M., ITSKAN, S. B., SANGER, R., FROTA-PESSOA, O. and SALDANHA, P. H. (1974) New linkage data for the X-linked types of muscular dystrophy and G6PD variants, color blindness and Xg blood groups. *Journal of Medical Genetics,* **11,** 321–327

14
Mitochondrial genes and neurological disease

A. E. Harding and I. J. Holt

ASPECTS OF MITOCHONDRIAL STRUCTURE AND FUNCTION

Mitochondria are cylindrical structures with a diameter of 0.5–1.0 μm. They consist of an outer membrane, an inner membrane and two internal compartments, the intermembranous space and the matrix. The outer membrane is permeable to ions and small proteins, whereas the highly convoluted inner membrane, which forms a series of infoldings into the matrix called cristae, has highly selective permeability.

Many mitochondrial enzymes are concentrated in the matrix, including those required for the oxidation of pyruvate and fatty acids, as well as acetyl CoA in the Kreb's cycle. The main products of these processes are carbon dioxide and NADH; the latter is the principal substrate for the enzymes of the respiratory chain which are situated in the inner mitochondrial membrane. Electrons from NADH (and $FADH_2$) pass along the electron transfer chain, gradually releasing energy which pumps protons across the inner membrane. This process utilizes oxygen and creates an electrochemical proton gradient across the inner membrane which is mainly used to drive the conversion of ADP and phosphate to ATP by the enzyme mitochondrial ATPase. The cell is thus dependent on mitochondria for aerobic glycolysis, which produces 36 molecules of ATP from each molecule of glucose, compared to only two from anaerobic glycolysis (Tzagoloff, 1982; Alberts *et al.*, 1983; Hatefi, 1985).

The respiratory chain consists of four major enzyme complexes which are embedded in the inner mitochondrial membrane (*Figure 14.1*). Most of the electron carriers in the chain contain metal atoms (haem or non-haem iron, and copper) which are tightly bound to a protein surface. The NADH–coenzyme Q (CoQ) reductase complex (complex I) consists of about 26 polypeptides, including eight iron sulphur (FeS) proteins, and accepts electrons from NADH before transferring them to complex III (ubiquinol-cytochrome *c* reductase) via a small lipid-soluble molecule, ubiquinone (CoQ). $FADH_2$ feeds into the chain at complex III via complex II (succinate–CoQ reductase). The 11 or so subunits of complex III include cytochromes *b* and c_1, and two FeS proteins. Electrons pass from ubiquinone to cytochrome *c* via complex III, and then to the cytochrome oxidase complex (complex IV) which consists of 13 subunits including cytochromes *a* and

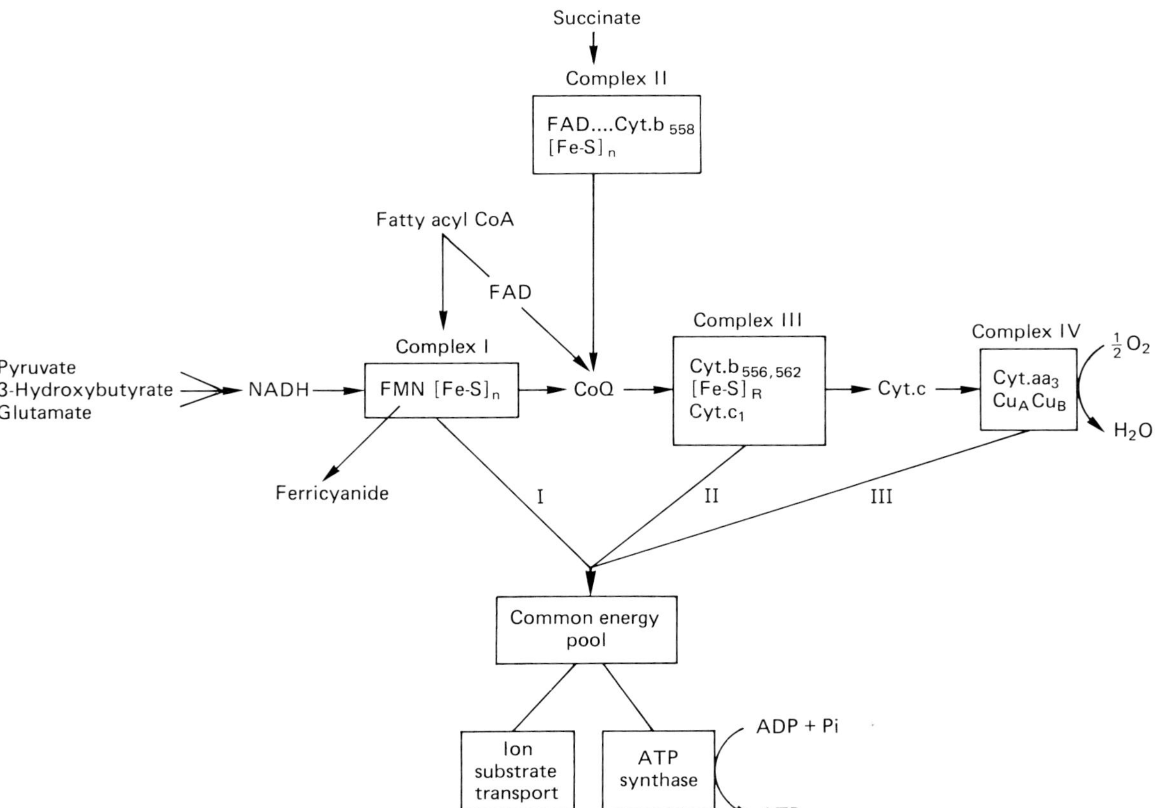

Figure 14.1 The mitochondrial respiratory chain and oxidative phosphorylation system

a_3. Cytochrome oxidase forms two water molecules from O_2 and four electrons transferred from cytochrome *c* (Hatefi, 1985).

The electrochemical proton gradient produced by electron transfer along the respiratory chain drives the production of ATP from ADP and phosphate by the inner membrane-bound enzyme complex ATP synthetase (complex V). The energy supplied by the gradient is also used to transport substances such as pyruvate and other mitochondrial enzyme substrates and mitochondrial proteins synthesized by the nuclear genetic system into the mitochondrial matrix (Alberts *et al.*, 1983; DiMauro *et al.*, 1985).

MITOCHONDRIAL GENETICS

The nucleus is not the only organelle which contains DNA in mammalian cells; mitochondria possess their own independently replicating genetic system. Each mammalian mitochondrion contains five to ten circular DNA molecules which are double-stranded and 16.6 kilobases (kb) in length (Alberts *et al.*, 1983; Attardi, 1986; Tzagoloff and Myers, 1986). Mitochondrial (mt) DNA differs from nuclear DNA in that it contains very little non-coding sequence, and to some extent in its genetic code which dictates, for example, that UGA reads tryptophan instead of a stop codon. Each strand of mtDNA is transcribed from a single promotor site, and then processed; the heavy (H) strand transcripts consist of two ribosomal RNAs, 14 tRNAs, and 12 protein coding sequences, whereas the L strand only codes for eight tRNAs and one protein coding sequence. The mitochondrial protein coding transcripts are polyadenylated with about 55 nucleotides at their 3′ end by mitochondrial poly-A polymerase, but they are not capped at their 5′ end (Anderson *et al.*, 1981; Attardi, 1981, 1986; Clayton, 1984).

Mitochondrial DNA encodes for 13 of the 67 or so subunits of the mitochondrial respiratory chain and oxidative phosphorylation system (*Figure 14.2*): seven

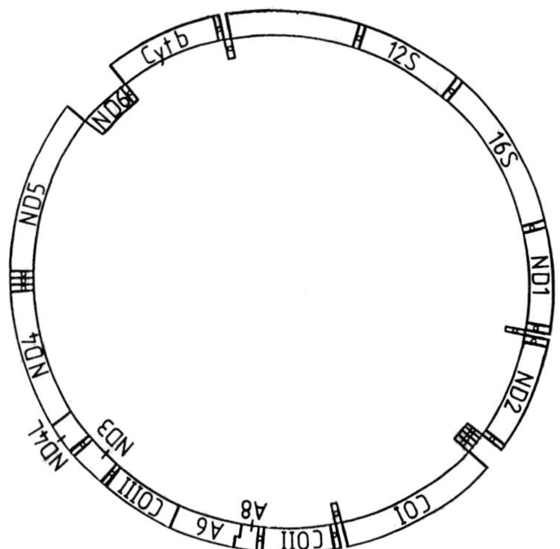

Figure 14.2 Map of the mitochondrial genome showing location of coding regions. ND = NADH dehydrogenase (or NADH–CoQ reductase; complex I), A = ATPase, CO = cytochrome oxidase, 12S and 16S = ribosomal RNAs; the symbol |○| indicates a tRNA

subunits of complex I; cytochrome *b* (complex III); subunits I, II, and III of cytochrome oxidase (complex IV); and subunits 6 and 8 of ATP synthetase (Chomyn *et al.*, 1985a, 1985b). The nuclear genome encodes the remaining polypeptides in the respiratory chain, and also controls their transport into the mitochondrion by synthesizing 'leader' peptides which appear to direct the proteins to sites of adhesion between the inner and outer mitochondrial membranes prior to transport into the matrix (Neupert and Schatz, 1981; Alberts *et al.*, 1983). Transcription and translation of the mitochondrial genome are also dependent on nuclear products such as the enzymes involved in mtDNA replication, RNA polymerase, and ribosomal proteins (Attardi, 1981).

It is not clear why the mitochondrion requires its own genetic system, but it has been suggested that a number of the mtDNA products are hydrophobic and therefore potentially difficult to import into the organelle from the cytoplasm. However, some of the mitochondrial proteins synthesized in the cytosol are very hydrophobic, and it may be that this complex situation merely represents a quirk of evolution; it is thought that mitochondria have evolved from prokaryocytes (such as bacteria) engulfed by eukaryocytic cells millions of years ago (Alberts *et al.*, 1983; Tzagoloff, 1982).

Mitochondria divide by binary fission, at a rate appropriate to that of division of their parent cells. Mitochondrial DNA molecules also replicate with every cell cycle, keeping the amount of organelle DNA per cell constant; mtDNA comprises about 1% of total cellular DNA. It has been clearly established that mtDNA is exclusively transmitted by females in many species, including man (Giles *et al.*, 1980; Lansman, Avise and Huettel, 1983). The cytoplasm of the ovum is an integral part of the developing embryo, and ova contain many more mitochondria than sperm. For example, it has been estimated that there are about 10^6 mitochondria in a cow's ovum, compared to about 70 in the mature spermatid of a bull (Hutchison *et al.*, 1974). Although a few paternal mitochondria may penetrate the ovum, these appear to degenerate subsequently (Fine, 1978). Thus each mammal should have the same mitochondrial genetic makeup as its mother and all matroclinal relatives, that is, any individual related only through females.

Maternal inheritance of mtDNA has been confirmed in recent years by means of restriction endonuclease analysis; mtDNA from all individuals in a single maternal line shows the same pattern of fragments after digestion with restriction endonucleases (Brown, 1980; Giles *et al.*, 1980). Nevertheless, just as inherited polymorphism is present in nuclear DNA, extensive nucleotide sequence divergence between different maternal lines occurs in mtDNA (Blanc *et al.*, 1983; Horai, Gojobori and Matsunaga, 1984; Cann, Stoneking and Wilson, 1987). The mutation rate of mtDNA is high; as it is inherited maternally and does not recombine, restriction mapping of mtDNA in different human populations can be used to trace their origins. It has been suggested that the current populations of Europe, North America, Asia, Australasia originated from one woman who is postulated to have lived in Africa about 200 000 years ago (Cann, Stoneking and Wilson, 1987).

MITOCHONDRIAL GENES AND DISEASE

If mutations of mtDNA cause disease, their manifestations would theoretically be expected in all the offspring of either sex of an affected female, but paternal transmission should not occur. It has been proposed that defects of the mitochondrial genome may give rise to two categories of disease in man, the mitochondrial myopathies (or encephalomyopathies) and Leber's optic atrophy.

Mitochondrial myopathies

The term mitochondrial myopathy (MM) is applied to a clinically and biochemically heterogeneous group of disorders which share the common feature of major mitochondrial structural abnormalities in skeletal muscle. Ragged red fibres containing peripheral and intermyofibrillar accumulations of abnormal mitochondria, seen with the modified Gomori trichrome stain (Olson *et al.,* 1972), are the major morphological hallmark of these diseases. They were first reported by Luft *et al.* (1962) in a patient with hypermetabolism due to defective coupling of oxidation and phosphorylation in muscle mitochondria; it later became apparent that they were common in patients presenting with syndromes of chronic progressive external ophthalmoplegia (CPEO) and/or proximal myopathy, often with weakness induced or enhanced by exertion. More recently they have been reported in children and adults with complex multisystem disorders predominantly or exclusively affecting the central nervous system (CNS), giving rise to clinical features such as psychomotor retardation, dementia, pigmentary retinopathy, ataxia, seizures, movement disorders, stroke-like episodes, deafness, and peripheral neuropathy in various combinations. Involvement of other systems, such as the heart, endocrine glands, and haemopoietic tissues, has also been reported. The protean manifestations of the mitochondrial myopathies have recently been reviewed by DiMauro *et al.* (1985) and Petty, Harding and Morgan-Hughes (1986).

A number of authors have suggested that cases of mitochondrial myopathy can be classified into distinct syndromes on clinical grounds, which include: the Kearns–Sayre syndrome, a combination of CPEO and pigmentary retinopathy developing before the age of 20 years, associated with ataxia, cardiac conduction defects, and elevation of the cerebrospinal fluid protein (Berenberg *et al.,* 1977); the syndrome of mitochondrial myopathy, encephalopathy, lactic acidosis and stroke-like episodes (MELAS) (Pavlakis *et al.,* 1984); and a third syndrome of myoclonus epilepsy with ragged red fibres (MERRF) (Fukuhara *et al.,* 1980). Petty, Harding and Morgan-Hughes (1986), reporting a series of 66 patients with mitochondrial myopathy, felt that these syndromes were not specific, even on clinical grounds, but merely represent combinations of some of the more striking features of these disorders.

It is possible to isolate human muscle mitochondria from large biopsy specimens in sufficient amounts for polarographic studies of oxidation and phosphorylation, and for the determination of cytochrome spectra (Morgan-Hughes *et al.,* 1977). These techniques have identified a variety of defects of the respiratory chain and oxidative phosphorylation system in patients with mitochondrial myopathy, all of which are associated with a wide range of clinical syndromes. Conversely, symptoms and signs may be similar in patients with different biochemical defects (Morgan-Hughes *et al.,* 1985; DiMauro *et al.,* 1985; Petty, Harding and Morgan-Hughes, 1986). These observations limit the usefulness of clinical classifications of the mitochondrial myopathies.

In vitro studies of mitochondrial metabolism in adult patients with mitochondrial myopathy most frequently show defects of the respiratory chain localized to complex I (NADH-CoQ reductase) (Morgan-Hughes *et al.*, 1985; Petty, Harding and Morgan-Hughes, 1986). Approximately half of the reported cases have had myopathy alone, and half a multisystem disorder involving muscle, the retina, and/or the CNS (Morgan-Hughes *et al.*, 1985). Complex I defects may also be seen in cases of infantile lactic acidosis with myopathy (Robinson *et al.*, 1986).

The clinical features observed in association with complex III defects in adults are also variable, but most commonly consist of proximal myopathy with CPEO although CNS involvement has been described (Morgan-Hughes *et al.*, 1985). Approximately 50% of patients with complex III defects have measurably low reducible cytochrome *b* (DiMauro *et al.*, 1985).

Several clinical syndromes occur in association with cytochrome *c* oxidase (complex IV) deficiency, including:

(1) fatal infantile mitochondrial myopathy with lactic acidosis, often associated with the de Toni–Fanconi–Debre renal syndrome and sometimes with cardiac or hepatic involvement;
(2) benign infantile mitochondrial myopathy resulting from reversible cytochrome *c* oxidase deficiency; and
(3) subacute necrotizing encephalomyelopathy (Leigh's disease) (DiMauro *et al.*, 1985).

Complex II deficiency is rare; it has been reported in some cases of Zellweger's disease, and two sibs with ataxia and myoclonus (Riggs *et al.*, 1984). Patients with mitochondrial myopathy and defects involving more than one complex have also been described, as has one example of the Kearns–Sayre syndrome with mitochondrial ATP synthetase deficiency (DiMauro *et al.*, 1985; Petty, Harding and Morgan-Hughes, 1986).

The majority of cases of mitochondrial myopathy are not familial, but kindreds containing more than one affected individual have been described. In the series of Harding, Petty and Morgan-Hughes (1988; *see also* Petty, Harding and Morgan-Hughes, 1986), 18% of index cases had similarly affected relatives. No consistent pattern of inheritance was evident for any of the clinical syndromes or identified defects of mitochondrial metabolism. The clinical and biochemical heterogeneity of the mitochondrial myopathies makes it difficult to draw any definite conclusions about their genetic basis. Some pedigrees suggest autosomal recessive or dominant inheritance. There is no evidence in favour of a significant contribution to the aetiology of mitochondrial myopathy from mutant genes on the X chromosome; male-to-male transmission occurs and there is no obvious difference in disease severity between males and females (Petty, Harding and Morgan-Hughes, 1986; Harding, Petty and Morgan-Hughes, 1988).

Data from both our series and a review of previously reported familial cases indicate that, when individuals are affected in more than one generation, maternal transmission to offspring is far more frequent than paternal transmission (in a ratio of approximately 9:1). Hudgson, Bradley and Jenkison (1972) first suggested that this could be explained on the basis of mitochondrial inheritance, and this hypothesis was later supported by Egger and Wilson (1983). In view of the coding functions of the mitochondrial genome, and the fact that the majority of patients with mitochondrial myopathy have biochemical defects localized to complex I, III, or IV of the respiratory chain, it is reasonable to suggest that these could result from mutations of mtDNA.

Nevertheless, if mitochondrial myopathy is mitochondrially inherited in pedigrees indicating maternal transmission, theoretically all the offspring of affected females should be affected, and only about half of them are (Harding, Petty and Morgan-Hughes, 1988). There are two possible explanations for this. One is that some individuals carrying the abnormal mitochondrial genotype do not express it clinically or histologically. The diagnosis of mitochondrial myopathy is sometimes difficult to confirm or exclude with certainty; clinically affected individuals with normal skeletal muscle biopsies and *in vitro* studies of mitochondrial metabolism have been described in some families (Rosing *et al.*, 1985; Petty, Harding and Morgan-Hughes, 1986). This is not surprising given that muscle is clinically unaffected in a significant proportion of patients.

An alternative explanation for reduced penetrance of an abnormal mitochondrial genotype in any single maternal line is that ova, collectively or individually, contain a heterogeneous population of mtDNA molecules. This hypothesis is also compatible with the partial deficiencies seen in patients with defects of the respiratory chain, and their variable expression in different tissues. It should be mentioned in this context that Robinson *et al.* (1986) cultured fibroblasts from three patients with infantile lactic acidosis and showed that the cells expressed the metabolic defects; two of the cases had NADH-CoQ reductase deficiency, and the third cytochrome *c* oxidase deficiency.

Although significant nucleotide divergence in a normal individual's mtDNA is not known to occur in man, it has been demonstrated in *Drosophila* (Solignac, Monnerot and Mounolou, 1983) and a single maternal line of Holstein cows (Hauswirth and Laipis, 1982; Hauswirth *et al.*, 1984). As has been mentioned, the high mutation rate of mtDNA leads to extensive polymorphism between different maternal lines (Cann, Stoneking and Wilson, 1987) and it would be rather surprising if mtDNA heteroplasmy did not occur.

Limited analysis of mitochondrial DNA in families with mitochondrial myopathy by means of restriction mapping has excluded major deletions of leukocyte mtDNA in patients, or any differences in restriction fragment patterns between normal and abnormal individuals in the same maternal line (Wallace *et al.*, 1985; Harding, Holt and Morgan-Hughes, 1987; Holt, Harding and Morgan-Hughes, 1988a; Poulton *et al.*, personal communication). Nevertheless, this approach clearly does not exclude the presence of small deletions or pathologically significant mutations outside restriction sites, and there is only about a 10% chance of detecting the latter using multiple restriction endonuclease analysis. It is difficult to assess the significance of previously unreported mtDNA polymorphisms in mitochondrial myopathy families in these studies; these may also be observed in controls and thus are of dubious relevance to the disease (Holt, Harding and Morgan-Hughes, 1988a).

More recently, Holt, Harding and Morgan-Hughes (1988b) have shown that some patients with mitochondrial myopathy have two populations of mtDNA in muscle, one of which is deleted (*Figure 14.3*). Deletions of muscle mtDNA, up to 7 kb in length, were observed in nine out of 25 cases, none of which had detectably abnormal leukocyte mtDNA. The proportion of abnormal mtDNAs in muscle ranged from 18 to 79%. In one patient with complex I deficiency, the deletion involved the two ribosomal RNAs and two subunits of NADH–CoQ reductase. All the others (four with complex I defects and two complex III deficiency), had deletions between about 7000 and 13 500, thus including three subunits of complex I, two each of cytochrome oxidase and mitochondrial ATP synthetase, and several tRNAs.

Figure 14.3 Hybridization patterns of ^{32}P labelled HeLa cell mtDNA to mtDNA from blood (B) and muscle (M) from a patient with mitochondrial myopathy. The DNA has been cut with the restriction endonuclease Pvu II, which cleaves the mitochondrial genome at one site, thus making it linear. In blood only one fragment (16.5 kb) is seen, but in muscle there are two. The lower band represents a population of mtDNAs bearing a deletion of about 6 kb

These observations demonstrate conclusively that mtDNA heteroplasmy can occur in man, and that human disease may be associated with defects of the mitochondrial genome. They also provide one explanation for the variable expression and distribution of respiratory chain defects in patients with mitochondrial myopathy, as mentioned above. The origin of the deletions is not known. Only one of the nine patients studied by Holt, Harding and Morgan-Hughes (1988b) had a family history suggesting maternal inheritance. The others were all single cases, and their mtDNA deletions presumably arose during oogenesis. Random partitioning of the two populations of mtDNA would be expected to occur during development; survival of the abnormal mtDNAs in muscle is compatible with the fact that the number of muscle fibres does not increase significantly after early fetal life. Conversely, genetically defective mitochondria might be selected against in cells, such as leukocyte precursors, which divide frequently.

On statistical grounds it is unlikely that all cases of mitochondrial myopathy are due to defective mitochondrial genes, as the nuclear genome codes for the majority of the respiratory chain subunits, as well as controlling their transport into the mitochondrion and subsequent assembly into functional enzyme complexes; transcription and translation of the mitochondrial genome are also dependent on the nucleus (Attardi, 1981). Evidence that mitochondrial myopathy may be caused by mutant nuclear genes is provided by Schapira *et al.* (1986) who showed that some patients with complex I defects have a specific deficiency of the 24 kD FeS protein which is a nuclear product.

Leber's optic atrophy

Leber's optic atrophy gives rise to acute or subacute visual loss, usually in males in the second or third decades of life. Patients often notice loss of vision in one eye a few days before the second. In the acute phase the optic discs are swollen with tortuous retinal arterioles, and telangiectases are present in the peripapillary small vessels (Nikoskelainen, Hoyt and Nummelin, 1983). There is some evidence that the appearances of capillary microangiopathy predate the onset of symptoms (Nikoskelainen, Hoyt and Nummelin, 1982). The visual field loss initially consists of an enlarged blind spot; this increases to involve central vision, producing a large centrocaecal scotoma. Optic atrophy is apparent after a few months. Visual loss is generally severe (6/60 or less), although limited improvement is sometimes observed (Carroll and Mastaglia, 1979; Nikoskelainen, Hoyt and Nummelin, 1983). There is a curious association between Leber's optic atrophy and cardiac pre-excitation syndromes (Nikoskelainen *et al.*, 1987).

Leber's optic atrophy shows a pattern of transmission which suggests cytoplasmic, or possibly mitochondrial, inheritance (Seedorf, 1970; Wallace, 1970; Nikoskelainen, 1984; Nikoskelainen *et al.*, 1987). About 85% of patients are male, and 18% of female carriers are affected. Paternal transmission of the disease to offspring has never been described. Between 70 and 100% of daughters of female carriers are also carriers; between 50 and 100% of the sons of carriers are affected.

Nikoskelainen *et al.* (1984, 1987) have suggested that Leber's optic atrophy may be a mitochondrial disease, partly on the basis of finding enlarged mitochondria with proliferation of cristae in skeletal muscle biopsies from patients. In view of the pattern of transmission, this is an interesting hypothesis. However, given that all the mitochondrial genes code for subunits of the mitochondrial respiratory chain and oxidative phosphorylation system, it is difficult to understand how mutations of mtDNA could give rise to Leber's optic atrophy.

Multiple restriction endonuclease analysis of leukocyte mtDNA from five families with the disease has not shown any differences in fragment patterns between normal and abnormal patients in any single maternal line, or any evidence of deletion of mtDNA in patients (Holt, Miller and Harding, unpublished data). A similar study of mtDNA in an unusual family, in which Leber's disease was seen in some members and maternally inherited dystonia in others (one female had both disorders), was also negative (Novotny *et al.*, 1986). Further patients with familial dystonia and a disorder resembling Leber's optic atrophy have been described by Marsden *et al.* (1986).

As is the case in mitochondrial myopathy, the finding of normal leukocyte mtDNA in patients with Leber's optic atrophy does not exclude the possibility of mitochondrial inheritance. It would, however, be difficult to examine relevant tissue for mtDNA heteroplasmy (*see above*) in Leber's disease.

There have been two reports of reduced rhodanese activity, in liver and rectal mucosa, in Leber's optic atrophy (Cagianut *et al.*, 1981; Poole and Kind, 1986), although others have reported normal activity in liver, brain (Wilson, 1965), and muscle (Nikoskelainen *et al.*, 1984). Rhodanese is a ubiquitous enzyme which may be important in cyanide detoxification, and Wilson (1965) suggested that Leber's optic atrophy was associated with defective cyanide metabolism. More recent evidence has indicated that rhodanese plays a role in the formation of iron sulphur proteins (Cerletti, 1986), which is of interest in relation to the hypothesis that the molecular basis of Leber's disease is one which would predictably be associated with defects of the mitochondrial respiratory chain. Intriguingly, Roger (1986) has recently reported a patient with the syndrome of myoclonus epilepsy with ragged red fibres (MERRF) who also has 'typical Leber's disease'.

References

ALBERTS, B., BRAY, D., LEWIS, J., RAFF, M., ROBERTS, K. and WATSON, J. D. (1983) *Molecular Biology of the Cell*. New York: Garland

ANDERSON, S., BANKIER, A. T., BARRELL, B. G. *et al.* (1981) Sequence and organisation of the human mitochondrial genome. *Nature (London)*, **290**, 457–465

ATTARDI, G. (1981) Organization and expression of the mammalian mitochondrial genome: a lesson in economy. *Trends in Biochemical Sciences*, **6**, 86–89, 100–103

ATTARDI, G. (1986) The elucidation of the human mitochondrial genome: a historical perspective. *BioEssays*, **5**, 34–39

BERENBERG, R. A., PELLOCK, J. M., DI MAURO, S. *et al.* (1977) Lumping or splitting? 'Ophthalmoplegia-plus' or Kearns–Sayre syndrome? *Annals of Neurology*, **1**, 37–54

BLANC, H., CHEN, K., D'AMOUR, M. A. and WALLACE, D. C. (1983) Amino acid change associated with the major polymorphic *Hinc II* site of oriental and Caucasian mitochondrial DNAs. *American Journal of Human Genetics*, **35**, 167–176

BROWN, W. M. (1980) Polymorphism in mitochondrial DNA of humans as revealed by restriction endonuclease analysis. *Proceedings of the National Academy of Sciences USA*, **77**, 3605–3609

CAGIANUT, B., RHYNER, K., FURRER, W. and SCHEBLI, H. P. (1981) Thiosulphate-sulphur transferase (rhodanese) deficiency in Leber's hereditary optic atrophy. *Lancet*, **i**, 981–982

CANN, R. L., STONEKING, M. and WILSON, A. C. (1987) Mitochondrial DNA and human evolution. *Nature (London)*, **325**, 31–36

CARROLL, W. M. and MASTAGLIA, F. L. (1979) Leber's optic neuropathy: a clinical and visual evoked potential study of affected and asymptomatic members of a six generation family. *Brain*, **102**, 559–580

CERLETTI, P. (1986) Seeking a better job for an under-employed enzyme: rhodanese. *Trends in Biochemical Sciences*, **11**, 369–372

CHOMYN, A., MARIOTTINI, P., CLEETER, M. W. J. *et al.* (1985a) Six unidentified reading frames of human mitochondrial DNA encode components of the respiratory chain NADH dehydrogenase. *Nature (London)*, **314**, 592–597

CHOMYN, A., MARIOTTINI, P., CLEETER, M. W. J. *et al.* (1985b) Functional assignment of the unidentified reading frames of human mitochondrial DNA. In: *Achievements and Perspectives of Mitochondrial Research, volume II: Biogenesis*, edited by E. Quagliariello *et al.*, pp. 259–275. Amsterdam: Elsevier Science Publishers

CLAYTON, D. A. (1984) Transcription of the mammalian mitochondrial genome. *Annual Review of Biochemistry*, **53**, 573–594

DiMAURO, S., BONILLA, E., ZEVIANI, M., NAKAGAWA, M. and DEVIVO, D. C. (1985) Mitochondrial myopathies. *Annals of Neurology*, **17**, 521–538

EGGER, J. and WILSON, J. (1983) Mitochondrial inheritance in a mitochondrially mediated disease. *New England Journal of Medicine*, **309**, 142–145

FINE, P. E. M. (1978) Mitochondrial inheritance and disease. *Lancet*, **ii**, 659–662

FUKUHARA, N., TOKIGUCHI, S., SHIRAKAWA, S. and TSUBAKI, T. (1980) Myoclonus epilepsy associated with ragged red fibres (mitochondrial abnormalities): disease entity or syndrome. *Journal of the Neurological Sciences*, **47**, 117–133

GILES, R. E., BLANC, H., CANN, R. M. and WALLACE, D. C. (1980) Maternal inheritance of human mitochondrial DNA. *Proceedings of the National Academy of Science USA*, **77**, 6715–6719

HARDING, A. E., HOLT, I. J. and MORGAN-HUGHES, J. A. (1987) Restriction enzyme analysis of mitochondrial DNA in patients with mitochondrial myopathy. *Journal of Medical Genetics*, **24**, 243

HARDING, A. E., PETTY, R. K. H. and MORGAN-HUGHES, J. A. (1988) Mitochondrial myopathy: a genetic study of 71 cases. *Journal of Medical Genetics* (in press)

HATEFI, Y. (1985) The mitochondrial electron transport and oxidative phosphorylation system. *Annual Review of Biochemistry*, **54**, 1015–1069

HAUSWIRTH, W. W. and LAIPIS, P. J. (1982) Mitochondrial DNA polymorphism in a maternal lineage of Holstein cows. *Proceedings of the National Academy of Science USA*, **79**, 4686–4690

HAUSWIRTH, W. W., VAN DE WALLE, M. J., LAIPIS, P. J. and OLIVO, P. D. (1984) Heterogeneous mitochondrial DNA D-loop sequences in bovine tissue. *Cell*, **37**, 1001–1007

HOLT, I, J., HARDING, A. E. and MORGAN-HUGHES, J. A. (1988a) Mitochondrial DNA polymorphism in mitochondrial myopathy. *Human Genetics* (in press)

HOLT, I. J., HARDING, A. E. and MORGAN-HUGHES, J. A. (1988b) Deletions of mitochondrial DNA in patients with mitochondrial myopathies. *Nature (London),* (in press)

HORAI, S., GOJOBORI, T. and MATSUNAGA, E. (1984) Mitochondrial DNA polymorphism in Japanese. I: Analysis with restriction enzymes of six base pair recognition. *Human Genetics,* **68,** 324–332

HUDGSON, P., BRADLEY, W. G., JENKISON, M. (1972) Familial mitochondrial myopathy. A myopathy with disordered oxidative metabolism in muscle fibres. Part 1. Clinical, electrophysiological and pathological findings. *Journal of the Neurological Sciences,* **16,** 343–370

HUTCHISON, C. A., NEWBOLD, J. E., POTTER, S. S. and EDGELL, M. H. (1974) Maternal inheritance of mitochondrial DNA. *Nature (London),* **251,** 536–538

LANSMAN, R. A., AVISE, J. C. and HUETTEL, M. D. (1983) Critical experimental test of the possibility of 'paternal leakage' of mitochondrial DNA. *Proceedings of the National Academy of Science,* **80,** 1969–1971

LUFT, R., IKKOS, O., PALMIERI, G., ERNSTER, L. and AFZELIUS, B. (1962) A case of severe hypermetabolism of non-thyroid origin with a defect in the maintenance of respiratory control: a correlated clinical, biochemical and morphological study. *Journal of Clinical Investigation,* **41,** 1776–1804

MARSDEN, C. D., LANG, A. E., QUINN, M. P., McDONALD, W. I., ABDALLAT, A., NIMRI, S. (1986) Familial dystonia and visual failure with striated CT lucencies. *Journal of Neurology, Neurosurgery and Psychiatry,* **49,** 500–509

MORGAN-HUGHES, J. A., DARVENIZA, P., KAHN, S. N. *et al.* (1977) A mitochondrial myopathy characterised by a deficiency in reducible cytochrome b. *Brain,* **100,** 617–640

MORGAN-HUGHES, J. A., HAYES, D. J., COOPER, M. and CLARK, J. B. (1985) Mitochondrial myopathies: deficiencies localised to complex I and complex III of the respiratory chain. *Biochemical Society Transactions,* **13,** 648–650

NEUPERT, W. and SCHATZ, G. (1981) How proteins are transported into mitochondria. *Trends in Biochemical Sciences,* **6,** 1–4

NIKOSKELAINEN, E. (1984) New aspects of the genetic, etiologic and clinical puzzle of Leber's disease. *Neurology,* **34,** 1482–1484

NIKOSKELAINEN, E. K., HOYT, W. F. and NUMMELIN, K. (1982) Ophthalmoscopic findings in Leber's hereditary optic neuropathy. Fundus findings in asymptomatic family members. *Archives of Ophthalmology,* **100,** 1597–1602

NIKOSKELAINEN, E. K., HOYT, W. F. and NUMMELIN, K. (1983) Ophthalmoscopic findings in Leber's hereditary optic neuropathy. The fundus findings in the affected family members. *Archives of Ophthalmology,* **101,** 1059–1068

NIKOSKELAINEN, E. K., HASSINEN, I. E., PALJARVI, L., LANG, H. and KALIMO, H. (1984) Leber's hereditary optic neuroretinopathy, a mitochondrial disease? *Lancet,* **ii,** 1474

NIKOSKELAINEN, E. K., SAVONTAUS, M. L., WANNE, O. P., KATILA, M. J. and NUMMELIN, K. U. (1987) Leber's hereditary optic neuroretinopathy, a maternally inherited disease. *Archives of Ophthalmology,* **105,** 665–671

NOVOTNY, E. J., SINGH, G., WALLACE, D. C. *et al.* (1986) Leber's disease and dystonia: a mitochondrial disease. *Neurology,* **36,** 1053–1060

OLSON, W., ENGEL, W. K., WALSH, G. O. and EINAUGLER, R. (1972) Oculocraniosomatic neuromuscular disease with 'ragged red' fibres. *Archives of Neurology,* **26,** 193–211

PAVLAKIS, S. G., PHILLIPS, P. C., DIMAURO, S., DEVIVO, D. C. and ROWLAND, L. P. (1984) Mitochondrial myopathy, encephalopathy, lactic acidosis, and strokelike episodes: a distinctive clinical syndrome. *Annals of Neurology,* **16,** 481–488

PETTY, R. K. H., HARDING, A. E. and MORGAN-HUGHES, J. A. (1986) The clinical features of mitochondrial myopathy. *Brain,* **109,** 915–938

POOLE, C. J. and KIND, P. R. N. (1986) Deficiency of thiosulphate sulphurtransferase (rhodanese) in Leber's hereditary optic neuropathy. *British Medical Journal,* **292,** 1229–1230

RIGGS, J. E., SCHOCHET, S. S., FAKADEJ, A. V. *et al.* (1984) Mitochondrial encephalomyopathy with decreased succinate-cytochrome *c* reductase activity. *Neurology,* **34,** 48–53

ROBINSON, B. H., WARD, J., GOODYER, P. and BAUDET, A. (1986) Respiratory chain defects in the mitochondria of cultured skin fibroblasts from three patients with lacticacidaemia. *Journal of Clinical Investigation,* **77,** 1422–1427

ROGER, J. (1986) Genetic transmission of myoclonus epilepsy with ragged red fibers. *Annals of Neurology,* **20,** 545

ROSING, H. S., HOPKINS, L. C., WALLACE, D. C. *et al.* (1985) Maternally inherited mitochondrial myopathy and myoclonic epilepsy. *Annals of Neurology,* **17,** 228–237

SCHAPIRA, A. H. V., COOPER, J. M., MORGAN-HUGHES, J. A. and CLARK, J. B. (1986) Evidence for molecular heterogeneity in human complex I deficiency. *Muscle and Nerve*, **9 (Suppl. 5)**, 183

SEEDORF, T. (1970) Leber's disease. *Acta Ophthalmologica*, **48**, 187–213

SOLIGNAC, M., MONNEROT, M. and MOUNOLOU, J.-C. (1983) Mitochondrial DNA heteroplasmy in *Drosophila mauritiana. Proceedings of the National Academy of Science USA*, **80**, 6942–6946

TZAGOLOFF, A. (1982) *Mitochondria.* New York: Plenum

TZAGOLOFF, A. and MYERS, A. M. (1986) Genetics of mitochondrial biogenesis. *Annual Review of Biochemistry*, **55**, 249–285

WALLACE, D. C. (1970) Leber's optic atrophy: a possible example of vertical transmission of a slow virus in man. *Australian Annals of Medicine*, **3**, 259–262

WALLACE, D. C., SINGH, G., HOPKINS, L. C. and NOVOTNY, E. J. (1985) Maternally inherited diseases of man. In: *Achievements and Perspectives of Mitochondrial Research, volume II: Biogenesis*, edited by E. Quagliariello *et al.*, pp. 427–436. Amsterdam: Elsevier Science Publishers

WILSON, J. (1965) Leber's hereditary optic atrophy: a possible defect of cyanide metabolism. *Journal of the Neurological Sciences*, **29**, 505–515

15
Molecular basis of retinoblastoma

Robert S. Sparkes

INTRODUCTION

Human retinoblastoma is the most common intraocular eye tumour of childhood with an occurrence of about 1 in 20 000. For many years it has been considered the prototype of the genetic developmental tumours.

In the majority of affected persons the tumour occurs sporadically and the family history is negative. About 10% of instances follow an autosomal dominant pattern of inheritance. But even with a negative family history, cases in which both eyes are affected are considered to represent germline mutations with the usual risk for autosomal dominant inheritance to children of the affected person. It has also been estimated that 15–20% of unilateral cases fall into this same category. Although the basis of the remaining sporadic cases is not known, most evidence would suggest that these represent a genetic disorder limited only to the somatic cells. It has also been estimated that less than 5% of cases have a constitutional chromosome deletion affecting band q14 of chromosome 13.

In all cases, the most clearly evident manifestation of the retinoblastoma gene is the retinoblastoma tumour itself. The tumour occurs in infancy in most cases, and is thought to be derived from a retinoblast cell which has the potential to develop into photoreceptor cells and neural cells of the retina. In keeping with other embryonic tumours, retinoblastoma occurs in a tissue which has a relatively late ongoing development and may still be partly immature at birth. Such tumours generally have a higher survival rate with spontaneous regression being more common.

The penetrance of the abnormal retinoblastoma gene is about 90%. Thus, some people with the gene will not show evidence of its clinical expression. Another benign manifestation may be retinoma (Connolly *et al.*, 1983). The retinoblastoma gene also appears to be related to the development of some pineoblastomas resulting in trilateral retinoblastoma (Bader *et al.*, 1982). Persons with the hereditary form of retinoblastoma appear to be at a greatly increased risk to develop other tumours, particularly osteogenic sarcoma (Kitchin and Ellsworth, 1974).

In recent years, retinoblastoma has received increasing attention from geneticists and molecular biologists. Much of this attention can be attributed to Knudson who in 1971 first formulated his two-mutation hypothesis for the development of retinoblastoma. His hypothesis attributes the development of retinoblastoma, both the hereditary and non-hereditary forms, to two mutational events. In the hereditary form, one mutation is inherited and only a second somatic cell mutation is required for the tumour to appear. Thus, such individuals who have inherited the gene are at an increased risk of developing the tumour and require only one additional event; in keeping with this theory, individuals with the inherited form of the disease tend to develop tumours earlier. In non-inherited cases, both mutations have to occur in the same cell and hence the tumour is more likely to be unilateral and to occur relatively late compared to the hereditary type. Most evidence to date supports this theory.

CHROMOSOMES IN RETINOBLASTOMA

Chromosome changes associated with retinoblastoma (Sparkes, 1984) have contributed significantly to the mapping of the retinoblastoma gene, and its apparent recent isolation as described below. Probably less than 5% of cases of retinoblastoma are associated with a constitutional chromosome deletion which, while it may vary in extent, always appears to involve band 13q14. Because chromosome changes are often associated with tumours, the retinoblastoma tumours themselves have been extensively analysed cytogenetically. These studies focused on chromosome 13 and in some instances chromosome changes, including deletions, were observed. However, most retinoblastomas have other chromosome abnormalities. One of the most consistent abnormalities appears to be an isochromosome of the short arm of chromosome 6 which is generally in addition to two normal chromosomes 6. Another chromosome abnormality which has been observed in relatively high frequency in retinoblastomas involves the long arm of chromosome 1 and generally leads to extra copies of at least part of this chromosome, especially band 1q23. The relationship of these chromosome changes to the development and progression of the retinoblastoma is unclear at this time.

ESTERASE D STUDIES

Esterase D is an enzyme found in most human tissues but its function is unclear. A recent report (Varki, Muchmore and Diaz, 1986) suggests that esterase D may be involved in the recycling of O-acetylated sialic acid molecules. Based upon its relative specificity for methylumbelliferyl esters and its different electrophoretic mobility, esterase D can be distinguished from other tissue esterases. Two relatively common alleles of esterase D have been identified as 1 and 2 (Hopkinson *et al.*, 1973).

Earlier somatic cell hybridization studies had assigned the gene for esterase D to human chromosome 13. Using a gene dosage evaluation in patients with deletions of human chromosome 13, it was possible to assign the locus for esterase D to band 13q14 (Sparkes *et al.*, 1979, 1980). This was the same band that appeared to be always involved in the constitutional chromosome deletion associated with retinoblastoma. It was therefore concluded that at least in the chromosome deletion form of retinoblastoma, a locus related to the development of the retinoblastoma is located in the band 13q14. Patients with the sporadic form of retinoblastoma and no demonstrable constitutional chromosome deletion were found to have normal esterase D activity suggesting that the esterase D locus is different from the retinoblastoma locus.

Two possible inconsistencies between esterase D activity and chromosomal constitution have been observed. One was a patient who had retinoblastoma with apparently normal chromosomes but approximately half normal esterase D activity; her parents had normal esterase D activity (Benedict *et al.*, 1983). It was concluded that the patient probably had a submicroscopic chromosome deletion which included both the retinoblastoma and esterase D. These studies indicate that these two loci are very close to one another. The retinoblastoma tumour cells from this patient demonstrated no esterase D activity and contained only one apparently normal chromosome 13. These results suggested that only the chromosome 13 with the presumed submicroscopic chromosome deletion was present in the tumour cells. Another patient had retinoblastoma with a constitutional chromosome deletion but normal esterase D activity (Sparkes *et al.*, 1984). This patient had sporadic retinoblastoma with a karyotype of 46,XX, del(13)(q14.1q22.3). It was suggested that the break in the band 13q14 had deleted the retinoblastoma locus but not the esterase D locus, thus indicating that the linear order of these genes is centromere–esterase D–retinoblastoma.

The next logical question was to determine whether the gene for the hereditary form of retinoblastoma was also located in 13q14. Using standard genetic linkage analysis in three informative families, it was possible to establish that the locus for the hereditary form of retinoblastoma is closely linked to esterase D and presumably at 13q14 (Sparkes *et al.*, 1983). This information also made it potentially possible to use genetic linkage analysis to determine in informative families who has the retinoblastoma gene even before the disease becomes manifest. Because of the limited genetic polymorphism for esterase D, there was considerable interest in identifying DNA markers for chromosome 13 close to the retinoblastoma gene. As discussed below, this was successful.

The esterase D studies gave the first insight into further genetic changes occurring within the tumour. The first indication that a recessive genetic mechanism may be operative within the tumour cells themselves came from the patient reported by Benedict *et al.* (1983). This is the patient described above in whom only one chromosome 13 was observed in the tumour which contained no esterase D activity. A recessive mechanism had been earlier postulated by Comings (1973).

Godbout *et al.* (1983) determined the electrophoretic types of esterase D in the red blood cells and tumours of patients with retinoblastoma. They found six patients who were heterozygous for the electrophoretic type of esterase D, and four of these expressed only one electrophoretic type in the tumours as had been found in the patient described earlier. The results were also in keeping with the recessive model. Elegant molecular studies have clarified this situation further.

MOLECULAR STUDIES OF RETINOBLASTOMA

Cavenee *et al.* (1983) used polymorphic DNA markers from chromosome 13 to evaluate retinoblastoma tumours in culture. They first showed that non-tumour cells from the affected individuals were heterozygous for one or more of the DNA markers. A study of tumours from the same individual showed that at least some of the tumours became homozygous for the chromosome 13 DNA markers. They suggested that at least three different chromosomal mechanisms were involved in the production of the homozygous state for the retinoblastoma gene, which resulted in the unmasking of the mutant allele. These included loss of one chromosome 13 leaving only a single chromosome 13 containing the retinoblastoma gene, as in the patient described above (Benedict *et al.*, 1983). The second mechanism seemed to be one of non-disjunction of chromosome 13 followed by reduplication of the chromosome 13 presumed to be carrying the retinoblastoma gene. The third mechanism was somatic cell mitotic recombination. Other possible explanations, which Cavenee *et al.* (1985) were not able to test directly, included gene conversion, gene deletion, and point mutation of the affected gene. Other studies by Dryja *et al.* (1984) of tumour tissue itself also demonstrated loss of heterozygosity for chromosome 13 in both sporadic and hereditary retinoblastomas. Because the retinoblastoma gene itself was not studied in these investigations, one can only infer that the retinoblastoma locus became homozygous. This may now be possible to test as it appears that the retinoblastoma gene may have been isolated, as discussed below. Further studies by Cavenee *et al.* (1985) demonstrated that the normal chromosome 13 is lost and that the chromosome 13 containing the mutant retinoblastoma allele is retained.

As noted above (p. 211), patients with the hereditary form of retinoblastoma show a greatly increased risk of developing osteogenic sarcoma later in life. This raised the possibility as to whether osteogenic sarcomas could also be related to chromosome 13. Hansen *et al.* (1985) evaluated several osteogenic sarcomas and found that chromosome 13 markers also became homozygous in these tumours. It was of interest that of the seven children studied by these authors, three had a history of retinoblastoma in infancy while four had no such history. Nevertheless, the results were the same in all the osteogenic sarcomas. Similar results were found by Dryja *et al.* (1986a).

Because of the apparent close genetic linkage between the gene for esterase D and the retinoblastoma locus noted earlier, there was much interest in the possibility of cloning the gene for esterase D. In the same issue of the *Proceedings of the National Academy of Science USA*, two laboratories reported this achievement (Lee and Lee, 1986; Squire *et al.*, 1986b). This was exciting for at least two reasons. First, it offered the possibility of finding DNA restriction fragment length polymorphisms (RFLPs) which would make the esterase D locus more informative than phenotypic enzyme electrophoretic analysis. Second, because it was presumed that the esterase D locus and the retinoblastoma locus are physically close to one another, it offered the possibility of 'chromosome walking' from the esterase D gene to the retinoblastoma gene.

Simultaneously, efforts were underway in many laboratories to isolate DNA from the 13q14 band. Dryja *et al.* (1986b) found that a sequence identified as H3-8 detected deletions in some retinoblastomas. These observations raised the possibility that the isolated DNA was close to or at the retinoblastoma locus. Follow up of this finding has led to the apparent isolation of the retinoblastoma gene (Friend *et al.*, 1986). A DNA segment close to the original H3-8 sequence was conserved between humans and mice, suggesting that this conserved fragment probably contained a functioning gene. This fragment also hybridized to RNA from a retinal cell line which had been transformed by adenovirus. However, it did not hybridize to some of the retinoblastomas and osteogenic sarcomas that were examined. It was then possible to isolate a cDNA complementary to the RNA transcript. Using this new cDNA, it was also possible to demonstrate that hybridization did not occur with some retinoblastomas or osteogenic sarcomas. Friend *et al.* (1986) extended their evaluation to 50 tumours and showed that about 30% had deletions or rearrangements affecting the gene. The authors admit that they have not definitively demonstrated that this gene is the retinoblastoma gene and that other studies as well as confirmation by their laboratories will be necessary to reach this determination. In an editorial in the same issue of *Nature*, Harris (1986) discusses these important observations. He raises the possibility that the authors may have identified a closely linked locus to retinoblastoma which is involved in more complex chromosome rearrangements. He also notes that DNA and protein sequence information should soon be forthcoming for the isolated gene and that this may give some clue as to the function of the retinoblastoma gene. Harris further indicates that he has not been convinced that there are dominant oncogenes, suggesting that the recessive mechanism may be more widespread than perhaps previously considered.

Friend *et al.* (1986) suggest that it may be important to determine that the gene that they have isolated can reverse the malignant properties of retinoblastoma and osteogenic sarcoma cells by introduction of the cloned sequences into these cells. However, while a positive result would be very suggestive, a negative result would be inconclusive because of the probable multiple steps involved in tumour formation and progression; by altering one part of this process, one may not get complete reversion.

Another laboratory (Lee *et al.*, 1987) has apparently isolated the same gene as Friend *et al.* (1986). Sequencing of this retinoblastoma gene by Lee *et al.* (1987) indicates that it should encode a hypothetical protein of 816 amino acids. A computer-assisted search revealed no closely associated proteins. Features of this hypothetical protein are interpreted by these authors to suggest that it may bind to DNA and hence could have a regulatory function.

Developments in molecular genetics have also led to the possibility of identifying persons at risk developing retinoblastoma before the disease has manifest. Cavenee *et al.* (1986) described the use of chromosome 13 RFLPs in five families predisposed to retinoblastoma. The esterase D electrophoretic types were more informative than the DNA markers in one family. Unfortunately, these studies were possible only in families in which there had been at least one other affected individual. Cavenee *et al.* (1986) felt that this approach could serve as a prototype model for at least some other human cancers and that it could be important for genetic counselling and clinical care of affected families. With the isolation of the retinoblastoma gene some of the limitations of using linked genetic markers will be overcome since the gene itself can be studied directly.

To evaluate the possible role of oncogenes in the pathogenesis of retinoblastoma, Lee, Murphree and Benedict (1984) studied the *N-myc* gene and found that it was amplified in two primary retinoblastomas and in the retinoblastoma cell line Y79. In most of the other retinoblastomas examined, there was elevated expression of this oncogene and the authors suggested that this may have a primary role in the development of retinoblastoma. In a follow-up study, Squire *et al.* (1986a) concluded that the *N-myc* in retinoblastoma tumours may reflect the origin of the tumour from an embryonic tissue which normally expresses the gene rather than being directly associated with the retinoblastoma gene. However, these studies still do not rule out the possibility that the gene at the retinoblastoma locus may control the expression of the *N-myc* oncogene or perhaps other oncogenes or growth factors; with the retinoblastoma mutation this control could be lost and these factors expressed, perhaps at elevated levels, leading to the development of the tumour.

CONCLUSIONS

The development of any cancer reflects a multiple step process and while it seems clear that the 'retinoblastoma locus' is an important, perhaps the most important, factor in the development of the tumour, it is still not clear at what step this acts.

Although it is not universally accepted that classical oncogenes act in a dominant fashion (Harris, 1986), there has been some speculation that, in contrast to a possible dominant mode of action for the oncogenes, the retinoblastoma locus acts in a recessive fashion, possibly as an 'anti-oncogene' (Knudson, 1986). This concept has to be speculative, at least until the normal function of the gene at the retinoblastoma locus is defined.

Because of the expression of retinoblastoma in infants and the development of second tumours in adults in specific tissues, it has been suggested that the retinoblastoma locus may be involved in differentiation (Sager, 1986; Knudson, 1986). Thus in this scheme, action of the retinoblastoma gene normally leads to tissue differentiation and, in its absence, these cells do not differentiate and possibly maintain their embryonic features, probably with uncontrolled growth.

In dealing with human problems, it is of interest to know whether animal models exist. To date, no recognized animal model for retinoblastoma has been found. However, as Knudson (1986) discusses, there is a hereditable renal carcinoma in rats in which the heterozygotes show no effect except for the development of the renal tumours. In this case, the homozygous state is lethal. This may serve as a useful model to gain better understanding of the carcinogenesis of retinoblastoma. In *Drosophila melanogaster*, at least 24 genes have been recognized as acting in a recessive fashion and lead to tissue specific tumours when mutated (Mechler, McGinnis and Gehring, 1985). The gene for one of these is the 'lethal (2) giant larvae' in which the heterozygotes appear to be normal, but the homozygotes develop neuroblastic tumours in the presumptive optic centres of the larval brain. The gene for this locus has now been isolated and it will be of interest to learn of its normal function, which is unknown (Mechler, McGinnis and Gehring, 1985).

Thus, while there have been major advances toward our better understanding of the retinoblastoma gene, a number of questions still need to be answered. The first concerns the product and function of the normal gene at this locus; the second is why retinoblasts and osteoblasts seem to be particularly affected by this locus. Also, it will be interesting to gain further insight into the mechanisms that lead to the loss of heterozygosity for chromosome 13 in the tumour cells, why this occurs and how widely it may occur in other tumours, perhaps for other chromosomes.

In conclusion, it is fair to say that the study of retinoblastoma has been most interesting and fruitful and has much broader implications for the study of human differentiation and development of tumours in general.

References

BADER, J. L., MEADOWS, A. T., ZIMMERMAN, L. E., RORKE, L. B., VOUTE, P. A., CHAMPION, L. A. A. and MILLER, R. W. (1982) Bilateral retinoblastoma with ectopic intracranial retinoblastoma: trilateral retinoblastoma. *Cancer Genetics and Cytogenetics*, **5**, 203–213

BENEDICT, W. F., MURPHREE, A. L., BANERJEE, A., SPINA, C. A., SPARKES, M. C. and SPARKES, R. S. (1983) Patient with 13 chromosome deletion: evidence that the retinoblastoma gene is a recessive cancer gene. *Science,* **219,** 973–975

CAVENEE, W. K., DRYJA, T. P., PHILLIPS, R. A. *et al.* (1983) Expression of recessive alleles by chromosomal mechanisms in retinoblastoma. *Nature (London),* **305,** 779–783

CAVENEE, W. K., HANSEN, M. F., NORDENSKJOLD, M. *et al.* (1985) Genetic origin of mutations predisposing to retinoblastoma. *Science,* **228,** 501–503

CAVENEE, W. K., MURPHREE, A. L., SHULL, M. M., BENEDICT, W. F., SPARKES, R. S., KOCK, E. and NORDENSKJOLD, M. (1986) Prediction of familial predisposition to retinoblastoma. *New England Journal of Medicine,* **314,** 1201–1207

COMINGS, D. E. (1973) A general theory of carcinogenesis. *Proceedings of the National Academy of Sciences USA,* **70,** 3324–3328

CONNOLLY, M. J., PAYNE, R. H., JOHNSON, G., GALLIE, B. L., ALLDERDICE, P. W., MARSHALL, W. H. and LAWTON, R. D. (1983) Familial, EsD-linked, retinoblastoma with reduced penetrance and variable expressivity. *Human Genetics,* **65,** 122–124

DRYJA, T. P., CAVENEE, W., WHITE, R., RAPAPORT, J. M., PETERSEN, R., ALBERT, D. M. and BRUNS, G. A. P. (1984) Homozygosity of chromosome 13 in retinoblastoma. *New England Journal of Medicine,* **310,** 550–553

DRYJA, T. P., RAPAPORT, J. M., EPSTEIN, J., GOORIN, A. M., WEICHSELBAUM, R., KOUFOS, A. and CAVENEE, W. K. (1986a) Chromosome 13 homozygosity in osteosarcoma without retinoblastoma. *American Journal of Human Genetics,* **38,** 59–66

DRYJA, T. P., RAPAPORT, J. M., JOYCE, J. M. and PETERSEN, R. A. (1986b) Molecular detection of deletions involving band q14 of chromosome 13 in retinoblastomas. *Proceedings of the National Academy of Sciences USA,* **83,** 7391–7394

FRIEND, S. H., BERNARDS, R., ROGELJ, S., WEINBERG, R. A., RAPAPORT, J. M., ALBERT, D. M. and DRYJA, T. P. (1986) A human DNA segment with properties of the gene that predisposes to retinoblastoma and osteosarcoma. *Nature (London),* **323,** 643–646

GODBOUT, R., DRYJA, T. P., SQUIRE, J., GALLIE, B. L. and PHILLIPS, R. A. (1983) Somatic inactivation of genes on chromosome 13 is a common event in retinoblastoma. *Nature (London),* **304,** 451–453

HANSEN, M. F., KOUFOS, A., GALLIE, B. L. *et al.* (1985) Osteosarcoma and retinoblastoma: a shared chromosomal mechanism revealing recessive predisposition. *Proceedings of the National Academy of Sciences USA,* **82,** 6216–6220

HARRIS, H. (1986) Malignant tumours generated by recessive mutations. *Nature (London),* **323,** 582–583

HOPKINSON, D. A., MESTRINER, M. A., CORTNER, J. and HARRIS, H. (1973) Esterase D: a new human polymorphism. *Annals of Human Genetics,* **37,** 119–137

KITCHIN, F. D. and ELLSWORTH, R. M. (1974) Pleiotropic effects of the gene for retinoblastoma. *Journal of Medical Genetics,* **11,** 244–246

KNUDSON, A. G. JR (1971) Mutation and cancer: statistical study of retinoblastoma. *Proceedings of the National Academy of Sciences USA,* **68,** 820–823

KNUDSON, A. G. JR (1986) Genetics of human cancer. *Annual Review of Genetics,* **20,** 231–251

LEE, W-H., MURPHREE, A. L. and BENEDICT, W. F. (1984) Expression and amplification of the *N-myc* gene in primary retinoblastoma. *Nature (London),* **309,** 458–460

LEE, E. Y.-H. P. and LEE, W.-H. (1986) Molecular cloning of the human esterase D gene, a genetic marker of retinoblastoma. *Proceedings of the National Academy of Sciences USA,* **83,** 6337–6341

LEE, W.-H., BOOKSTEIN, R., HONG, F., YOUNG, L.-J., SHEW, J.-Y. and LEE, E. Y.-H. P. (1987) Human retinoblastoma susceptibility gene: cloning, identification, and sequence. *Science,* **235,** 1394–1399

MECHLER, B. M., McGINNIS, W. and GEHRING, W. J. (1985) Molecular cloning of lethal (2) giant larvae, a recessive oncogene of *Drosophila melanogaster. EMBO Journal,* **4,** 1551–1557

SAGER, R. (1986) Genetic suppression of tumor formation: a new frontier in cancer research. *Cancer Research,* **46,** 1573–1580

SPARKES, R. S. (1984) Cytogenetics of retinoblastoma. *Cancer Surveys,* **3,** 479–496

SPARKES, R. S., TARGUM, S., GERSHON, E., SENSABAUGH, G. F., SPARKES, M. C. and CRIST, M. (1979) Evidence for a null allele at the esterase D (EC 3.1.1.1) locus. *Human Genetics,* **46,** 319–323

SPARKES, R. S., SPARKES, M. C., WILSON, M. G., TOWNER, J. W., BENEDICT, W., MURPHREE, A. L. and YUNIS, J. J. (1980) Regional assignment of genes for human esterase D and retinoblastoma to chromosome band 13q14. *Science,* **208,** 1042–1044

SPARKES, R. S., MURPHREE, A. L., LINGUA, R. W., SPARKES, M. C., FIELD, L. L., FUNDERBURK, S. J. and BENEDICT, W. F. (1983) Gene for hereditary retinoblastoma assigned to human chromosome 13 by linkage to esterase D. *Science,* **219,** 971–973

SPARKES, R. S., SPARKES, M. C., KALINA, R. E., PAGON, R. A., SALK, D. J. and DISTECHE, C. M. (1984) Separation of retinoblastoma and esterase D loci in a patient with sporadic retinoblastoma and del (13)(q14.1q22.3). *Human Genetics,* **68,** 258–259

SQUIRE, J., GODDARD, A. D., CANTON, M., BECKER, A., PHILLIPS, R. A. and GALLIE, B. L. (1986a) Tumour induction by the retinoblastoma mutation is independent of *N-myc* expression. *Nature (London)*, **322,** 555–558

SQUIRE, J., DRYJA, T. P., DUNN, J. *et al.* (1986b) Cloning of the esterase D gene: a polymorphic gene probe closely linked to the retinoblastoma locus on chromosome 13. *Proceedings of the National Academy of Sciences USA*, **83,** 6573–6577

VARKI, A., MUCHMORE, E. and DIAZ, S. (1986) A sialic acid-specific O-acetylesterase in human erythrocytes: possible identity with esterase D, the genetic marker of retinoblastomas and Wilson disease. *Proceedings of the National Academy of Sciences USA*, **83,** 882–886

16
Detection of viral genes in neurological disease

R. F. Itzhaki

INTRODUCTION

This chapter is concerned with persistent and latent viral infections, excluding those due to RNA tumour viruses. These topics are both fascinating and challenging but are far less well understood than acute infections. Definitions of persistence and latency have not been standardized. Persistence is sometimes equated with chronic or dynamic latency and usually includes the category of slow infection; latency is sometimes treated as synonymous with persistence but is occasionally classed as a subcategory of the latter. However, in general, and in this chapter, persistent or chronic infection is defined as that where there is continuous production, although at a very low level, of the virion. In contrast, in latent infection, no virion is produced and production of viral messenger RNA (mRNA) and of viral protein is low or zero, but the viral genome is maintained in the host cell, either integrated within the cell genome or in episomal form.

INTERACTION OF PERSISTENT VIRUS AND BRAIN

What are the special features of virus–brain interaction, particularly with respect to viral persistence? These have been discussed by Wolinsky and Johnson (1980) in their excellent review on the role of viruses in chronic neurological disease. The features reflect:

(1) The lack of vascular permeability and tightly packed parenchymal structure, which impede invasion by viruses but also impede their removal once they have entered.
(2) The lack, normally, of an immune system; this may retard clearance and also prevent immunological detection of foreign antigens. Possibly, low levels of antibody may even aid the virus by removing surface antigens but not causing cell lysis, so that viral nucleic acid and protein are preserved within the cell.
(3) There is striking diversity of cell types and of viral receptors; some viruses can affect all cell types whereas others can infect only specific types, and such specificity can differ between species and vary with age of the individual.

(4) Ease of transport by cell-to-cell contacts and across extracellular gaps, along axons, dendrites or via mobile leucocytes.
(5) Stability and lack of regenerative ability. Thus, in brain the virus may be provided with a secure home for life. In contrast, in infected tissues which normally divide rapidly, infected cells that do not lyse divide more slowly and so eventually die out, replaced by uninfected cells that divide at their normal rate.
(6) The extremely high metabolic rate of brain results in a greater interference with cell function than would be the case in infected cells of lower metabolic rate.

How are viruses able to persist despite the host's defence mechanisms? Evasion can occur through a low rate of production of viral protein (due sometimes, but not always, to a low or zero rate of production of viral messenger (m) RNA), so that host antibodies are not elicited, or there may be formation of an incomplete set of products. This is believed to be the case with measles virus proteins, in particular the matrix (M) protein, which stabilizes the envelope glycoproteins and directs the alignment of nucleocapsids to allow budding. Also in the case of measles virus, cyclic nucleotides have been implicated in the establishment and presumably maintenance of latency, and these may well be relevant to other persistent or latent viruses. A study of the conversion of acute measles virus infection to a latent state in mouse neuroblastoma cell cultures showed that agents which can enhance neural differentiation, and decrease the breakdown of cyclic AMP (cAMP), lead to decreased production of infectious virus and of M protein (Miller and Carrigan, 1982). The effects were reversible on removal of the agent or by addition of cGMP, and were specific for neural cells. The authors suggested that viral replication depends on host factors, and pointed out that levels of cAMP vary with the state of growth and differentiation of the cell, being relatively high in non-cycling cells.

Another mechanism involves antigenic variation through mutation of the viral nucleic acid; this is especially true of RNA viruses since they lack the 'proofreading' enzymes that ensure the correctness of DNA replication (Holland *et al.*, 1982). Thus, in the lentiviruses, visna and human immunodeficiency virus (HIV), there is continuous antigenic drift, particularly in the region coding for the envelope glycoproteins against which neutralizing antibody is directed. In influenza virus, a more major change, genetic shift, occurs, involving recombination between different RNA molecules. In yet another mechanism, cells involved in the immune system, such as macrophages and lymphocytes, are infected; this has been described by Haase (1986) as the 'Trojan Horse' effect, leading to infiltration by the virus into regions of the host such as the brain, cerebrospinal fluid (CSF) and lungs. Finally, there is another mechanism that operates mainly, though not exclusively, in RNA viruses and interferes with infectious virus replication. This is caused by the generation of non-infectious, recombinant subgenomic nucleic acid in virions containing normal protein, called defective interfering (DI) particles that compete for the replicase: the DI particles have to replicate via enzymes and proteins encoded by the wild type virus since they lack 3′ terminal sequences and so cannot act as template for mRNA. Examples of these mechanisms will be discussed in the sections below on specific viruses.

THE ROLE OF IMMUNOSUPPRESSION

Although it is well known that latent viruses can be reactivated during immunosuppression, in humans especially the sites of latency and the mechanisms whereby the CNS is affected are rarely known. Further, the immune status of humans cannot of course be manipulated, nor is it always possible to obtain information about previous infection by the virus under investigation. Saldanha *et al.* (1986) conducted a study which combined serological investigation with molecular hybridization in immunosuppressed humans. They asked the question as to whether latent herpes simplex virus type 1 (HSV1) genomes affect host cells or whether damage occurs only after the virus is reactivated. As a first step they decided to establish if, in humans, reactivation occurs after immunosuppression as it does in mice (Kastrukoff and Koprowski, 1981). The authors investigated patients with acute leukaemia whose illness and treatment involved immunosuppression. The patients were tested for evidence of previous infection with HSV1 by standard virological techniques. In those coming to autopsy, *in situ* hybridization (*see below*, p. 222) was used with a HSV1 DNA probe to look for viral DNA in brain specimens (temporal lobe).

Two groups were examined: group A comprised seven patients with acute leukaemia, all of whom had been immunocompromised and two of whom had received whole body irradiation (another patient in this group had HSV encephalitis); group B consisted of three individuals who had not been immunocompromised and who were not suffering from neuropsychiatric disease; virological data could not be obtained for this group.

The leukaemia patients were found to have HSV antibody levels within the normal range for healthy adults. In specimens taken from peripheral sites, HSV was recovered in tissue culture or identified by electronmicroscopy, or both. Thus, these patients had had a latent HSV infection which was capable of reactivation. In two other patients with evidence of past infection, no virus reactivation was detected, and in two others there was no evidence of past infection. Virus was not detected in any post-mortem brain tissue by inoculation into tissue culture or by electronmicroscopy.

In the three individuals from group B, no virological data were available but since 80–90% of humans have antibodies to HSV1, some or all of these must have been infected with this virus.

In situ hybridization showed labelled cells in some brain specimens; the cells were mainly neurons but also, in two cases, endothelial cells. We compared the specimens from different patients by counting in each case the grains in at least 400 nuclei, but not in cytoplasm since in the case of neurons it was too poorly defined. Cumulative frequency values were then plotted against number of grains per nucleus. From these curves we deduced that HSV sequences were present in brain from the encephalitis patient, from three others of group A with serological evidence of past infection and with reactivation of virus in the oropharynx, and from one other with serological evidence alone. No sequences were detected in two patients with no virological evidence of past infection, nor in the three brains from group B. From this study, we concluded that immunosuppression of humans results in reactivation of HSV1, leading to an amplification of viral sequences (and also infection of endothelial cells) such that they were then detectable by the author's group's assay. Our results are consistent with those of Taylor *et al.* (1985), who searched for cytomegalovirus (CMV) DNA by dot blot hybridization in 83 brain samples from normal individuals and from patients with Huntington's disease and schizophrenia. In only one case was hybridization detected, a normal individual who had previously been immunocompromised.

EFFECTS OF PERSISTENT AND LATENT VIRUSES ON HOST CELL METABOLISM

Few studies have been made on this most important aspect, probably because of experimental difficulties in working *in vivo*. The main system used is that of lymphocytic choriomeningitis virus (LCM), which has been studied in mice as well as in mouse cell cultures, and is particularly useful since infection is easily established and the type can be varied by altering the age or strain of mouse and strain of virus. Oldstone *et al.* (1982) have found that LCM replicates in growth hormone producing cells in mouse pituitary gland, resulting in a decreased synthesis of growth hormone and consequently in retarded growth and hypoglycaemia. However, there is no evidence of cell necrosis or inflammation. This is consistent with previous studies by Oldstone, Holstoen and Welsh (1977) with mouse neuroblastoma cells in culture; these indicated that although morphological appearance, growth, and cloning rates remained normal after LCM infection, there was an effect on the 'luxury', that is, differentiated function, namely a reduced synthesis of acetylcholine, choline acetyl transferase and acetylcholine transferase.

Broadly analogous types of results were obtained with other viruses by Koschel and Münzel (1980) and Koschel *et al.* (1986). In the first study, it was found that persistent infection of rat glioma cell cultures with canine distemper virus reduced the level of β-adrenergic receptors to 50% of normal value. In the second study, persistent infection of mouse neuroblastoma–rat glioma hybrids with rabies virus led to impaired response to prostaglandins and total loss of opiate receptor function; response to acetylcholine remained normal.

DETECTION METHODS FOR VIRAL GENOME

Molecular hybridization is the most direct and sensitive method for detecting virus and can be carried out using either radiolabelled or fluorescent probes. Only radiolabelling will be referred to here since it has been used far more frequently. Probe is prepared using either whole viral DNA or specific cloned sequences, incorporating the (labelled) precursor (^{3}H or ^{32}P) by nick translation or by random primed oligonucleotide labelling.

The three most commonly used hybridization methods are *in situ*, dot blot and Southern blot. As the name implies, *in situ* hybridization is carried out with whole cells, smears or tissue sections on slides, rather than with extracted DNA. The DNA in the cells is denatured by immersion of the slides in alkali, and after restoration of the pH to neutrality, the labelled denatured probe is added to the slides and left overnight. ^{3}H is used rather than ^{32}P, to ensure high resolution of the autoradiograph for subsequent microscopy. After removal of excess probe by washing, the slide is coated with photographic emulsion. Exposure is carried out for an appropriate time and the slide is then developed and fixed, and the cells are stained for viewing by light microscopy. Using this method, it is possible to determine the cell type in which silver grains, corresponding to sites of hybridization, are located. If viral sequences are present in only a very small percentage of cells, the task of locating them is laborious, but correspondingly, the sensitivity is very high; neither of the other two methods that use extracted DNA is capable of detecting specific sequences if present in such low proportions. The detectability in all three methods will depend, of course, on the size of the viral genome, and in these two methods on the *average* number of viral genomes per cell. Clearly though, in the *in situ* method, to ensure that the occasional cell detected with grains in excess of background is not an artefact, very large numbers of cells need to be screened. A further disadvantage is that the amount of tissue which can be screened even on a number of slides is extremely small.

In the dot blot and Southern blot methods, the first step is extraction of DNA from cells and tissues – a process necessitating an extra couple of days work; 1 g of human brain yields up to 700 μg DNA (Saldanha, Gannicliffe and Itzhaki, 1984). In the dot method, an appropriate amount of DNA, such as to contain a detectable level of viral DNA, is placed on a filter; in the Southern blot method, an appropriate amount is digested with a restriction endonuclease, then subjected to electrophoresis, denatured and finally transferred to a nitrocellulose or nylon filter. After hybridization of the filter with the ^{32}P labelled probe, autoradiography is performed. Blackened regions on the X-ray film denote the presence of viral sequences. Obviously, the dot method can easily be quantified by applying known amounts of viral DNA as standards on the filter and comparing their degree of blackening.

The advantage of dot and Southern hybridization over *in situ* hybridization is that they enable large numbers and/or amounts of specimens to be screened, although as a consequence of this, their sensitivity is correspondingly less than that of *in situ* hybridization. However, the Southern blot method, unlike the other two methods, can yield information about the structure of the viral genome. For example, it can sometimes distinguish, if viral DNA termini are used as probes, between integrated and non-integrated viral DNA; in the former the size of the terminal fragments after restriction endonuclease digestion will be larger, because of their linkage to host cell fragments, than that of purified viral DNA. The presence or absence of specific bands would then reveal if integration sites are random or non-random. However, there is some ambiguity since larger than normal bands containing terminal sequences could also be produced by circularization of the viral genome. These types of problems are exemplified by studies of HSV DNA conformation (see below, p. 224).

Various problems can arise with these techniques. First, there is the needle-in-a-haystack difficulty of searching for viral genomes in samples which, to provide adequate sensitivity and some degree of sampling, must each necessarily be small, from a brain which weighs well over 1 kg. However, such difficulties might be overcome in future by use of combined macroscopic and microscopic hybridization techniques such as those devised by Southern, Blount and Oldstone (1984) and Haase *et al.* (1985a).

Second, hybridization of probe could occur to any sequences which are homologous to the viral genome in normal, uninfected cells. According to Peden, Mounts and Hayward (1982) and Puga, Cantin and Notkins (1982), some human middle repetitive (G + C)-rich DNA sequences are homologous to the repeated sequences (the joint regions and the ends) of HSV DNA. However, on the basis of blocking experiments with added poly (U, G), Jones and Hyman (1983) maintain that the above findings are artefactual, due to hybridization of G-rich probe (portions of the joint region are over 80% (G + C)) to C-rich human DNA (see also Patient, 1984). It appears that it is advisable to use as probes only those sequences (cloned) of a viral genome which are known not to show any homology (whether real or apparent) with human DNA.

VIRUSES OF SPECIAL INTEREST

Herpes simplex virus type 1 (HSV1)

HSV is well known for its ability to give rise to latent infection and for its capacity for reactivation, although mechanisms of establishment and maintenance of latency, and of reactivation, are not yet elucidated. Over 50 years ago, it was proposed that latent infection occurs in sensory ganglia (Goodpasture, 1929). Subsequently, preliminary molecular hybridization studies suggested that HSV1 sequences were present in the brains of patients with chronic but not acute psychiatric disease, and also of mice 6 months after infection with HSV1 (Sequiera *et al.*, 1979). These results were confirmed by other studies of human (Fraser *et al.*, 1981) and mouse brain (Cabrera *et al.*, 1980). The role of immunosuppression in reactivation of HSV1 in the brain was referred to earlier in this chapter.

More recently it has been found that the conformation of the viral DNA differs in latent infection from that in acute infections. Rock and Fraser (1983) used Southern blot hybridization to examine DNA from experimental latently infected mice and found that most, if not all, of the HSV genome was present; however, they were unable to detect terminal sequences, indicating that the genome was not in linear form. Puga *et al.* (1984), also working with latently infected mice, concluded that structural changes had occurred, reflecting either extensive rearrangement of viral fragments, or integration of linear sequences into host cell DNA at preferred sites. Alternatively, most of the fragments detected could have been part of defective genomes, thus concealing any molecules which were non-defective. As to human brain, Fraser *et al.* (1981) found HSV1 DNA sequences in six out of 11 specimens, obtained from both normal individuals and from patients with multiple sclerosis. In some cases the terminal fragments were detected, indicating that the DNA was in linear, non-integrated form. Similarly, Wigdahl *et al.* (1984) found evidence of HSV genome predominantly in linear non-integrated, and non-concatameric form, in latent infections of cultures of human diploid fibroblasts and rat fetal neurons.

Efstathiou *et al.* (1986) have questioned whether the above results using animal models are biologically significant – in other words, if the observed DNA structures are capable of being reactivated, or if they are merely artefacts arising from the very high doses needed to establish latency. To examine this and also the stability of the HSV DNA structure with time, and to find if the conformation is similar to that detected in latently infected humans, Efstathiou *et al.* probed both latently infected mice and human trigeminal ganglia. They found in the former, in agreement with Rock and Fraser (1983), that the viral DNA is 'endless', that all four isomeric forms of the genome are present and that the conformation is stable. In the human specimens, they found the viral DNA organization to be similar to that in the mice but in lower concentrations.

As to expression of the HSV DNA, Galloway *et al.* (1979) used *in situ* hybridization with HSV2 DNA as probe (the genome has considerable homology with that of HSV1) to examine human sensory ganglia and found that a very small proportion (0.04–8%) of the neurons expressed HSV RNA. In a later paper (Galloway, Fenoglio and McDougall, 1982), it was shown that transcripts from the left hand 30% of the viral genome were detectable in all HSV-positive cases; transcripts from other sequences of the long component were less often present and none was detected from the short component. This indicated that specific transcription occurs in latently infected human ganglia. Broadly similar results, showing very few neurons expressing RNA at low concentrations, were found on studying trigeminal ganglia of latently infected mice (Puga *et al.*, 1978) and guinea pigs (Tenser *et al.*, 1982).

Human immunodeficiency virus (HIV)

It is estimated that at least one-third of patients with AIDS suffer from dysfunction of the CNS, usually due to an encephalitis. Originally, most cases were thought to be the result of an opportunistic infection such as progressive multifocal leucoencephalopathy, or of cytomegalovirus (CMV) or herpes simplex infection, but it is now known that they are due to direct infection of the brain by HIV. An important clinical observation supporting this is the fact that infected children born to infected mothers do not show opportunistic infection, being too young to have been exposed to any extent (Johnson and McArthur, 1986), yet they show arrested brain growth and fail to thrive. At autopsy, over three-quarters of their brains show neurological changes.

The majority of patients with encephalitis develop a subacute encephalopathy with progressive dementia and cerebral atrophy. HIV RNA and DNA have been identified in affected brain of adults and children by *in situ* hybridization and Southern blot analysis, and viral sequences are present in larger quantities than in spleen, lymph nodes, liver or lung (Shaw *et al.*, 1985). The sequences are both chromosomally integrated and unintegrated, the latter form suggesting a replication intermediate for virus within the CNS.

Very recent work using cocultivation for virus isolation, *in situ* hybridization, immunocytochemistry and transmission electronmicroscopy has shown that the infected cells are mononucleated and multinucleated macrophages (Koenig *et al.*, 1986). In another study using *in situ* hybridization and immunocytochemistry (Wiley *et al.*, 1986), nine out of 12 patients were found to have significant HIV infection of the CNS; in all cases white matter was affected more than grey matter and the infection was restricted to capillary endothelial cells, mononuclear inflammatory cells and giant cells. In one case only, a low level infection was detected in some astrocytes and neurons; however, as Wiley *et al.* (1986) pointed out, the assay was such as to detect only high levels of viral DNA (10–100 copies of the viral genome per cell), so that a latent HIV infection might have been present in neuronal cells but undetectable by their assay.

Prior to the above studies it was known that the lymphocytes of AIDS patients harboured HIV. The dual lymphotropism and neurotropism of HIV has since been explained as due to expression of the T4 gene in brain as well as in lymphoid cells. The T4 glycoprotein appears to act as receptor on the cell membrane for HIV. To examine this further, human cells of a type not normally expressing T4 (and resistant to HIV) were experimentally transformed so as to express this protein on their surface; they then became susceptible to infection by HIV. In contrast, transformation of mouse cells led to T4 expression but not to infectibility, suggesting that the cells could not support virus entry (Maddon *et al.*, 1986).

The pathological mechanisms leading to CNS dysfunction in AIDS patients remain to be elucidated. It has been suggested that the neuropsychiatric symptoms reflect an infection of the endothelial cells, leading to damage to the blood–brain barrier and thence to fluctuations in fluid and electrolyte levels, or cell injury consequent to activation of infected microphages (Wiley *et al.*, 1986).

HIV is known to be a lentivirus, and its mechanism of persistence resembles that of visna, another member of the same subfamily, which affects sheep. Both viruses are slow, that is, have long incubation periods ranging from months to years, both involve the CNS and both show genetic drift, with mutations occurring in the part of the genome coding for the envelope glycoproteins. This mechanism enables these viruses to evade an immune response.

Unconventional agents

These are responsible for the slow, degenerative encephalopathies: scrapie in sheep and goats; kuru, Creutzfeld–Jakob disease (CJD) and Gerstmann–Straussler (GS) syndrome in man. The name reflects the unusual characteristics of these agents: they have not been visualized by electronmicroscopy; they do not cause cytopathic effects in cell cultures or non-neural tissues; they do not evoke an immune response; they show unusually high resistance to physical and chemical agents; and, most startlingly of all, they appear not to have a nucleic acid genome.

In their extensive studies on scrapie, Prusiner and colleagues (*see*, for example, Prusiner, 1984; and for a refutation, Carp *et al.*, 1985) have proposed that this disease, and also kuru, Creutzfeldt–Jakob disease and Gerstmann–Straussler syndrome, are caused by a proteinaceous infectious particle which they have named 'prion'. The partially purified agent, extracted from infected hamster brain (which has frequently been used as a model system) was found to consist of a sialoglycoprotein of molecular weight 27 000–30 000, called a prion protein. Great interest was aroused by the apparent association of the extract with infectivity, despite the absence of any detectable nucleic acid (consistent with the early findings of Alper *et al.*, 1967). Another reason for the great interest engendered by this agent was the resemblance of the aggregated scrapie-associated fibrils seen in the disease to the amyloid observed in brain with patients with Alzheimer's disease.

The extraction procedures used by Prusiner and colleagues involved digestion with proteinase K under conditions in which most of the host protein was degraded, and indeed no 27 000–30 000 protein was detected in uninfected hamster brain processed in this way. Subsequently, it was suggested that prion protein might be an altered pathological protein normally residing in, or derived from, cell membranes rather than an infectious agent and this has recently been substantiated; part of the protein has been sequenced and a set of oligonucleotides representing possible coding sequences to that part has been constructed (Oesch *et al.*, 1985; Chesebro *et al.*, 1985). Using the latter as a probe, a scrapie prion protein-specific mRNA has been found in both scrapie-infected and uninfected hamster brain, and cDNA prepared from the former has been cloned.

Further, in order to study the human counterpart of the prion protein, homologous clones have been isolated using human brain, the primary structure of the protein has been deduced and the chromosomal location mapped using dot blot hybridization on separated individual chromosomes (Liao *et al.*, 1986). Homology between rodent and primate prion sequences was found to be 89% and the gene was located on human chromosome 20, as a single copy per haploid genome. Measurement of mRNA levels showed that expression was the same in both infected and uninfected hamster brain and occurred at various levels in several other tissues such as placenta and kidney (Oesch *et al.*, 1985). It was also found that highly purified prions do not contain related nucleic acid sequences, indicating that they are not encoded by a nucleic acid carried within the infectious particles, and that the protein is present in both normal and infected brain. However, in the former, it is not resistant to proteinase K action.

Despite these great advances in our knowledge of this infection, due mainly to the use of molecular biological techniques, mechanisms of infection and pathogenicity have still not been elucidated.

DISEASES OF KNOWN VIRAL AETIOLOGY

Subacute sclerosing panencephalitis (SSPE)

This is a rare, usually fatal, inflammatory disease of humans caused by persistent infection with measles virus. A high proportion of patients are known to have had acute measles virus infection before the age of 2 and then to have been normal for some 5 or 6 years before the development of neurological symptoms. Because of this long time interval, SSPE is termed a slow infection.

Molecular hybridization studies, which combined the *in situ* technique with macroautoradiographic screening of large regions of tissue, using cDNA clones specific for the matrix (M) protein gene and the nucleocapsid protein gene of the virus, showed that at the early stage of the disease there was a global restriction in the synthesis of negative strand and positive strand RNAs, and in production of antigen. In the terminal stage, most infected cells still had somewhat low levels of RNA and only nucleocapsid antigens were detected; there appeared to be a specific block in virus replication in that the M protein was lacking. This explains why attempts to isolate virus were unsuccessful until the techniques of cocultivation and explanation were used, accounting for the interval of over 30 years between the time a viral aetiology was first proposed (in the 1930s) and the actual attribution to a specific virus. The absence of M protein was attributed to a decrease in transcription and in translation of the corresponding mRNA, and possibly to a decrease in stability of the protein (Haase *et al.*, 1985b).

Very recently it has been shown that M protein may in fact be present in SSPE brain tissue. It has been suggested that since the clinical course of the disease and the brain pathology can vary, the state of viral persistence might vary between different cases or even between different regions of one brain (Liebert *et al.*, 1986). Data obtained by the latter authors with four SSPE patients, using immunofluorescence with monoclonal antibodies, indicated that nucleocapsid protein and phosphoprotein were present in every case and in every brain region examined, haemagglutinin protein was present in two cases, fusion protein in three and M protein in only one case. Within any one brain, there was some variation in viral protein type, but in general the same proteins were either present or absent, regardless of the neuropathology.

In this disease, and probably in other slow infections, the low levels of viral antigen enable the infected cell to evade detection and destruction by the immune system. Presumably the progression of the disease is due to the gradual accumulation of tissue damage caused by infection in those few cells which produce complete virus.

The pathogenesis of the disease is not yet understood. However, by use of *in situ* hybridization with a cloned measles virus DNA probe specific for the nucleocapsid protein gene, viral RNA sequences have been detected in peripheral blood lymphocytes of SSPE patients (Fournier *et al.*, 1985). It is not known whether lymphocytes are infected and then carry virus to the brain, whether they are infected by virus initially in brain, or if each cell type is infected separately.

Progressive multifocal leucoencephalopathy (PML)

Progressive multifocal leucoencephalopathy (PML) is a rare, subacute demyelinating disease and was the first chronic disease in humans known to be of viral aetiology, the usual agent responsible being the polyoma virus, JC; 75% of adults carry antibodies to the virus, and primary infection in childhood results in long-term persistence. Whether PML occurs during immunosuppression, after primary infection, or after virus reactivation is uncertain. However, the disease is rare even in groups prone to reactivation, such as renal transplant recipients.

Grinnell, Padgett and Walker (1983b), using dot blot hybridization with probes of cloned JC and also BK virus (whose genome has up to 20% homology with that of JC), examined tissues from nine PML patients. They found that viral DNA was present in the kidneys of seven patients, at levels of 0.2–10 copies/cell genome equivalent, and in diseased areas of brain at 600–4000 copies/cell. Southern blot analysis gave no evidence of viral integration into the host genome.

A study by Chesters, Heritage and McCance (1983) using cloned JC and BK virus DNA probes and Southern blot hybridization, indicated that in healthy individuals, BK persists in the kidney after primary infection and may reactivate when the immune response is impaired. No JC or BK virus DNA was detected in normal brain tissue. Possibly, therefore, the kidneys can act as a source of infection which, during severe immunosuppression, can spread to the brain. The route would presumably be via the blood, and consistent with this, Grinnell, Padgett and Walker (1983b) showed that in some PML cases, viral DNA was present in liver, lung, lymph nodes and spleen as well as in kidney and brain.

Whether the pathogenesis is related to strain differences in JC virus is unknown. To address this question, Grinnell, Padgett and Walker (1983a) cloned JC virus DNA from brain of PML patients and compared the DNA by restriction endonuclease mapping. The patterns obtained suggested that there are two subtypes of JC virus circulating in the population.

SOME DISEASES OF POSSIBLE VIRAL AETIOLOGY

Multiple sclerosis (MS)

Evidence for viral involvement in the demyelinating disease MS has been discussed in detail by Johnson (1982). Epidemiological studies suggest that the disease depends on an environmental factor(s) of childhood. As with many infections (and other diseases), it shows a unimodal, age-specific onset, being rare before 15 years or after 50 years, with a peak at about 30 years of age. The geographical distribution provides evidence suggestive of a viral aetiology; there is a relationship to latitude, the disease being more prevalent in the northerly regions of the northern hemisphere and in the more southerly regions of the southern hemisphere. The birthplace risk is retained in those who migrate from a high to a low risk area before the age of 15. There are a few areas of extremely high prevalence; also, in strong support of an infectious agent, a relatively large number of cases were recorded in the Faroe islands between 1943 and 1960, after a period of British occupation. Another argument which points to an infectious slow agent is that although the period of risk apparently occurs before puberty, the peak age of onset occurs many years later. Lastly, the viral hypothesis is supported by the fact that in MS there are remissions and exacerbations, and multifocal areas of demyelination; these are characteristic of viral diseases such as HSV infection and also visna, a demyelinating CNS disease. Significantly, another demyelinating disease, PML (*see above*, p. 228), is known to be caused by viral infection.

A plethora of viruses have been implicated in MS, usually by immunological and morphological studies; these include measles (CSF measles antibody titres are elevated in MS), which *in situ* hybridization studies have detected in the brain of one of four patients (Haase *et al.*, 1981), and Epstein-Barr, which has been detected by serum antibodies to the virus (Sumaya *et al.*, 1985). HSV2, rabies, parainfluenza and many others have also been implicated.

Roos (1983) has discussed mechanisms of demyelination by viruses as well as experimental models, such as canine distemper virus infection of germ-free dogs. However, reports of transmission of MS to primates and other animals, and of recovery of viruses after inoculation into animals, have not been substantiated. Thus, as Johnson (1982) puts it, a viral cause for MS remains only an attractive hypothesis, not an established fact.

Amyotrophic lateral sclerosis (ALS)

One of the main arguments in favour of a viral cause of the degenerative neurological disease ALS is that target cells in paralytic poliomyelitis and ALS are the same; it has also been suggested that individuals infected with poliovirus show a higher incidence of ALS. Unfortunately, neither epidemiological data, immunological studies nor transmission experiments give any clear evidence of a viral cause. Brahic *et al.* (1985), using *in situ* hybridization with complementary (c) DNA to the RNA of the picornaviruses, polio or Theiler's virus, have recently examined brain and spinal cord from 15 ALS patients and six controls. The detection limit was estimated to be 30 copies of viral RNA per cell. RNA sequences were found in several sections, including many of a few hundred copies per cell, in only one ALS patient and one control. The authors pointed out that the presence of picorna sequences could be underestimated because autolysis time periods were 10 or more hours in many cases – though only 2 h or less in the positives. They mentioned that, unfortunately, no records of vaccination with live polio vaccine could be obtained for any of the cases.

Epilepsy

A study by Gannicliffe *et al.* (1985), in which post-mortem and operative specimens from patients with temporal lobe epilepsy were examined by dot blot hybridization using HSV1 DNA as probe, showed that hybridizing sequences were present in the temporal lobe of three patients, though at a very low level in all but one case. The significance of these findings is unclear.

Alzheimer's disease (AD)

This topic is dealt with in a detailed review by Esiri (1987) and will therefore only be discussed briefly here. Evidence in favour of a viral aetiology includes the distribution of lesions, especially of neurofibrillary tangles (NFT), which are most prevalent in parts of the limbic system, in neurons near incoming fibres of the olfactory system. Generally, the involvement of the olfactory pathways, and regions nearby, in the pathology of Alzheimer's disease would be compatible with the presence of an environmental agent such as a virus reaching the brain by the olfactory route; certainly there is evidence that rabies, some togaviruses and HSV use this route. Further, NFT resemble the pathological changes seen in some viral diseases, such as HSV encephalitis. Esiri (1987) also cites an electronmicroscope study of a particularly well-prepared cerebral biopsy from an Alzheimer's disease patient at an early stage of the disease; this showed abnormal tubular structures as well as paired helical filaments within some neurons, the structures resembling certain viruses such as rhabdoviruses. Lastly, in Alzheimer's disease patients, previous head injury is more common than in age-matched controls and it may be relevant that axonal trauma can reactivate HSV. Arguments against a viral aetiology include the inability to transmit Alzheimer's disease to animals and the lack of evidence of viral expression or multiplication in AD brain; the latter would not preclude a role for latent viruses.

As to hybridization studies, in the author's previous laboratory, examination of six post-mortem brain specimens from Alzheimer's disease patients showed three positives by dot blot hybridization to HSV1 sequences, one of the three being a possible positive by *in situ* hybridization. Using an SV40 DNA probe (as a partial homologue for JC virus), one of the six was positive by the dot blot, and none by the *in situ* method. In eight age-matched controls, three HSV positives and one possible SV40 positive were found (Gannicliffe, Sutton and Itzhaki, unpublished data). Taylor *et al.* (1984), using dot blot hybridization, did not detect any HSV DNA sequences in brain specimens from eight Alzheimer's disease patients and nine age-matched controls, and no evidence of HSV in Alzheimer's disease brain was found by immunocytochemistry (Roberts *et al.*, 1986). The question of a viral aetiology in this disease is thus unresolved.

CONCLUSIONS

Clearly, before a role can be ascribed to viruses in certain neurological diseases, much more information is needed about persistence and latency, in particular the effect on host cells and the mechanisms of reactivation of latent genome at the molecular level. Of all the results so far obtained, one of the most intriguing is the detection of viral sequences in normal brain as well as in certain diseased brains. In the case of HSV, this is perhaps not surprising as about one-third of the population suffer reactivation due to factors such as stress, sunlight and menstruation. Possibly many other people harbour HSV or other viral genomes but can keep the virus quiescent either because of a more efficient immune system or through genetic or environmental factors unfavourable to reactivation. In contrast, others may be particularly prone to repeated bouts of reactivation, and if especially susceptible regions of brain are affected, these could cumulatively produce neurological dysfunction. Further, HSV has an interesting characteristic in that it resembles a transposable genetic element (Ritchie and Timbury, 1980). Whether part of its genome, if integrated, could act as a mobile element, and even if not itself expressed, could affect neighbouring host genes, is a matter for surmisal. Similarly, the role of the scrapie gene in normal brain provides fuel for speculation.

Acknowledgements

The review was written while the author was holding grants from the Mental Health Foundation and the Sir Jules Thorn Charitable Trust.

References

ALPER, T., CRAMP, W. A., HAIG, D. A. and CLARKE, M. C. (1967) Does the agent of scrapie replicate without nucleic acid? *Nature (London)*, **214**, 764–766

BRAHIC, M., SMITH, R. A., GIBBS JR., C. J., GARRUTO, R. M., TOURTELLOTTE, W. W. and CASH, E. (1985) Detection of picornavirus sequences in nervous tissue of amyotrophic lateral sclerosis and control patients. *Annals of Neurology*, **18**, 337–343

CABRERA, C. V., WOHLENBERG, C., OPENSHAW, H., REY-MENDEX, M., PUGA, A. and NOTKINS, A. L. (1980) Herpes simplex virus DNA sequences in the CNS of latently infected mice. *Nature (London)*, **288**, 288–290

CARP, R. I., MERZ, P. A., KASCSAK, R. J., MERZ, G. S. and WISNIEWSKI, H. M. (1985) Nature of the scrapie agent: current status of facts and hypotheses. *Journal of General Virology*, **66**, 1357–1368

CHESEBRO, B., RACE, R., WEHRLY, K. *et al.* (1985) Identification of scrapie prion protein-specific mRNA in scrapie-infected and uninfected brain. *Nature (London)*, **315**, 331–333

CHESTERS, P. M., HERITAGE, J. and McCANCE, D. J. (1983) Persistence of DNA sequences of BK virus and JC virus in normal human tissues and in diseased tissues. *Journal of Infectious Diseases*, **147**, 676–684

EFSTATHIOU, S., MINSON, A. C., FIELD, H. J., ANDERSON, J. R. and WILDY, P. (1986) Detection of herpes simplex virus-specific DNA sequences in latently infected mice and in humans. *Journal of Virology*, **57**, 446–455

ESIRI, M. (1987) Typical and atypical viruses in the aetiology of senile dementia of the Alzheimer type. In: *Histology and Histopathology of the Ageing Brain*, edited by J. Ulrich. Basel: Karger, in press

FOURNIER, J.-G., TARDIEU, M., LEBON, P. *et al.* (1985) Detection of measles virus RNA in lymphocytes from peripheral-blood and brain perivascular infiltrates of patients with subacute sclerosing panencephalitis. *New England Journal of Medicine*, **313**, 910–915

FRASER, N. W., LAWRENCE, W. C., WROBLEWSKA, Z., GILDEN, D. H. and KOPROWSKI, H. (1981) Herpes simplex type 1 DNA in human brain tissues. *Proceedings of the National Academy of Science USA*, **78**, 6461–6465

GALLOWAY, D. A., FENOGLIO, C., SHEVCHUCK, M. and McDOUGALL, J. K. (1979) Detection of herpes simplex RNA in human sensory ganglia. *Journal of Virology*, **95**, 265–268

GALLOWAY, D. A., FENOGLIO, C. M. and McDOUGALL, J. K. (1982) Limited transcription of the herpes simplex virus genome when latent in human sensory ganglia. *Journal of Virology*, **41**, 686–671

GANNICLIFFE, A., SALDANHA, J. A., ITZHAKI, R. F. and SUTTON, R. N. P. (1985) Herpes simplex viral DNA in temporal lobe epilepsy. *Lancet*, **i**, 214–215

GRINNELL, B. W., PADGETT, B. L. and WALKER, D. L. (1983a) Comparison of infectious JC virus DNAs cloned from human brain. *Journal of Virology*, **45**, 299–308

GRINNELL, B. W., PADGETT, B. L. and WALKER, D. L. (1983b) Distribution of nonintegrated DNA from JC papovavirus in organs of patients with progressive multifocal leukoencephalopathy. *Journal of Infectious Diseases*, **147**, 669–675

GOODPASTURE, E. W. (1929) Herpetic infection with especial reference to involvement of nervous system. *Medicine*, **8**, 223–243

HAASE, A. T ., VENTURA, P., GIBBS JR, C. J. and TOURTELLOTTE, W. W. (1981) Measles virus nucleotide sequences: detection by hybridization in scrapie. *Science*, **212**, 672–675

HAASE, A. T., GANTZ, D., BLUM, H. *et al.* (1985a) Combined macroscopic and microscopic detection of viral genes in tissues. *Journal of Virology*, **140**, 201–206

HAASE, A. T., GANTZ, D., EBLE, B. *et al.* (1985b) Natural history of restricted synthesis and expression of measles virus genes in subacute sclerosing panencephalitis. *Proceedings of the National Academy of Science USA*, **82**, 3020–3024

HAASE, A. T. (1986) The pathogenesis of slow virus infections: molecular analyses. *Journal of Infectious Diseases,* **153,** 441–443

HOLLAND, J., SPINDLER, K., HORODYSKI, F., GRABAU, E., NICHOL, S. and VANDEPOL, S. (1982) Rapid evolution of RNA genomes. *Science,* **215,** 1577–1585

JOHNSON, R. T. (1982) *Viral Infections of the Nervous System.* New York: Raven

JOHNSON, R. T. and McARTHUR, J. C. (1986) AIDS and the brain. *Trends in Neurosciences,* **9,** 91–93

JONES, T. R. and HYMAN, R. W. (1983) Specious hybridization between herpes simplex virus DNA and human cellular DNA. *Journal of General Virology,* **131,** 555–560

KASTRUKOFF, L. and KOPROWSKI, H. (1981) Isolation of virus from brain after immunosuppression of mice with latent herpes simplex. *Nature (London),* **291,** 432–433

KOENIG, S., GENDELMAN, H. E., ORENSTEIN, J. M. *et al.* (1986) Detection of AIDS virus in macrophages in brain tissue from AIDS patients with encephalopathy. *Science,* **233,** 1089–1093

KOSCHEL, K. and MÜNZEL, P. (1980) Persistent paramyxovirus infections and behaviour of β-adrenergic receptors in C-6 rat glioma cells. *Journal of General Physiology,* **47,** 513–517

KOSCHEL, K., MUNZEL, P., HALBACH, M. and METZNER, R. (1986) Persistent rabies virus infection of neuronal cells causes defects in neuroreceptor regulation. In *Molecular Aspects of Neurobiology,* edited by R. L. Montalcini, P. Calissano, E. R. Kandel and A. Maggi, pp. 158–161. Berlin: Springer-Verlag

LIAO, Y-C. J., LEBO, R. V., CLAWSON, G. A. and SMUCKLER, E. A. (1986) Human prion protein cDNA: molecular cloning, chromosomal mapping, and biological implications. *Science,* **233,** 364–367

LIEBERT, U. G., BACZKO, K., BUDKA, H. and TER MEULEN, V. (1986) Restricted expression of measles virus proteins in brains from cases of subacute sclerosing panencephalitis. *Journal of General Virology,* **67,** 2435–2444

MADDON, P. J., DALGLEISH, A. G., McDOUGAL, J. S., CLAPHAM, P. R., WEISS, R. A. and AXEL, R. (1986) The T4 gene encodes the AIDS virus receptor and is expressed in the immune system and the brain. *Cell,* **47,** 333–348

MILLER, C. A. and CARRIGAN, D. R. (1982) Reversible repression and activation of measles virus infection in neural cells. *Proceedings of the National Academy of Science, USA,* **79,** 1629–1633

OESCH, B., WESTAWAY, D., WÄLCHLI, M. *et al.* (1985) A cellular gene encodes scrapie PrP 27–30 protein. *Cell,* **40,** 735–746

OLDSTONE, M. B. A., HOLMSTOEN, J. and WELSH, R. M. JR (1977) Alterations of acetylcholine enzymes in neuroblastoma cells persistently affected with lymphocytic choriomeningitis virus. *Journal of Cell Physiology,* **91,** 459

OLDSTONE, M. B. A., SINHA, Y. N., BLOUNT, P. *et al.* (1982) Virus-induced alterations in homeostasis: alterations in differentiated functions of infected cells *in vivo. Science,* **218,** 1125–1127

PATIENT, R. (1984) DNA hybridization – beware. *Nature (London),* **308,** 15–16

PEDEN, K., MOUNTS, P. and HAYWARD, G. S. (1982) Homology between mammalian cell DNA sequences and human herpes virus genomes detected by a hybridization procedure with high-complexity probe (1982). *Cell,* **31,** 71–80

PRUSINER, S. B. (1984) Prions. *Scientific American,* **251,** 48–57

PUGA, A., ROSENTHAL, J. D., OPENSHAW, H. and NOTKINS, A. L. (1978) Herpes simplex virus DNA and mRNA sequences in acutely and chronically infected trigeminal ganglia of mice. *Journal of Virology,* **89,** 102–111

PUGA, A., CANTIN, E. M. and NOTKINS, A. L. (1982) Homology between murine and human cellular DNA sequences and the terminal repetition of the S component of herpes simplex virus type 1 DNA. *Cell,* **31,** 81–87

PUGA, A., CANTIN, E. M., WOHLENBERG, C., OPENSHAW, H. and NOTKINS, A. L. (1984) Different sizes of restriction endonuclease fragments from the terminal repetitions of the herpes simplex virus type 1 genome latent in trigeminal ganglia of mice. *Journal of General Virology,* **65,** 437–444

RITCHIE, D. A. and TIMBURY, M. C. (1980) Herpes virus and latency: possible relevance of the structure of the viral genome. *FEMS Microbiology Letters,* **9,** 67–72

ROBERTS, G. W., TAYLOR, G. R., CARTER, G. I. *et al.* (1986) Herpes simplex virus; a role in the aetiology of Alzheimer's disease? *Journal of Neurology, Neurosurgery and Psychiatry,* **49,** 216

ROCK, D. L. and FRASER, N. W. (1983) Detection of HSV-1 genome in central nervous system of latently infected mice. *Nature (London),* **302,** 523–525

ROOS, R. P. (1983) Viruses and demyelinating disease of the central nervous system. *Neurologic Clinics,* **1,** 681–695

SALDANHA, J., GANNICLIFFE, A. and ITZHAKI, R. F. (1984) An improved method for preparing DNA from human brain. *Journal of Neuroscience Methods,* **11,** 275–279

SALDANHA, J., SUTTON, R. N. P., GANNICLIFFE, A., FARRAGHER, B. and ITZHAKI, R. F. (1986) Detection of HSV1 DNA by *in situ* hybridisation in human brain after immunosuppression. *Journal of Neurology, Neurosurgery and Psychiatry,* **49,** 613–619

SEQUIERA, L. W., CARRASCO, L. H., CURRY, A., JENNINGS, L. C., LORD, M. A. and SUTTON, R. N. P. (1979) Detection of herpes simplex viral genome in brain tissue. *Lancet,* **ii,** 609–612

SHAW, G. M., HARPER, M. E., HAHN, B. H. *et al.* (1985) HTLV III infection in brains of children and adults with AIDS encephalopathy. *Science,* **227,** 177–182

SOUTHERN, P. J., BLOUNT, P. and OLDSTONE, M. B. A. (1984) Analysis of persistent virus infections by *in situ* hybridization to whole-mouse sections. *Nature (London),* **312,** 555–558

SUMAYA, C. V., MYERS, L. W., ELLISON, G. W. and ENCH, Y. (1985) Increased prevalence and titer of Epstein-Barr virus antibodies in patients with multiple sclerosis. *Annals of Neurology,* **17,** 371–377

TAYLOR, G. R., CROW, T. J., MARKAKIS, D. A., LOFTHOUSE, R., NEELEY, S. and CARTER, G. I. (1984) Herpes simplex virus and Alzheimer's disease: a search for virus DNA by spot hybridisation. *Journal of Neurology, Neurosurgery and Psychiatry,* **47,** 1061–1065

TAYLOR, G. R., CROW, T. J., HIGGINS, T. and REYNOLDS, G. (1985) Search for cytomegalovirus in postmortem brain tissue from patients with Huntington's chorea and other psychiatric disease by molecular hybridization using cloned DNA. *Journal of Neuropathology and Experimental Neurology,* **44,** 176–184

TENSER, R. B., DAWSON, M., RESSELL, S. J. and DUNSTAN, M. E. (1982) Detection of herpes simplex virus mRNA in latently infected trigeminal ganglion neurons by *in situ* hybridization. *Annals of Neurology,* **11,** 285–291

WIGDAHL, B., SCHECK, A. C., ZIEGLER, R. J., DE CLERCQ, E. and RAPP, F. (1984) Analysis of the herpes simplex virus genome during *in vitro* latency in human diploid fibroblasts and rat sensory neurons. *Journal of Virology,* **49,** 205–213

WILEY, C. A., SCHRIER, R. D., NELSON, J. A., LAMPERT, P. W. and OLDSTONE, M. A. (1986) Cellular localization of human immunodeficiency virus infection within the brains of acquired immune deficiency syndrome patients. *Proceedings of the National Academy of Science, USA,* **83,** 7089–7093

WOLINSKY, J. S. and JOHNSON, R. T. (1980) Role of viruses in chronic neurological diseases. In: *Comprehensive Virology 16,* edited by H. Fraenkel-Conrat and R. R. Wagner, pp. 257–296. New York: Plenum

17
Immunogenetics: genetic polymorphism and susceptibility to neurological disease

D. A. S. Compston and R. J. Swingler

INTRODUCTION

Classical methods have established the importance of genetic factors in various neurological diseases, and in some cases disease associations with phenotypic markers of polymorphic genetic systems involved in regulating immune response have since been demonstrated. But it has been necessary to propose incomplete penetrance, polygenic inheritance, disease heterogeneity and multifactorial aetiology in order to account for the pattern of associations observed to date; it seemed possible that difficulties arising in the interpretation of the mechanism of genetic susceptibility would be resolved with the application of molecular genetics, but despite improved understanding of the structure and organization of immunogenetic systems, these expectations have not been realized.

GENETICS OF IMMUNE RESPONSE

Immune response involves a sequence in which complex antigen is absorbed, processed and presented to T-helper lymphocytes which activate suppressor subpopulations, cytotoxic T cells and B lymphocytes. These processes are regulated by highly polymorphic genes, grouped within the major histocompatibility complex (MHC) – in man the HLA system – that encode proteins determining recognition events at cell surfaces. The HLA system lies on the short arm of chromosome 6 in the distal portion of the 6p21.3 band and is 1–2 cM (1–2 $\times$ 10^6 base pairs) long (*Figure 17.1*).

Class 1 (HLA-A, B and C) and class 2 (HLA-DR, DP and DQ amongst others) antigens regulate immune function by restricting T-cell activation and effector cell responses; class 3 products are involved in the classical (C2 and C4) and alternate (Bf) complement pathways. Serological methods have been used to distinguish 23 HLA-A, 47 HLA-B, 8 HLA-C, 16 HLA-DR and 3 HLA-DQ phenotypes, but the variety of responses encountered in functional tests indicates that other alleles exist. Cellular methods have been used to define a range of more complex specificities which reflect integrated function of class 2 antigens and are assigned to the HLA-D locus; this approach has been extended to define 6 HLA-DP alleles by

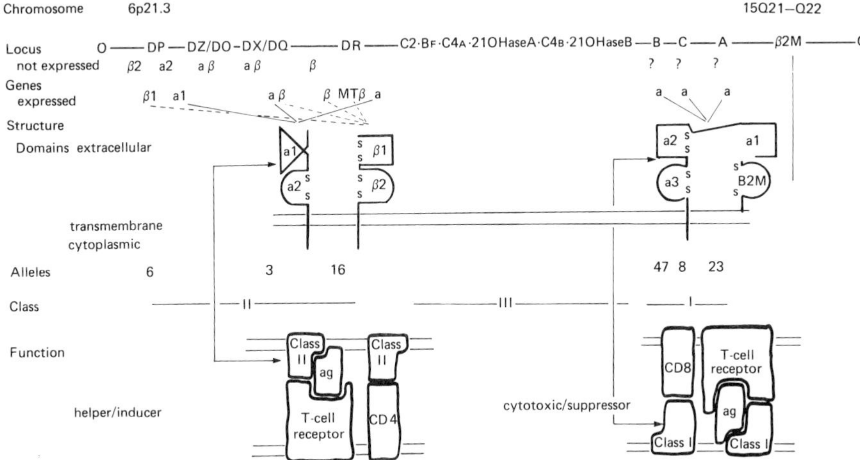

Figure 17.1 Diagram of the HLA region showing the class I, II and III loci, the polymorphic genes and non-expressed pseudogenes encoded at each locus, the structure of their products and the number of designated phenotypes; recognition events at the cell surface involving antigen, class I or II HLA gene product, T-cell receptor and the accessory CD4 or CD8 molecules are shown. For details, see text

primed lymphocyte typing. The expression of antigen at corresponding loci on each chromosome 6 is codominant and because recombination occurs in only 1–2% of meioses, alleles are usually inherited together as a haplotype. Recent advances in molecular biology and protein chemistry have shown that these phenotypic analyses conceal further complexities of the MHC loci and their genes.

Class 1 gene products

Class 1 antigens consist of an invariable polypeptide chain (β_2-microglobulin; molecular weight 12 000) encoded on chromosome 15, inserted into the cell membrane (Strachan, 1987). This is associated with a larger chain (molecular weight 45 000) which is the product of either an HLA-A, B or C locus gene; this has cytoplasmic and transmembrane portions and three extracellular domains (α_1, α_2 and α_3). In fact, although only three class 1 HLA genes are expressed, up to 17 may exist, many of which code for sequences that do not produce messenger (m) RNA or result in protein synthesis (pseudogenes). Most polymorphism in class 1 HLA genes is due to variations in amino acid structure of the α_1 and α_2 domains; α_3 is more conserved.

Class 2 gene products

The class 2 genes are more complicated and occupy approximately 1100 kb of the HLA region (*see Figure 17.1*; and Trowsdale, 1987). Class 2 antigens are dimers consisting of 33 000 molecular weight α chains and 28 000 molecular weight β chains encoded by genes within class 2 loci, each of which therefore contains a minimum of one α and one β gene. Both polypeptide chains have cytoplasmic, transmembrane and two extracellular domains named α_1 and α_2 or β_1 and β_2, respectively.

The DP locus has two α and two β genes but one pair (DP α_2 and DP β_2) is not expressed. Polymorphisms arise mostly from variations in the β_1 domain. The DQ-DX locus has two α and two β chains; DQ α and DQ β are expressed, antigenic polymorphism arising from variations in the α_1 and β_1 domains respectively. DX α and DX β genes are not expressed. The DZ-DO locus may be artefactual in that the distance separating the two identified genes, DZ α and DO β, is considerable and neither are expressed as protein products. The DR locus has one α gene only and this is conserved so that antigenic polymorphism depends on variations in the two β genes, DR β and DR (MT) β either of which can form a heterodimer with the DR α chain. There are additional DR β pseudogenes that are not expressed. Most variation of amino acid sequence is found in the DR β_1 domain. Although the class 2 products that are not expressed (DP α_2 and DP β_2, DZ α, DO β, DX α, DX β and some DR β genes) seem capable of making mRNA that could lead to protein synthesis, they may be incipient pseudogenes.

The status of HLA-D is less certain; for example, correlations between specificities defined serologically and by cellular techniques allow subdivision of DR2 into five HLA-D subtypes (HLA DR2-Dw2, Dw12, WJR, AZH and tb24; Nepom *et al.*, 1984). Using two dimensional electrophoresis the DR2 proteins separate into two regions, implying structural differences; one is found in all DR2 subtypes but the other is only found associated with HLA DR2–Dw2 and Dw12.

Class 3 gene products

There are two polymorphic C4 genes (C4A and C4B) both of which are expressed; there is less diversity of amino acid structure and polymorphism in C2 and Bf than is found elsewhere in the human MHC (Campbell, 1987; Carroll and Alper, 1987). C4 is polymorphic, although one allele is found in over 90% of individuals.

Biology of MHC gene products

Molecules encoded by the MHC act as determinants that are recognized in association with foreign antigen by T-cell receptors. The T-cell receptor consists of a two-chain glycoprotein dimer encoded by one α and one β gene found on 14q11.2 and 7q35 respectively. Each gene consists of variable, joining, and constant regions which undergo rearrangements during ontogeny to provide the diversity which characterizes the T-cell receptor through a variety of transcription mechanisms (Dembic, von Boehmer and Steinmetz, 1986). The α and β genes encode the serologically defined T-cell receptor polypeptide chains, but chains encoded by the γ and δ genes have also been identified. It seems probable that both α and β chains participate in recognition of antigen and MHC products by T cells.

Class 1 MHC products are involved in interactions of T-cytotoxic cells; recognition by the T-cell receptor occurs in combination with the CD8 molecule on the T-cell membrane. Conversely, T-helper/inducer interactions involve class 2 MHC products and the CD4 molecule. Antigen presenting cells can only activate the CD4/T cell receptor complex on helper/inducer cells if the foreign antigen is presented with a class 2 molecule of the same allelic specificity. Similarly, the CD8/T cell receptor complex on cytotoxic/suppressor T lymphocytes only responds to foreign antigen on target cells if it is found in association with a complementary class 1 molecule. There is therefore a variable (antigen and T-cell receptor rearrangement) and constant (MHC product and CD4 or 8 molecule) component to cell presentation and recognition (*see Figure 17.1*).

Population genetics

HLA antigens are not evenly distributed throughout or between populations (Bodmer *et al.*, 1987) and certain haplotypes tend to be population specific; A1, B8 and DR3 is the most frequent haplotype in northern Europeans but A24 Dw52 DR2 is commonest in Japanese. There is a gradient from north to south Europe for the haplotype A3, B7 and DR2 with higher frequencies occurring in Scandinavian, English, German and Swiss than in Celtic, Dutch, French, Italian, Spanish and Austrian populations. Certain alleles (such as A3, B7, DR2, DQw1) occur together more often than expected by chance (linkage disequilibrium), but the antigens in each linkage group can differ between populations so that, for example, DR2 is linked to DQw1 in northern Europeans and DQw3 in South American Indians. Since recombination should abolish linkage disequilibrium, these haplotypes are presumably maintained by natural selection or non-equilibrium population conditions resulting from stratification or migration.

HLA AND DISEASE

Products of these MHC loci provide obvious candidate genes for the study of neurological diseases which are thought to have an autoimmune basis. Occasionally they also make excellent markers for linked susceptibility genes in non-immunological conditions. Although the phenotypic MHC associations have been weak, the early studies have been repeated in the hope that genetic polymorphisms might reveal closer markers of disease susceptibility. HLA antigen associations have been demonstrated by comparing antigen frequency in patients and controls of the same ethnic group, but almost without exception these are incomplete, weak and often differ between populations.

Several explanations have been offered for these findings based first on the assumption that HLA substances are themselves the relevant factors in the mechanism of disease (Batchelor and McMichael, 1987). Autoreactive responses may result from cross-reactivity between HLA antigen and pathogen; HLA substances could act as receptors for pathogens which thereby gain entry into target cells, and the associations may reflect aberrations in the appropriate rate or degree of response to pathogens through physical interactions of antigen, HLA substance, T-cell receptor and CD4/8 molecules on the cell membrane. Although stronger or more complete associations may be found with rare, but as yet unidentified alleles, it is necessary to suggest significant environmental contributions, polygenic inheritance, disease heterogeneity and a multifactorial aetiology in order to account for the existing associations in terms of a direct effect of HLA substances themselves.

An alternative explanation is that phenotypes conceal complexities of molecular sequence which are more critical for the pathogenesis of disease. The suggestion has been made that class 2 α and β genes might recombine not with partners from the appropriate locus, but with complementary genes on adjacent loci or the other 6th chromosome to form heterodimers which are not normally expressed. If proteins are then synthesized in response to viral infection and α interferon production, self-tolerance would be lost resulting in autoimmune responses. It is also possible that conserved molecular sequences shared between phenotypes and not affecting specificity could determine disease susceptibility accounting for the association of non-linked genes with the same disease within, and specifically different genes between, populations. The HLA associations might depend on linkage to functionally related (such as class 3) or unrelated genes which happen to be located within the MHC (such as 21 hydroxylase deficiency and one form of autosomal dominant spinocerebellar degeneration). But there are probably no non-HLA genes within the HLA-D region, and for those disorders in which the closest associations are found with alleles of these loci, linkage to disease susceptibility genes is an unlikely explanation.

MULTIPLE SCLEROSIS

The aetiology of multiple sclerosis

The aetiology of multiple sclerosis (MS) is probably polygenic and multifactorial involving genetic, exogenous and immunological factors. Evidence for an environmental influence comes from epidemiological studies (Acheson, 1985). While the disease appears to have a predilection for caucasians, the frequency may vary by a factor of 10 according to their place of residence. In addition, migrants from high to low-risk areas appear further to reduce their chances of getting the disease if they move early in life, and an increased risk may be present for those migrating early from low to high-risk regions. The offspring of migrant populations tend to take on the risk of the indigenous population and this may vary considerably from that seen in their parental generation. There is also circumstantial evidence that individuals from high-risk areas could have introduced an infectious agent which caused small epidemics in isolated populations previously free of the disease (Kurtzke, 1986). The assumption that these epidemiological clues reflect involvement of a virus in the cause of MS is not substantiated by direct evidence from population serology, viral culture or *in situ* hybridization. Evidence from case control studies suggests that respiratory and other infections may act as trigger factors and in a recent prospective survey, 27% of relapses were preceded by an infectious illness, although overall viral infections were less common in MS patients than controls (Sibley, Bamford and Clark, 1985).

Immunological abnormalities have consistently been observed in MS patients. The earliest pathological change consists of perivenous lymphocyte, plasma cell and macrophage infiltration, followed by myelin breakdown. Later, lesions are more sharply demarcated and there is astrocytosis with reduced oligodendrocyte and axon numbers. There are raised levels of immunoglobulin (mainly IgG subclass 1) in cerebrospinal fluid; much of this appears to be synthesized in the central nervous system and immunoglobulin is deposited in plaques. IgG_1 is complement fixing and there is evidence that complement activation occurs in the CNS of patients with MS (Compston *et al.*, 1986a). There are both phenotypic and functional abnormalities in peripheral blood from patients with MS, indicating a defect in T-suppressor cells and B-lymphocyte overactivity (Hughes and Compston, unpublished data).

HLA phenotype associations

Caucasoids appears to be more susceptible to MS than individuals of African or Mongolian descent irrespective of their place of residence. Up to 17% of cases are familial; concordance rates are higher in monozygotic twins, one of whom is already known to have MS, than in dizygotic pairs or other siblings (Ebers *et al.*, 1986).

Studies of north European MS patients have shown progressively stronger associations with HLA-A3, B7, DR2 and Dw2 carrying relative risks of 2.9, 4.1, 5.2 and 5.5 respectively (Compston, 1986). The association with HLA-DR2 becomes progressively weaker through southern Europe, but can still be detected in France and Hungary. HLA-A3 and B7 confer susceptibility in Greeks and the association with B7 has also been detected in Iranians. The HLA-DR2 association is still found in northern European migrants to North America and south-east Australia, in coloured South Africans and black Americans. The prevalence of MS is very high in the Orkney Islands and north-east Scotland, but there is no association with HLA-A3, B7, DR2 or Dw2. This arises from the fact that the control frequency of DR2/Dw2 is much higher in these islands than in other northern European populations, whereas the frequency of these antigens in Orcadian patients is similar to individuals with MS elsewhere in northern Europe. More recently, a stronger association has been reported with DQw1 than the linked allele DR2 in a study from north-east Scotland (Francis *et al.*, 1987), but this finding has not been confirmed in south-east Wales (Swingler *et al.*, 1987) or Holland (van Lambalgen, van Sanders and D'Amaro, 1986). A weak association has been described with DPw4 in French patients (Moen *et al.*, 1984). The distribution of MS in the United Kingdom correlates with variations in the normal frequency of DR2 (Swingler and Compston, 1986) and this may also be true for other parts of the world. However, MS is rare in Hungarian gypsies, black Africans and Japanese, in each of whom DR2 has a normal frequency which is over 30%.

This simple hypothesis also fails to explain the absent and specifically different associations found in non-European populations. For example, no relationship has been detected in Israeli Jews; HLA-B18 is increased in Spanish patients with MS, whereas B12 is associated with the disease in the Canary Islands. There are associations with DR2, DRw53 and DR4 in Mediterranean, Gulf and Jordanian Arabs respectively. An increased frequency of DR4 and DR2 has been reported in Italian patients. Studies of Indians have shown an association with B12 but a subsequent study in the same area showed no associations. DR6 was initially shown to be associated with MS in Japan, but later reports showed a slight, but not significant excess of DR2 and DR4. Lastly, in European populations where DR2 is the main risk factor, there is also a subsidiary association with DR3.

Class 3 HLA gene polymorphisms have been investigated in MS patients. Schroder *et al.* (1983) found a significant association between MS and the C4 haplotype A4/B2 in twin-pair studies, and Stewart, Basten and Kirk (1979) demonstrated an association between all BfS alleles and MS; both of these observations may be explained by linkage disequilibrium between these antigens and HLA-DR2/Dw2. Fielder *et al.* (1981) reported a reduced frequency of all BfF alleles in patients with definite MS but not isolated demyelinating lesions and suggested that DR2 might act as a marker for susceptibility, whereas BfF determined subsequent course of the disease, but these observations have not been confirmed in a sample of MS cases from a different region selected retrospectively for extremes of severity (Francis *et al.*, 1987). Susceptibility may be conferred by haplotypes, rather than individual alleles and the linkage group A3, B7, DR2, DQw1, C4 A4/B2 and BfS may have significance for pathogenesis in its function as a unit.

Immunoglobulin heavy chain (Gm) allotypes

Immunoglobulins consist of two paired heavy and light chains; the heavy chain constant and variable regions are encoded on chromosome 14 by adjacent genes including those that code for structural differences determining the separate classes and subclasses together with minor alterations in the constant region which give rise to the *Gm* allotypes. Other genes regulate the switch from synthesis of one heavy chain class to another (Schanfield and van Loghem, 1986). Pandey, Goust and Salier (1981) found the *G1m*1, 17; *G3m*21 phenotype to be increased in 70 MS patients compared with controls. Propert, Bernard and Symons (1982) also reported elevated levels of *G1m*1 in MS patients, whereas *G1m*3; *G3m*5, 13, 14 appeared protective. However Stewart and Kirk (1980) found an increase in the *G1m*1, 17; *G3m*21, 26 haplotype in only some groups of Australian patients and overall the study populations were not in Hardy–Weinberg equilibrium. Salier *et al.* (1986) and Francis *et al.* (1986) failed to demonstrate a primary *Gm* association with MS, but both authors reported an interaction between HLA-DQw1 or DR2 and the *Gm* haplotypes *G1m*1 and *G1m*3:*G3m*5, 10, 11, 13, 14 in determining susceptibility respectively. As in the *Bf* studies the difference in results seen between these investigations may have resulted from sampling errors, but racial and ethnic variations could also be relevant. There is a gradient of increasing frequency of *G1m*1, 17:*G3m*21 and *G1m*1, 2, 17:*G3m*21 and fall in *G1m*3:*G3m*5, 10, 11, 13, 14 from south-eastern to northern Europe, but although variations in the normal distribution of DR2 may account for the north–south gradient of MS in the United Kingdom, *Gm* haplotype frequencies are similar in the north-east of Scotland and southern parts of the United Kingdom (Brazier and Goldsmith, 1968).

Restriction fragment length polymorphisms (RFLPs) and multiple sclerosis

Since restriction enzymes cleave DNA at preferential sites, enzyme digestion will produce DNA fragments of varying size due to small alterations in nucleotide sequence; these can be detected by hybridization with a radiolabelled complementary DNA probe after electrophoretic separation. Since these polymorphisms are inherited they can be used to type individuals within families. These techniques have been applied to MS in the hope that one or other RFLP would show a complete and invariable association with MS between populations.

Using a cDNA DQ β gene probe and the *Eco RI* restriction enzyme, Cohen *et al.* (1984) identified a 2.2 kb fragment that segregated in association with the DR2 phenotype in patients with MS and controls, but not DR2-positive patients with insulin dependent diabetes. Using restriction enzymes which cleave DNA at other sites, fragments were produced which also showed different typing patterns from serologically defined phenotypes in all individuals. Using this combined serological and genotypic approach, Marcadet *et al.* (1985a) showed that after digestion with *Eco RV* and hybridization with another cDNA DQ β gene probe, DR2-positive individuals had either a 2.6 kb fragment (designated DQR1) or a 5.2 kb fragment (designated DQR2.6). DQR1 is so named because it segregates with all serologically defined DR1 but also some DR2 and DRw6 positive individuals, whereas DQR2.6 segregates with the remaining DR2 and DRw6 individuals only. There is a disturbance in the distribution of these DQ beta clusters in MS patients (DQR2.6, 97%; DQR1, 16%) compared with controls (DQR2.6, 63%; DQR1, 56%) suggesting molecular differences between phenotypically identical DR2 positive MS patients and controls. These may be analogous to the HLA-Dw subtypes since there is a good correlation between RFLPs and HLA-Dw typing patterns. The DQR1 cluster may recognize the HLA-DR2/WJR D-locus subtype, whereas DQR2.6 identifies HLA-DR2/Dw2 positive individuals. The differences between RFLP and cellular typing patterns in DR positive individuals who are serotypically indistinguishable raises the further possibility that the DR2 gene has a unique structure in MS patients, but RFLP studies cannot distinguish between DR2/Dw2 MS cases and controls (Jacobsen *et al.*, 1986) and sequence data have not yet demonstrated that the DR beta chains in MS patients differ from those in controls. A disease specific defect in transcription or translation cannot be excluded but two-dimensional gel electrophoresis has so far failed to demonstrate structural differences in the polymorphic chains of MS patients and controls (Sriram *et al.*, 1985).

Restriction fragment length polymorphisms (RFLPs) have been identified using cDNA probes for fragments of the three subclasses of immunoglobulin G heavy chains and this approach has considerably extended the range of recognizable *Gm* allotypes. Gaiser *et al.* (1987), using DNA digested with *Bst EII* and a cDNA probe that identified RFLPs marking $_1$, $_2$, $_3$ and pseudo heavy chain regions, reported that the frequency of a 5.9 kb band was reduced and a 4.1 kb band increased in MS patients from France and California compared with controls. The 5.6 kb band segregated with the *G3m*21 and the 4.1 kb band with the *G3m*5, 10, 11, 13 phenotype. The strength of the association is comparable but the specificity different from previous phenotypic studies.

IgG subclass 3 has not previously been implicated in MS – the oligoclonal bands consisting mainly of IgG_1 – and this association cannot be explained by linkage disequilibrium. The possibilities of environmental or aetiological heterogeneity, genetic stratification and sampling error have again to be considered as possible explanations.

HLA linked immune responsiveness

HLA linked serological abnormalities have been sought especially for measles, but the available studies do not demonstrate conclusively whether the finding of increased serum titres is an effect of HLA type over and above that attributable to the disease itself, or whether immune reactivity is altered in all individuals possessing the HLA antigens which increase susceptibility to MS; taken together no consistent pattern of altered reactivity has emerged from these investigations. Compston *et al.* (1986b) compared serological and historical evidence for viral infection in 177 patients and 164 controls. Differences in age of infection and/or antibody titre to rubella, measles, mumps and parainfluenza I were identified especially in comparisons of HLA-DR2 positive cases and controls. HLA-DR3 positive patients had increased titres of varicella zoster and adenovirus antibodies; Lamoureux, Lapierre and Ducharme (1983) also found an association between the presence of HLA-DR3 and a history of upper respiratory tract infections in childhood.

In a subsequent study (Hughes *et al.*, 1987), peripheral blood CD8 cell phenotypes were correlated with serological evidence for exposure to 15 infectious agents in longitudinal studies involving 13 patients with MS, 13 of their siblings, nine spouses and 13 unrelated controls; exposure to no single infectious agent could account for the observed serial changes in CD8 cells, but there was an association between the presence of HLA-DR2 and periodic reductions in CD8 cells irrespective of clinical status. Taken with previously reported serial observations in patients and cohabiting relatives, this finding provides indirect evidence for an interplay between environmental and genetic factors in determining CD8 cell phenotypes in MS.

There is a greater difference in interferon synthesis by mitogen stimulated peripheral blood mononuclear cells between Dw2 positive patients and controls than occurs in comparisons of individuals without this antigen, but interferon production in response to viral antigen is reduced in all Dw2 positive individuals (Salonen *et al.*, 1982; Vervliet *et al.*, 1983). In an attempt to identify a DNA polymorphism determining an MHC controlled immunological function, Jacobsen *et al.* (1986) distinguished an HLA-DR2/Dw2 associated RFLP pattern segregating with expression of one of two distinct DR β polypeptide chains (DR β 2) and correlating with HLA class 2 restriction of T4 positive measles virus-specific cytotoxic T-cell clones.

The concentration of CSF IgG and the presence of oligoclonal bands have been correlated with the presence of HLA-Dw2, and CSF complement component concentrations have also been related to the presence of MHC alleles; Trouillas *et al.* (1976) found low levels of C3, total complement haemolytic activity and factor B in 30% of MS patients, abnormalities which were linked to HLA-B18 and BfF1. Delasnerie-Laupretre, Suet-Hubert and Marcelli-Barge (1982) found lower C2 levels in B18 positive individuals, and Bertrams (1982) reported C2 hypocomplementaemia in Dw2 positive patients.

Polygenic inheritance

Even if the general hypothesis linking immune function, the HLA system and susceptibility to disease is correct, strong associations would not be expected if the aetiology is multifactorial and polygenic. Taken together the available evidence suggests that no one risk factor appears necessary or is itself sufficient for the development of MS, but it is their occurrence in the same individual perhaps in a special sequence or at critical times which determines development of the disease. The influence of each factor will depend on the absolute risk conferred as well as its distribution in the at risk population. Disease associations are based on differences in frequency between cases and controls, but these do not necessarily reflect biological significance; they become obscured in any region where the frequency in controls approximates to that in cases. In a population where a major risk factor is common and a minor one rare, prevalence of the disease will be high, but associations with the major risk factor may disappear and the distribution of MS within this population will then be determined more by the presence in individuals of less common complementary genes or exogenous factors.

It is therefore possible that the class 2 HLA associations in the Orkneys and North-east Scotland may be more important than their weak associations suggest, for although the relative risks are small, these genes are present in a high proportion of the population and may be making a large contribution to the incidence of disease. In areas where all susceptibility gene frequencies are low in the normal population, the chance that an individual will possess each risk factor and be exposed to environmental conditions necessary for the development of MS will also be low, so that incidence of the disease will fall. Under these circumstances the gene that confers the greatest absolute risk will show the strongest association and more detailed studies involving large numbers will always be required to demonstrate the weaker associations.

The implications of this multifactorial hypothesis, in which individual risk factors confer different susceptibility and are not evenly distributed, is that in some studies environmental conditions and in others genetic factors will appear to be of exclusive aetiological importance, giving the impression of greater aetiological heterogeneity than may in fact be the case. Unless multiple sclerosis as presently defined is the expression of many different diseases, or there is an as yet unidentified unique RFLP association, only the polygenic model can account for the pattern of genetic associations observed within and between populations.

MYASTHENIA GRAVIS

Circulating acetylcholine receptor antibody (AChR) is present in most patients with generalized and penicillamine-induced myasthenia gravis (MG), in fewer ocular cases, but not in patients with congenital myasthenia, usually a familial condition which should be distinguished from neonatal myasthenia due to placental transfer of IgG antibody or rare cases of myasthenia gravis where there is a family history. Genetic studies have nevertheless provided evidence for disease heterogeneity in MG (Compston *et al.*, 1980).

The peak age of onset in myasthenia gravis patients with thymoma is in the 5th decade, whereas there is a bimodal distribution in the remainder. Preoperatively, patients with thymoma have high levels of circulating AChR antibody and prevalence of striated muscle but not other autoantibodies; older onset non-thymoma cases have lower levels of receptor antibody, but striated muscle antibody and other autoantibodies are commonly present in these predominantly male and often ocular cases. Younger onset female patients have intermediate levels of AChR antibody but do not show the same high prevalence of all other autoantibodies. The postsynaptic membrane receptor is structurally complex, consisting of at least five subunits and after immunization with purified antigen, monoclonal antibodies can be raised to several epitopes. There is a considerable difference in the titre of antibodies detected against these separate antigens in myasthenia gravis patients, and significant differences in their distribution are seen in the three categories of patients defined by the presence of thymoma and age of onset (Whiting, Vincent and Newsom Davis, 1986).

HLA phenotypes in myasthenia gravis

This evidence for immunological and pathological heterogeneity is complemented by HLA studies. In caucasians there are independent associations between A3, B7 and DR2 and male sex, onset over the age of 40, low mean AChR antibody titre and prevalence of striated muscle antibody; conversely A1, B8 and DR3 are associated with female sex, onset below the age of 40, higher mean titres of receptor antibody and thymic hyperplasia. Penicillamine-induced myasthenia is associated in caucasians with Bw35 and DR1 (Garlepp, Dawkins and Christiansen, 1983).

Although detailed comparisons have not been carried out in non-caucasian patients, it is clear that specifically different associations exist especially with clinical subgroups. In northern India there are associations with Bw21, Bw35 and B8 in patients with thymic hyperplasia (Mehra *et al.*, 1983). HLA-B12 is increased in young Japanese myasthenic females with thymic hyperplasia and B5 in thymoma cases (Yoshida *et al.*, 1977). In Chinese populations, there are associations with Aw26, Bw46 and DR4 (Chee *et al.*, 1978; Chan-Lui, Leung and Lau, 1984; Lee *et al.*, 1984), whereas Chinese childhood onset cases living in Hong Kong, who have low or absent AChR antibody and a high prevalence of ocular disease, show a Bw46 association which is not seen in adult Chinese from the same area (Hawkins *et al.*, 1984, 1986). In a study of predominantly young female American blacks with generalized myasthenia, there was an increase in frequency of A1, B8 and DR5 but not DR3; Aw30 was decreased and Aw33 increased in this group. A similar pattern of association between A1, B8 but not DR3 has been observed in Thais (Christiansen *et al.*, 1984).

The situation is similar to that found in multiple sclerosis, namely incomplete associations with one or more alleles of the same or linked loci, differing between populations.

Restriction fragment length polymorphisms and myasthenia gravis

Molecular techniques have therefore been applied to this problem. Bell *et al.* (1986) identified a DQw2 associated 15 kb band after digestion with *Hinc II* and hybridization with a cDNA DQβ probe which was present in seven out of 16 DR3 positive myasthenic patients but only one out of 44 DR3 positive individuals who were normal or suffering from other DR3 phenotype associated diseases. An identical band appeared to be present in most DR7 positive individuals irrespective of clinical status; this band was also detected in two DR3 negative myasthenia gravis patients but not in 35 DR3 negative controls. This study demonstrates a stronger RFLP association with myasthenia gravis than is seen with B8, DR3 or DQw2; but whether the 15 kb fragment is directly involved in the disease process or merely includes an epitope which is more critical for the pathogenesis than DR3 itself will become clear once the results of DQβ gene sequence analysis between phenotypically identical myasthenia gravis and other individuals are available. It is probable that this finding will only increase precision in categories of myasthenia gravis which are DR3 associated and not the disease as a whole; it should also be noted that in most studies the B8 phenotype shows a closer association than DR3, an observation which invites caution in concluding that the HLA association in myasthenia gravis is necessarily related to its autoimmune basis.

Immunoglobulin heavy chain (*Gm*) allotypes and myasthenia gravis

There is evidence that *Gm* associations exist with some forms of myasthenia gravis but these also differ between populations. In Japanese patients there is an increased frequency of *G1m1*, 2:G3 m21, especially in patients with thymoma and severe generalized myasthenia (Nakao *et al.*, 1980). *G1m1* is decreased in Swedish patients with severe myasthenia gravis but not Finns, although in the latter AChR antibody titres are higher in *G1m1* positive than negative individuals (Smith *et al.*, 1983, 1984). In a comparison between Chinese and caucasian patients with myasthenia, Chiu *et al.* (1987) found that *G2m23* was significantly increased in Chinese female patients with high AChR antibodies, and in caucasians with ocular myasthenia or no detectable AChR antibody.

Gaiser *et al.* (1987) studied 47 patients with myasthenia gravis and found no alteration in control frequencies of three RFLPs detected using an Ig γ_1 probe marking the γ_{1}, $_{2}$ or $_{3}$ and pseudod γ heavy chain loci.

LAMBERT–EATON MYASTHENIC SYNDROME

There is evidence that the Lambert–Eaton syndrome, which results from failure of acetylcholine release, is due to the presence of a circulating IgG autoantibody directed against presynaptic calcium channels (Lang *et al.*, 1981). Lambert–Eaton syndrome is associated with HLA-B8 but not DR3 and with the *Gm* allotypes *G1m1*, 2:G3*m21* in caucasians from northern Europe, but there have been no other immunogenetic studies of this rare syndrome (Willcox *et al.*, 1985). There is an increased frequency of RFLPs marking the μ and α_1 switch regions of Ig heavy chain constant region genes in patients with this disease (Demaine *et al.*, 1987).

THE NARCOLEPTIC SYNDROME

The narcoleptic syndrome is characterized by excessive daytime sleepiness, cataplexy, sleep paralysis and presleep dreams usually of an unpleasant character. The diagnosis is easily made in the typical case, but difficulties sometimes arise in deciding whether or not patients with only daytime sleepiness are abnormal. The syndrome has to be distinguished from neurological diseases which are complicated by pathological drowsiness. Pedigree analysis is difficult in the absence of a diagnostic test, but approximately 30% of patients with the narcoleptic syndrome have an affected first-degree relative; a dominant mode of inheritance with incomplete penetrance has been proposed, although monozygotic twins discordant for narcolepsy have been described and other investigators have since suggested that the condition is multifactorial (Parkes, 1985).

Narcolepsy and the HLA system: phenotypes and restriction fragment length polymorphisms

HLA associations were first described with Bw35 and Cw3 in Japanese patients with a reduced frequency of Bw52 and B7 (Honda *et al.*, 1983); conversely B7 was increased in French patients (Seignalet and Billiard, 1984). This situation changed dramatically when a complete association was demonstrated between narcolepsy and HLA-DR2 and DQw1 in studies from England, France and Japan (Juji *et al.*, 1984; Langdon *et al.*, 1984; Billiard and Seignalet, 1985). Other sleep disorders show no DR2 association. The finding that rapid eye movement (REM) sleep latencies are shorter in all DR2 individuals irrespective of clinical status has not been confirmed in a further sample analysed by the original investigators (Schulz *et al.*, 1987).

In Japan, all narcoleptic patients studied so far have the DR2/Dw2 subtype compared with 10% of unaffected controls, consistent with the suggestion that this may be necessary for the development of narcolepsy in any population. Studies of DRβ gene polymorphisms have failed to identify a narcolepsy-associated RFLP but DQ polymorphisms do discriminate between these patients and controls. Using the *Eco RI* restriction enzyme and a cDNA DQβ gene probe, the 5.2 kb fragment characterizing the DQR2.6 cluster has been identified in all narcolepsy patients, although this is also present in up to 95% of controls (Marcadet *et al.*, 1985b; Andreas-Zietz *et al.*, 1986). In MS patients this genotype associates with the DR2/Dw2 subtype. A few narcoleptic patients negative for the DR2 phenotype and DQR2.6 cluster have been described (Andreas-Zietz *et al.*, 1986).

The DR2 association has also been further investigated by recombinant DNA techniques in Japanese patients; using *Eco RI* and a cDNA DQβ probe, a 2.4 kb band was present in all (DR2 positive) narcoleptic patients, but only 38% of controls; using *Bam H1* a 2.9 kb band was present in all patients and 55% of controls and using *Pst 1* an informative fragment was present in all patients and 83% of controls (Honda *et al.*, 1986). These RFLP studies have confirmed the DR2/Dw2 association at the genotypic level, but are no more discriminating than combined serological and cellular typing methods, as would be expected given the almost complete association of the phenotypic markers with narcolepsy. Sequence data should resolve whether an epitope of the DR2 or DQw1 molecule confers absolute specificity for narcolepsy but studies of the DRβ gene have so far failed to detect a sequence specific mutation.

Any interpretation of these findings must accommodate the facts that the association with DR2/Dw2 (and the DR2.6 cluster) is virtually complete; the same association is seen in all populations despite the susceptibility marker being found in different linkage groups, no non-HLA genes are known to exist in the HLA-D region, and there is no other reason for believing that narcolepsy is an autoimmune disease. It is still possible that DR2 is linked to a narcolepsy susceptibility gene or that the disease is autoimmune, but a further explanation is that the DR2 molecule acts as receptor for an extrinsic factor which interferes with transmission in the central nervous system pathways concerned with wakefulness.

CONCLUSIONS

The techniques of molecular and cellular biology have contributed considerably to understanding the HLA region. Cloning, sequencing and restriction studies have shown that genetic organization is much more complex than previously thought and even the known loci are more polymorphic than phenotypic studies had indicated. Investigation using two-dimensional electrophoresis and monoclonal antibodies is beginning to elucidate the molecular basis of serological and functional polymorphisms. While these developments are of considerable interest they have so far failed to account for the observed disease associations. DNA studies of the HLA region have verified the genetic basis of phenotypic associations, and in some cases revealed subgroups at greater risk than was previously apparent, but no disease-specific RFLPs have yet been identified. It is possible that DNA sequencing may reveal a specific susceptibility gene for narcolepsy where the phenotypic risk is extraordinarily high, but it seems unlikely that diagnostic markers will be found in autoimmune conditions such as multiple sclerosis and myasthenia gravis. Indeed, the incomplete and heterogeneous associations found to date probably reflect the multifactorial and polygenic nature of these disorders.

References

ACHESON, E. D. (1985) The epidemiology of multiple sclerosis. In: *McAlpine's Multiple Sclerosis,* edited by W. B. Matthews, pp. 3–46. Edinburgh: Churchill Livingstone

ANDREAS-ZIETZ, A., SCHOLZ, S., ROTH, B. *et al.* (1986) DR2 negative narcolepsy. *Lancet,* **ii,** 684–685

BATCHELOR, J. R. and McMICHAEL, A. J. (1987) Progress in understanding HLA and disease associations. *British Medical Bulletin,* **43,** 156–183

BELL, J., SMOOT, S., NEWBY, C. *et al.* (1986) HLA-DQ beta-chain polymorphism linked to myasthenia gravis. *Lancet,* **i,** 1058–1060

BERTRAMS, J. (1982) Factor B alleles and multiple sclerosis. *Lancet,* **i,** 288

BILLIARD, M. and SEIGNALET, J. (1985) Extraordinary association between HLA-DR2 and narcolepsy. *Lancet,* **ii,** 226–227

BODMER, J. G., KENNEDY, L. J., LINDSAY, J. and WASIK, A. M. (1987) Applications of serology and the ethnic distribution of three locus HLA haplotypes. *British Medical Bulletin,* **43,** 94–121

BRAZIER, D. M. and GOLDSMITH, K. L. G. (1968) Frequency of certain *Gm* and *Inv* factors in the United Kingdom. *Nature (London),* **219,** 193

CAMPBELL, R. D. (1987) The molecular genetics and polymorphism of C2 and factor B. *British Medical Bulletin,* **43,** 37–49

CARROLL, M. C. and ALPER, C. A. (1987) Polymorphism and molecular genetics of human C4. *British Medical Bulletin,* **43,** 50–65

CHAN-LUI, W. U., LEUNG, N. K. and LAU, T. T. Y. (1984) Myasthenia gravis in Chinese children. *Developmental Medicine and Child Neurology,* **26,** 717–724

CHEE, C.-Y., CHEN, R.-C., HUNG, T.-P., CHEN, T.-S., CHEN, G. H. and DENG, J.-S. (1978) Immunological studies in Chinese myasthenia gravis patients. *Journal of the Formosan Medical Association,* **77,** 621–628

CHIU, H.-C., deLANGE, G. G., WILLCOX, N. *et al.* (1987) Immunoglobulin allotypes in caucasian and Chinese myasthenia gravis: differences from Japanese patients. *Journal of Neurology, Neurosurgery and Psychiatry,* in press

CHRISTIANSEN, F. T., POLLACK, M. S., GARLEPP, M. J. and DAWKINS, R. L. (1984) Myasthenia gravis and HLA antigens in American blacks and other races. *Journal of Neuroimmunology,* **7,** 121–129

COHEN, D., COHEN, O., MARCADET, A. *et al.* (1984) Class II HLA-DC β chain DNA restriction fragments differentiate among HLA-DR2 individuals in insulin-dependent diabetes and multiple sclerosis. *Proceedings of the National Academy of Science, USA,* **81,** 1774–1778

COMPSTON, D. A. S., VINCENT, A., NEWSOM-DAVIS, J. and BATCHELOR, J. R. (1980) Clinical pathological HLA antigen and immunological evidence for disease heterogeneity in myasthenia gravis. *Brain,* **103,** 579–601

COMPSTON, D. A. S. (1986) Genetic factors in the aetiology of MS. In: *Multiple Sclerosis: BIMR Neurology 6,* edited by W. I. McDonald and D. Silberberg, pp. 56–73. London: Butterworths

COMPSTON, D. A. S., MORGAN, B. P., OLEESKY, D., FIFIELD, R. and CAMPBELL, A. K. (1986a) Cerebrospinal fluid C9 in demyelinating disease. *Neurology,* **36,** 1503–1506

COMPSTON, D. A. S., VAKARELIS, B. N., MCDONALD, W. I., BATCHELOR, J. R. and MIMS, C. A. (1986b) Viral infection in patients with multiple sclerosis and HLA-DR matched controls. *Brain,* **109,** 325–344

DELASNERIE-LAUPRETRE, N., SUET-HUBERT, C. and MARCELLI-BARGE, A. (1982) Cerebrospinal fluid C2 and HLA system in multiple sclerosis. *Tissue Antigens,* **19,** 79–84

DEMAINE, A. G., WILLCOX, N., WELSH, K. I. and NEWSOM-DAVIS, J. (1987) Susceptibility to Lambert-Eaton myasthenic syndrome associated with IgH region alleles. *Journal of Neurology, Neurosurgery and Psychiatry* (in press)

DEMBIC, S., von BOEHMER, H. and STEINMETZ, M. (1986) The role of T cell receptor α and β genes in MHC-restricted antigen recognition. *Immunology Today,* **7,** 308–310

EBERS, G. C., BULMAN, D. E., SADOVNICK, A. D. *et al.* (1986) A population-based study of multiple sclerosis in twins. *New England Journal of Medicine,* **315,** 1638–1642

FIELDER, A. H. L., BATCHELOR, J. R., VAKARELIS, B. N., COMPSTON, D. A. S. and McDONALD, W. I. (1981) Optic neuritis and multiple sclerosis: do factor B alleles influence progression of disease? *Lancet,* **ii,** 1246–1248

FRANCIS, D. A., BRAZIER, D. M., BATCHELOR, J. R., McDONALD, W. I., DOWNIE, A. E. and HERN, J. E. C. (1986) *Gm* allotypes in multiple sclerosis influence susceptibility in HLA-DQw1 positive patients from the north-east of Scotland. *Clinical Immunology and Immunopathology,* **41,** 409–416

FRANCIS, D. A., BATCHELOR, J. R., McDONALD, W. I. *et al.* (1987) Multiple sclerosis in north-east Scotland: an association with HLA-DQw1. *Brain,* **110,** 181–196

GAISER, C. N., JOHNSON, M. J., de LANGE, G., RASSENTI, L., CAVALLI-SFORZA, L. L. and STEINMAN, L. (1987) Susceptibility to multiple sclerosis associated with an immunoglobulin gamma-3 restriction fragment length polymorphism. *Journal of Clinical Investigation,* **79,** 309–313

GARLEPP, M. J., DAWKINS, R. L. and CHRISTIANSEN, F. T. (1983) HLA antigens and acetylcholine receptor antibodies in penicillamine induced myasthenia gravis. *British Medical Journal,* **286,** 338–340

HAWKINS, B. R., CHAN-LUI, W. Y., CHOI, E. K. K. and HO, A. Y. (1984) Strong association of HLA Bw46 with juvenile onset myasthenia gravis in Hong Kong Chinese. *Journal of Neurology, Neurosurgery and Psychiatry,* **47,** 555–557

HAWKINS, B. R., MSM, I. P., LAM, K. S. L., CHAN-LUI, W. Y., YEUNG, R. T. T. and DAWKINS, R. L. (1986) HLA antigens and acetylcholine receptor antibody in the subclassification of myasthenia gravis in Hong Kong Chinese. *Journal of Neurology, Neurosurgery and Psychiatry,* **49,** 316–319

HONDA, Y., ASAKA, A., TANAKA, Y. and JUJI, T. (1983) Discrimination of narcoleptic patients by using genetic markers and HLA. *Sleep Research,* **12,** 254

HONDA, Y., JUJI, T., MATSUKI, K. *et al.* (1986) HLA-DR2 and Dw2 in narcolepsy and in other disorders of excessive somnolence without cataplexy. *Sleep,* **9,** 133–142

HUGHES, P. J., KIRK, P. F., DYAS, J., MUNRO, J. A., WELSH, K. I. and COMPSTON, D. A. S. (1987) Factors influencing circulating CD8 cell phenotypes in patients with multiple sclerosis. *Journal of Neurology, Neurosurgery and Psychiatry,* **50,** 1156–1159

JACOBSEN, S., SORRENTINO, R., NEPOM, G. T., McFARLIN, D. E. and STROMINER, J. L. (1986) DNA restriction fragment length polymorphism of HLA-DR2: correlation with HLA-DR2-associated functions. *Journal of Neuroimmunology,* **12,** 195–203

JUJI, T., SATAKE, M., HONDA, Y. and DOI, Y. (1984) HLA antigens in Japanese patients with narcolepsy – all the patients were DR2 positive. *Tissue Antigens,* **24,** 316–319

KURTZKE, J. F. (1986) Multiple sclerosis in the Faroe Islands. II. Clinical update transmission and the nature of MS. *Neurology,* **36,** 307–328

LAMBALGEN, R., van SANDERS, E. A. C. M. and D'AMARO, J. (1986) Sex distribution, age of onset and HLA profiles in two types of multiple sclerosis. *Journal of the Neurological Sciences,* **76,** 13–21

LAMOUREUX, G., LAPIERRE, Y. and DUCHARME, G. (1983) Past infectious events and disease evolution in multiple sclerosis. *Journal of Neurology,* **230,** 81–90

LANG, B., NEWSOM-DAVIS, J., WRAY, D., VINCENT, A. and MURRAY, N. (1981) Autoimmune aetiology for myasthenic (Eaton-Lambert) syndrome. *Lancet,* **ii,** 224–226

LANGDON, N., WELSH, K. I., DAM, M. V., VAUGHAN, R. W. and PARKES, D. (1984) Genetic markers in epilepsy. *Lancet,* **ii,** 1178–1780

LEE, T. D., ZHAO, T. M., BU, K. J., O'DONNELL, M. and SANDLER, S. G. (1984) HLA-DR4 with myasthenia gravis in the Chinese. *Tissue Antigens,* **23,** 127–129

MARCADET, A., MASSART, C., SEMANA, G. *et al.* (1985a) Association of class II HLA DQ β chain DNA restriction fragments with multiple sclerosis. *Immunogenetics,* **22,** 93–96

MARCADET, A., GEBUNRER, L., BETUEL, H. *et al.* (1985b) DNA polymorphism related to HLA-DR2 Dw2 in patients with narcolepsy. *Immunogenetics,* **22,** 679–683

MEHRA, N. K., AHUJA, G. K., TANEJA, V. and VAIDYA, M. C. (1983) HLA antigens and myasthenia gravis in North India. *Journal of Neurology, Neurosurgery and Psychiatry,* **46,** 361–364

MOEN, T., STIEN, R., BRATLIE, A. and BONDEVIK, E. (1984) Distribution of HLA-SB antigens in multiple sclerosis. *Tissue Antigens,* **24,** 126–127

NAKAO, Y., MIYAZAKI, T., OTA, K. *et al.* (1980) *Gm* allotypes in myasthenia gravis. *Lancet,* **i,** 677–680

NEPOM, G. T., NEPOM, B. S., WILSON, M., MICKELSON, E., ANTONELLI, P. and HANSEN, J. A. (1984) Multiple 'la-like' molecules characterise HLA-DR2 haplotypes which differ in HLA-D. *Human Immunology,* **10,** 143–151

PANDEY, J. R., GOUST, J.-M. and SALIER, J.-P. (1981) Immunoglobulin G heavychain (*Gm*) allotypes in multiple sclerosis. *Journal of Clinical Investigation,* **67,** 1797–1800

PARKES, J. D. (1985) In: *Sleep and its Disorders,* pp. 282–285. London: W. B. Saunders

PROPERT, D. N., BERNARD, C. C. A. and SIMONS, M. J. (1982) *Gm* allotypes and multiple sclerosis. *Journal of Immunogenetics,* **9,** 359–361

SALIER, J.-P., SESBOUE, R., MARTIN-MONDIERE, C. *et al.* (1986) Combined influences of *Gm* and HLA phenotypes upon multiple sclerosis susceptibility and severity. *Journal of Clinical Investigation,* **78,** 533–538

SALONEN, R., IONEN, J., REUNANEN, M. and SALMI, A. (1982) Defective production of interferon-α associated with HLA-Dw2 antigen in stable multiple sclerosis. *Journal of the Neurological Sciences,* **55,** 197–206

SCHANFIELD, M. S. and van LOGHEM, E. (1986) Human immunoglobulin allotypes. In: *Handbook of Experimental Immunology,* volume 3, edited by D. M. Weir, Chapter 94. Oxford: Blackwell

SCHULZ, H., GEISLER, P., POLLMAECHER, T. *et al.* (1987) HLA-DR2 and rapid eye-movement sleep latency: failure to replicate. *Lancet,* **i,** 627

SCHRODER, R., ZANDER, H., ANDREAS, A. and MAUFF, G. (1983) Multiple sclerosis: immunogenetic analyses of sib-pair double case families. II. Studies on the association of multiple sclerosis with C2, C4, Bf, C3, C6 and GLO polymorphisms. *Immunobiology,* **164,** 160–170

SEIGNALET, J. and BILLIARD, M. (1984) Possible association between HLA B7 and narcolepsy. *Tissue Antigens,* **23,** 188–189

SIBLEY, W. A., BAMFORD, C. R. and CLARK, K. (1985) Clinical viral infections and multiple sclerosis. *Lancet,* **i,** 1313–1315

SMITH, C. I. E., GRUBB, R., HAMMERSTROM, L. and MATELL, G. (1983) *Gm* allotypes in Swedish myasthenia gravis patients. *Journal of Immunogenetics,* **10,** 1–9

SMITH, C. I. E., GRUBB, R., HAMMARSTROM, L. and PIRSKANEN, R. (1984) *Gm* allotypes in Finnish myasthenia gravis patients. *Neurology,* **34,** 1604–1605

SRIRAM, S., STEWART, G. J., BUHLER, M., GRUMET, G. and ENGLEMAN, E. (1985) HLA-DR antigens in multiple sclerosis: two-dimensional gel electrophoresis. *Neurology,* **35,** 248–251

STEWART, G. J., BASTEN, A. and KIRK, R. L. (1979) Strong linkage disequilibrium between HLA-Dw2 and BfS in multiple sclerosis and in the normal population. *Tissue Antigens,* **14,** 86–97

STEWART, G. J. and KIRK, R. L. (1983) The genetics of multiple sclerosis: the HLA system and other genetic markers. In: *Multiple Sclerosis*; edited by J. F. Hallpike, C. W. M. Adams and W. R. Tourtellotte, pp. 97–128. London: Chapman and Hall

STRACHAN, T. (1987) Molecular genetics and polymorphism of class I HLA antigens. *British Medical Bulletin,* **43,** 1–14

SWINGLER, R. J. and COMPSTON, D. A. S. (1986) The distribution of multiple sclerosis in the United Kingdom. *Journal of Neurology, Neurosurgery and Psychiatry,* **49,** 1115–1124

SWINGLER, R. J., KIRK, P. F., DARKE, C. and COMPSTON, D. A. S. (1987) HLA and multiple sclerosis in South-East Wales. *Journal of Neurology, Neurosurgery and Psychiatry,* **50,** 1153–1155

TROUILLAS, P., BERTHOUX, F., BETUEL, H., BOUISSON, D., AIMARD, G. and DEVIC, M. (1976) Hypocomplementaemic MS. Heterozygous C2 deficiency linked to HLA A10, B18. *Lancet,* **ii,** 1023

TROWSDALE, J. (1987) Genetics and polymorphism: class II antigens. *British Medical Bulletin,* **43,** 15–36

VERVLIET, G., CLAEYS, H., van HAVER, H. *et al.* (1983) Interferon production and natural killer (NK) activity in leukocyte cultures from multiple sclerosis patients. *Journal of the Neurological Sciences,* **60,** 137–150

WHITING, P. J., VINCENT, A. and NEWSOM-DAVIS, J. (1986) Myasthenia gravis: monoclonal acetylcholine receptor antibody used to analyse antibody specificity and response to treatment. *Neurology,* **36,** 612–617

WILLCOX, N., DEMAINE, A. G., NEWSOM-DAVIS, J., WELSH, K. I., ROBB, S. A. and SPIRO, S. G. (1985) Increased frequency of IgG heavychain marker *G1m2* and of HLA B8 in Lambert-Eaton myasthenic syndrome with and without associated lung cancer. *Human Immunology,* **14,** 29–36

YOSHIDA, T., TSUCHIYA, M., ONO, A., YOSHIMATSU, H., SATOYOSHI, E. and TSUJI, K. (1977) HLA antigens and myasthenia gravis in Japan. *Journal of the Neurological Sciences,* **32,** 195–201

18
A neurological gene map

A. E. Harding and Roger N. Rosenberg

INTRODUCTION

The main purpose of this chapter is to summarize the 'neurological gene map' as it stands in the latter part of 1987. Undoubtedly, further gene loci associated with neurological disease will have been mapped to specific chromosomal regions by the time this book is published. *Tables 18.1* and *18.2* show the chromosomal localization of genes encoding proteins, enzymes, and transmitters involved in normal neurological function, as well as loci for mutations causing neurological disease.

The techniques used in gene mapping include: genetic linkage studies using conventional (such as blood group antigens and protein polymorphisms) and DNA (restriction fragment length polymorphism, RFLP) markers; somatic cell hybridization; *in situ* hybridization to specific chromosomes; and analysis of specific chromosomes isolated by fluorescence-activated sorting (McKusick, 1986). These and other methods have been described in *Chapter 1*, as have the purposes and prospects of gene mapping. The most immediate clinical application is the ability to diagnose genetic disorders preclinically and prenatally.

A number of the gene loci shown in *Tables 18.1* and *18.2* have been referred to in more detail in this book, including the following: hereditary motor and sensory neuropathy (HMSN) type 1 (chromosome (chr) 1 in some families), X-linked HMSN, familial amyloid polyneuropathy (chromosome 18), and manic depressive psychosis (chromosome 11) (*Chapter 1*); von Recklinghausen's and bilateral acoustic neurofibromatosis (chromosomes 17 and 22) (*Chapters 1* and *8*); those encoding the acetylcholine receptors (chromosomes 2, 17), phenylalanine hydroxylase (chromosome 12), and tyrosine hydroxylase (chromosome 11) (*Chapter 4*); dominant ataxias (chromosome 6) (*Chapter 11*); Huntington's disease (chromosome 4) (*Chapter 12*); Alzheimer's disease (chromosome 21) (*Chapters 1* and *10*); Duchenne, Becker, Emery–Dreifuss (X chromosome) and myotonic (chromosome 19) dystrophies (*Chapter 13*); various oncogenes and growth factors (*Chapters 5, 6* and *8*); retinoblastoma (chromosome 13) (*Chapter 15*); and genes involved in the immune system (chromosomes 6 and 14) (*Chapter 17*).

Table 18.1 A neurological gene map: autosomes

Chromosome/ region	*Gene* (disease) *locus*	*Reference***
1p	(Dominant retinitis pigmentosa*)	Yijian HGM2
1pter-p31	(Neuroblastoma*)	
1p36.2	Atrial natriuretic factor*	
1p34	α-L-fucosidase-1 (Fucosidosis)	
1p34	Uroporphyrinogen decarboxylase (Porphyria cutanea tarda)	
1p22.1	β-nerve growth factor	
1p21	Joseph disease*	
1p21-qter	Actin, α chains	
1cen-q32	Muscle phosphofructokinase* (Glycogenosis VII)	
1q21	Glucocerebrosidase (Gaucher's disease)	
1q2	(Hereditary motor and sensory neuropathy I; some families)	
1q31-q32.1	(Microcephaly*)	
1q	(Xeroderma pigmentosum A*)	
2p25	Pro-opiomelanocortin	
2p24-p23	*N-myc* oncogene	
2p	Carbamoyl phosphate synthetase (CPS) I* (CPS I deficiency)	
2p	(Dominant optic atrophy*; ? one type)	
2q	Muscle nicotinic ACHR, γ and δ subunits*	
2	Glutamate decarboxylase	Sparkes HGM644
3p21-cen	β-galactosidase I (GM_1 gangliosidosis)	
3q21-q25	Caeruloplasmin	
3q21-qter	Rhodopsin*	
3q28	Somatostatin	
4p16.3	(Huntington's disease)	
4p16	(Dominant diabetes insipidus*)	
4p15.3	Dihydropteridine reductase (deficiency)	Brown HGM8
4cen-q31	(Usher's syndrome*)	Daiger HGM577
4q11-q13	α-fetoprotein	
4q25-q27	Epidermal growth factor	
5q11	(Treacher-Collins syndrome*)	
5q11-q13	Arylsulphatase B (Maroteaux-Lamy syndrome)	
5q13	Hexosaminidase B (Sandhoff disease)	
5q31-q32	Platelet derived growth factor receptor*	
5	β-adrenergic receptor, surface*	
5	GM_2 activator protein* (GM_2 gangliosidosis, AB variant)	
6pter-p21	Tubulin, β*	
6pter-p23	Mitochondrial malic enzyme*	
6p23-q12	Prolactin	
6p21.3	Major histocompatibility complex	
6p21.3	Neuraminidase I (Sialidosis*)	
6p21.2-q12	(Dominant ataxia; some families)	
6q23	Arginase* (Argininaemia)	
7pter-q22	Actin, cytoskeletal β*	
7pter-q22	Neuropeptide Y*	
7p21.3-p21.2	(Craniosynostosis; some cases)	
7p21-q22	Argininosuccinate lyase (Argininosuccinicaciduria)	
7p13-p11	Epidermal growth factor receptor	
7q11.2-q22	β-glucuronidase (Mucopolysaccharidosis VII)	
8q23-q24	Proencephalin	Litt HGM203
9p13	Galactose-1-phosphate uridyltransferase (Galactosaemia)	
9q	(Tuberous sclerosis)	Fryer *et al.* (1987)
9q34	Argininosuccinate synthetase (Citrullinaemia)	
9q34	Dopamine-β-hydroxylase*	
9	Coproporphyrin oxidase* (Coproporphyria)	

Table 18.1 Continued

Chromosome/ region	*Gene* (disease) *locus*	*Reference***
10q23-q24	Glutamate dehydrogenase*	
10	Glycoprotein neuraminidase* (? Sialidosis)	
11p15.5	Tyrosine hydroxylase*	
11p14.2-p12	Calcitonin/gene related peptide	Rassool HGM592
11p	(Manic depressive psychosis, Amish type)	Egeland *et al.* (1987)
11q13-qter	Muscle glycogen phosphorylase* (McArdle's disease)	
11q23.2-qter	Porphobilinogen deaminase (Acute intermittent porphyria)	
11q	Pyruvate carboxylase*	
12q13-q14.3	*Gli* oncogene	Kinzler HGM156
12q24.1	Phenylalanine hydroxylase (Phenylketonuria)	
13q14	(Wilson's disease)	
13q14.1	(Retinoblastoma)	
14q32	Brain creatine kinase	
14q32.33	Immunoglobulin heavy chain gene cluster	
15q11	(Dominant dyslexia*)	
15q22-q25.1	Hexosaminidase A (Tay-Sachs disease)	
17p	Muscle nicotinic ACHR, β subunit*	
17p13	Myosin, heavy chain cluster	
17q21-q22	Neuroblastoma or glioblastoma derived oncogene*	
17q21-q22	Growth hormone	
17q21-q22	Nerve growth factor receptor	
17q23	Acid α-glucosidase (Acid maltase deficiency)	
17	(von Recklinghausen's neurofibromatosis)	Barker *et al.* (1987); Seizinger *et al.* (1987)
17	Acid lysosomal sphingomyelinase (Niemann-Pick disease)	Konrad HGM879
18q11.2-q12.1	Transthyretin (Familial amyloid polyneuropathy)	Sparkes *et al.* (1987)
18q22-q23	Myelin basic protein	Kamholz HGM161
19p13.2-q13.2	Lysosomal α-D-mannosidase B (Mannosidosis)	
19q13	Muscle creatine kinase*	
19cen-q13.2	(Myotonic dystrophy)	Brook HGM532
19cen-q13.3	Poliovirus sensitivity gene	Siddique HGM101
19q	(Paramyotonia congenita*)	Chitayat HGM424
19	Myelin associated glycoprotein	Barton HGM450
20pter-p12	Prion protein	
20pter-p12	Prodynorphin	Litt HGM203
20	Neuraminidase/β-galactosidase expression (Galactosialidosis)	
21q21-q22.1	Cystathione β-synthase (Homocystinuria)	
21q11-q22	(Familial Alzheimer's disease)	St George-Hyslop *et al.* (1987)
21q	Amyloid core plaque protein	
22q11	α-L-iduronidase* (Hurler, Scheie syndromes)	
22q13.31-qter	Arylsulphatase A (Metachromatic leucodystrophy)	
22q	(Bilateral acoustic neurofibromatosis)	Rouleau HGM363

p = short arm, q = long arm, ter = end of short or long arm, cen = centromere; for example 6q23 = band 23 on the long arm of chromosome 6 (band numbers increase moving in either direction away from the centromere); 3p21-cen = between band 21 on the short arm of chromosome 3 and the centromere, and 5 on its own implies localized to chromosome 5, but not to any specific region.

* provisional assignment only.

** From McKusick (1986) unless indicated; citations followed by HGM give first author and abstract number from the Ninth Human Gene Mapping conference, Paris, September 1987.

Table 18.2 A neurological gene map: the X chromosome

Chromosome/ region	*Gene* (disease) *locus*	*Reference**
Xp22	(Aicardi syndrome)	
Xp22	(Ocular albinism; two types)	
Xp22	(Retinoschisis)	
Xp22.2-22.1	(Coffin-Lowry syndrome*)	Hanauer HGM517
Xp21.1	Ornithine transcarbamylase (OTC deficiency)	
Xp21.2	(Duchenne muscular dystrophy)	
Xp21.2	(Becker muscular dystrophy)	
Xp21-p11	Monoamine oxidase A	Ozelius HGM361
Xp11.3	(Norrie's disease)	
Xp11.3	(X-linked retinitis pigmentosa)	
Xp11	(Incontinentia pigmenti)	
Xp11-q11	(Menkes' disease)	
Xq13-q21	(X-linked HMSN)	Fischbeck *et al.* (1986a)
Xq13-q21	(Choroideraemia)	
Xq13-q26	(Albinism and deafness*)	Litvak HGM316
Xq21.3-q22	(Bulbospinal muscular atrophy)	Fischbeck *et al.* (1986b)
Xq22	α-galactosidase A (Fabry's disease)	
Xq22	Myelin proteolipid protein (Pelizaeus–Merzbacher disease)	
Xq25	(Lowe oculocerebral syndrome)	
Xq26-q27.2	Hypoxanthine-guanine phosphoribosyltransferase (Lesch–Nyhan syndrome)	
Xq26-q28	(Hunter's disease)	
Xq27.3	(Fragile X mental retardation syndrome)	
Xq28	(Adrenoleukodystrophy, adrenoleukomyeloneuropathy)	
Xq28	(Deutan and protan colour blindness)	
Xq28	(Emery–Dreifuss muscular dystrophy)	
Xq28	(X-linked spastic paraplegia; one type)	Kenwrick *et al.* (1986)
Xq (proximal)	(X-linked spastic paraplegia; one type)	Leppert HGM328
Xq (distal)	(Myotubular myopathy)	Thomas HGM398

* From McKusick (1986), unless otherwise indicated; citations followed by HGM give first author and abstract number from the Ninth Human Gene Mapping conference, Paris, September 1987. Abbreviations as in *Table 18.1*.

References

BARKER, D., WRIGHT, E., NGUYEN, K. *et al.* (1987) Gene for von Recklinghausen neurofibromatosis is in the pericentric region of chromosome 17. *Science,* **236,** 1100–1102

EGELAND, J. A., GERHARD, D. S., PAULS, D. L. *et al.* (1987) Bipolar affective disorders linked to DNA markers on chromosome 11. *Nature (London),* **325,** 783–787

FISCHBECK, K. H., AR-RUSHDI, N., PERICAK-VANCE, M., ROZEAR, M., ROSES, A. D. and FRYNS, J. P. (1986a) X-linked neuropathy: gene localization with DNA probes. *Annals of Neurology,* **20,** 527–532

FISCHBECK, K. H., IONASESCU, V., RITTER, A. W. *et al.* (1986b) Localization of the gene for X-linked spinal muscular atrophy. *Neurology,* **36,** 1595–1598

FRYER, A. E., CONNOR, J. M., POVEY, S., YATES, J. R. W., CHALMERS, A. and FRASER, I. (1987) Evidence that the gene for tuberous sclerosis is on chromosome 9. *Lancet,* **i,** 659–661

KENWRICK, S., IONASESCU, V., IONASESCU, G. *et al.* (1986) Linkage studies of X-linked recessive spastic paraplegia using DNA probes. *Human Genetics,* **73,** 264–266

McKUSICK, V. A. (1986) *Mendelian Inheritance in Man,* 7th edn. Baltimore: Johns Hopkins University Press

ST GEORGE-HYSLOP, P. H., TANZI, R. E., POLINSKY, R. J. *et al.* (1987) The genetic defect causing familial Alzheimer's disease maps on chromosome 21. *Science,* **235,** 885–890

SEIZINGER, B. R., ROULEAU, G. A., OZELIUS, L. J. *et al.* (1987) Genetic linkage of von Recklinghausen neurofibromatosis to the nerve growth factor receptor gene. *Cell,* **49,** 589–594

SPARKES, R. S., SASAKI, H., MOHANDAS, T. *et al.* (1987) Assignment of the prealbumin (PALB) gene (familial amyloidotic polyneuropathy) to human chromosome region 18q11.2-q12.1. *Human Genetics,* **75,** 151–154

Index